Extreme Weather Events in a Warming Climate

Extreme Weather Events in a Warming Climate

Guest Editor

Masoud Rostami

MDPI

Basel • Beijing • Wuhan • Barcelona • Belgrade • Novi Sad • Cluj • Manchester

Guest Editor
Masoud Rostami
Earth System Analysis
Potsdam Institute for Climate
Impact Research (PIK)
Potsdam
Germany

Editorial Office
MDPI AG
Grosspeteranlage 5
4052 Basel, Switzerland

This is a reprint of the Special Issue, published open access by the journal *Atmosphere* (ISSN 2073-4433), freely accessible at: https://www.mdpi.com/journal/atmosphere/special_issues/7K180X502M.

For citation purposes, cite each article independently as indicated on the article page online and as indicated below:

Lastname, A.A.; Lastname, B.B. Article Title. *Journal Name* **Year**, *Volume Number*, Page Range.

ISBN 978-3-7258-5215-4 (Hbk)
ISBN 978-3-7258-5216-1 (PDF)
https://doi.org/10.3390/books978-3-7258-5216-1

Cover image courtesy of Masoud Rostami

Contents

About the Editor

Masoud Rostami

Masoud Rostami is a Senior Scientist at the Potsdam Institute for Climate Impact Research (PIK). His research focuses on Earth and planetary atmospheric science, geophysical fluid dynamics, and Earth system modeling. He is particularly known for his work on the dynamics of the Madden–Julian Oscillation, equatorial modons, and planetary-scale vortices. He developed Aeolus 2.0, a state-of-the-art atmospheric model that will serve as one of the atmospheric components of the Potsdam Earth Model (POEM). He has also made significant contributions to advancing theories on moist-convective dynamics, equatorial adjustment, and planetary atmospheres. His interdisciplinary research spans climate change, Earth and planetary sciences, and human–environment interactions. He holds a PhD from Sorbonne University (formerly UPMC), where he conducted research at the Laboratoire de Météorologie Dynamique (LMD). He continues to serve as a visiting scientist at both LMD and the Shenzhen International Center for Mathematics at SUSTech University, focusing on theoretical advances in the numerical modeling of geophysical flows.

Preface

As climate change intensifies, the frequency, duration, and severity of extreme weather and climate events have increased worldwide. These phenomena, ranging from heatwaves and floods to tropical cyclones and prolonged droughts, are exerting significant pressure on ecological systems, human health, and socio-economic structures. Understanding these events in both regional and global contexts is crucial for predicting future patterns and informing adaptation and mitigation strategies.

This Reprint brings together ten peer-reviewed articles originally published in the *Atmosphere* Special Issue titled "Extreme Weather Events in a Warming Climate". Spanning multiple continents and climate regimes, these articles offer diverse case studies and innovative methodologies to assess and interpret extreme climatic behavior. Below is a brief summary of each contribution.

1. El Niño-Induced Drought Impacts on Reservoir Water Resources in South Africa by Fhumulani I. Mathivha, Lufuno Mabala, Selelo Matimolane and Nkanyiso Mbatha

This study investigates the impact of El Niño-induced droughts on reservoir water levels in the Albasini Dam Catchment in South Africa's Limpopo Province. Using standardized drought indices (SPI and SSI), along with trend and correlation analyses (Mann–Kendall tests and wavelet coherence), the authors found a strong relationship between El Niño events and declining water levels, particularly at longer timescales (12 months). The results highlight that droughts driven by El Niño significantly reduce water availability, stressing the need for integrated, multi-scale water management strategies to enhance regional climate resilience.

2. Research on the Terrain Characteristics of Changbai Mountain and Their Impact on Precipitation and Wind Distribution by Li Liang, Wanxiu Ai, Xiaodan Yang and Luqiang Zhao

This study examines how the terrain of the Changbai Mountain region in northeastern China influences precipitation and wind patterns, especially during summer. Using two-dimensional discrete cosine transform spectral analysis, the researchers quantified topographic features and their relationship with atmospheric flows across three selected domains. Results show that terrain-induced atmospheric effects vary by region. The findings highlight that the large-scale terrain and minor terrain both play a crucial role in atmospheric uplift and the occurrence and development of summer rainfall.

3. Vegetation and Evapotranspiration Responses to Increased Atmospheric Vapor Pressure Deficit across the Global Forest by Rihong Wen, Meiou Qin, Peng Jiang, Feiyun Yang, Bin Liu, Mengyuan Zhu, Yuan Fang, Yichen Tian and Bo Shang

This study explores how rising atmospheric vapor pressure deficit (VPD), a key factor in plant stress, affects global forest ecosystems, particularly vegetation growth and evapotranspiration (ET). Using long-term satellite data and partial correlation analysis, the researchers found that over half of global forest regions, especially in North America and Eastern Europe, exhibit a negative correlation between VPD and both leaf area index (LAI) and ET. Areas like Eastern Europe and western Russia are becoming drier, with vegetation more strongly tied to soil water content. The results highlight the growing influence of VPD on forest health and water cycles under climate change, pointing to increased vulnerability of forest carbon sinks.

4. Estimating Concurrent Probabilities of Compound Extremes: An Analysis of Temperature and Rainfall Events in the Limpopo Lowveld Region of South Africa by Caston Sigauke and Thakhani Ravele

This study applies copula-based statistical models to estimate the joint probability of compound extreme events, specifically high temperatures and low rainfall, in South Africa's Limpopo Lowveld region. Using elevation-stratified data and indicators for drought severity, the researchers found that mild droughts have a 64–66% likelihood, moderate droughts 36–39%, and severe droughts 16–19% across elevation zones. Logistic regression models incorporating the Southern Oscillation Index produced comparable results. The approach is adaptable to other regions and offers valuable insights for disaster risk management and climate resilience planning.

5. Spatiotemporal Patterns of Typhoon-Induced Extreme Precipitation in Hainan Island, China, 2000–2020, Using Satellite-Derived Precipitation Data by Mengyu Xu, Yunxiang Tan, Chenxiao Shi, Yihang Xing, Ming Shang, Jing Wu, Yue Yang, Jianhua Du and Lei Bai

This study analyzes typhoon-induced extreme precipitation on Hainan Island from 2000 to 2020 using satellite data and cyclone track information. While typhoon frequency slightly declined, extreme precipitation events have intensified, especially in the eastern and central regions. Different typhoon paths exert region-specific impacts, and typhoon duration and wind speed are strongly correlated with precipitation extremes. High sea surface temperatures and ENSO-related changes, particularly in the Niño 1+2 region, are also key influencing factors. These findings can inform region-specific disaster prevention and mitigation strategies for Hainan Island.

6. Projection of Extreme Summer Precipitation over Hubei Province in the 21st Century by Abrar Mubark, Qian Chen, Mohamed Abdallah, Awad Hussien and Monzer Hamadalnel

This study projects future trends of extreme summer precipitation in Hubei Province, China, throughout the 21st century under the RCP8.5 scenario using the RegCM4 regional climate model driven by HadGEM2-ES. Model validation via five rainfall indices indicates a significant increase in extreme rainfall events, especially in southeastern regions such as Wuhan, Xiantao, and Qianjiang. All indices, except for consecutive dry days, show rising intensity and frequency of rainfall extremes, suggesting heightened flood risks in the region as global warming intensifies.

7. Assessment of the Drought Risk in Constanța County, Romania by Carmen Elena Maftei, Alina Bărbulescu and Amela Osman

This study evaluates drought risk across six sites in Constanța County, Romania, using the de Martonne aridity index, the Standardized Precipitation Index (SPI), and derived drought hazard and risk indices. Findings show climate conditions ranging from slightly arid to semi-arid, with locations like Adamclisi and Cernavodă experiencing notable climate variability. Drought Risk Index classifications vary by site, with some facing low to moderate risk, while others show high to very high risk. Drought intensity ranged widely, and prolonged dry periods exceeded 99 months across all stations, indicating significant vulnerability in this agriculturally important region.

8. Temperature and Precipitation Extremes in the Brazilian Legal Amazon: A Summary of Climatological Patterns and Detected Trends by Wanderson Luiz-Silva, Anna Carolina Fernandes Bazzanela, Claudine Pereira Dereczynski, Antonio Carlos Oscar-Júnior and Igor Pinheiro Raupp

This study analyzes trends in 20 climate extreme indices related to temperature and precipitation in the Brazilian Legal Amazon (BLA) using observed data from 1961 to 2020. The results reveal significant increases in heat-related extremes, with warm days and nights rising by 11 days per decade and heatwaves extending from 10 to 20, to 30 to 40 days. Simultaneously, extreme precipitation has declined, especially in the western BLA, while consecutive dry days are increasing in the east/south, by 1.5 days per decade, delaying the rainy season. These climatic shifts are reducing humidity convergence and may impact broader regional weather systems in central-southern Brazil.

9. Characteristics and Driving Mechanisms of Heatwaves in China During July and August by Jinping Liu and Mingzhe Li

This study analyzes the frequency, duration, and causes of heatwaves in China from 2013 to 2023 using the Excess Heat Factor (EHF). Results show regional contrasts; southeastern coastal areas face frequent but short heatwaves, while the arid northwest experiences both frequent and prolonged events. The Tibetan Plateau has milder heatwaves. Key drivers include subtropical highs, low-level wind anomalies, and soil moisture–temperature feedbacks, which vary by region. While atmospheric factors dominate in some areas, soil moisture plays a significant role in others. The study notes limitations due to the EHF's lack of humidity consideration and potential uncertainties in reanalysis data used.

10. Advancing Heat Health Risk Assessment: Hotspot Identification of Heat Stress and Risk Across Municipalities in Algiers, Algeria by Dyna Chourouk Zitouni, Djihed Berkouk, Mohamed Elhadi Matallah, Mohamed Akram Eddine Ben Ratmia and Shady Attia

This study assesses long-term heat-related risks in Algiers from 2001 to 2023 using satellite, meteorological, and census data. The Heat Health Risk Index (HHRI), based on IPCC's hazard-exposure-vulnerability framework, and the Universal Thermal Climate Index (UTCI) were analyzed together to identify hotspots of heat stress. Results show a significant increase in HHRI and UTCI, with "very-low" risk areas decreasing by 93% and UTCI rising by 6 °C , indicating "very strong heat stress". Coupling HHRI and UTCI revealed broader and different risk zones than using HHRI alone, emphasizing the need for targeted interventions in central and eastern Algiers and demonstrating the value of remote sensing for urban heat resilience planning.

Collectively, these ten studies provide a nuanced understanding of the evolving patterns and impacts of extreme weather and climate events across diverse geographic and climatic contexts. From droughts in South Africa and Romania to heatwaves in China and Algeria, and from precipitation extremes in China and Brazil to compound events in Southern Africa, the research highlights the multifaceted nature of climate risks intensified by global warming. The integration of advanced observational datasets, climate modeling, and innovative statistical approaches underscores the importance of interdisciplinary and multi-scale analyses for capturing the dynamics of these phenomena. Moreover, the regional specificity of findings emphasizes the need for localized adaptation and mitigation strategies tailored to distinct environmental and socio-economic conditions. This volume thus serves as a resource for researchers, policymakers, and practitioners seeking to enhance climate resilience and inform sustainable management in the face of accelerating climatic changes.

I extend my sincere gratitude to the editorial team of *Atmosphere* and the MDPI publishing platform for their invaluable support in facilitating the dissemination of this work. Special thanks are due to the peer reviewers, whose insightful and constructive feedback significantly enhanced the quality and rigor of the articles included in this volume.

I also wish to acknowledge the support of all participating institutions and funding agencies, without which this collective effort would not have been possible. The dedication and collaboration of numerous researchers, data providers, and technical teams from around the world have been essential to the successful completion of this research.

It is my hope that this reprint will serve as a meaningful resource for communities most vulnerable to climate extremes, contributing scientific knowledge that supports equitable and effective resilience and adaptation strategies.

Masoud Rostami
Guest Editor

atmosphere
MDPI

Article

El Niño-Induced Drought Impacts on Reservoir Water Resources in South Africa

Fhumulani I. Mathivha [1,*], Lufuno Mabala [2], Selelo Matimolane [3] and Nkanyiso Mbatha [4]

1 Department of Water and Sanitation, University of Limpopo, Sovenga 0727, South Africa
2 Department of Earth Sciences, University of Venda, Thohoyandou 0950, South Africa; 17008568@mvula.univen.ac.za
3 Equitable Education and Economies, Human Sciences Research Council, Pretoria 0002, South Africa; smatimolane@hsrc.ac.za
4 Department of Geography and Environmental Sciences, University of Zululand, KwaDlangezwa 3886, South Africa; mbathanb@unizulu.ac.za
* Correspondence: fhumulani.mathivha@ul.ac.za

Abstract: The ENSO phenomenon is associated with below average rainfall and influences the climate regime of southern Africa. With the advent of climate change, drought frequencies and magnitudes have worsened in the developing world and this in turn negatively impacts the natural environment and communities' livelihoods. This study evaluated the relationship between El Niño-induced drought and reservoir water levels over the Albasini Dam Catchment (ADC) areas in Limpopo Province, South Africa. Standardised indices (i.e., SPI and SSI) were used to define drought events over the study area. Mann–Kendall and Sequential Mann–Kendall were used for trends analysis as well as correlation and wavelet coherence to evaluate the relationship between variables of interest. There exists a relationship between El Niño-induced drought event and reservoir water levels. This was shown by the correlation between drought indices and reservoir water levels with the coefficient of determination being stronger at the 12th timescale (i.e., 0.743 and 0.59) compared to the 6th timescale (i.e., 0.07 and 0.44) for both precipitation and streamflow indices, respectively. Wavelet analysis further showed that there existed a phased relationship between the two variables. Although there are other factors that may affect reservoir water resources, these study findings show that El Niño-induced drought also negatively affect water resources. Therefore, this study recommends the development of multidimensional and multiscale management strategies to minimise drought impacts and adaptation in the region.

Keywords: drought; water resource management; water levels; wavelet analysis; streamflow; trends

1. Introduction

The El Niño–Southern Oscillation (ENSO) influences the global climatic regime [1] and is characterised by a cool (La Nina) and warm phase (El Niño). The phenomenon results from the fluctuating equatorial Pacific Ocean temperatures and large-scale air pressure changes [2,3]. The ENSO phenomenon impacts the southern Africa climatic regime, with regional and international events [1]. It is associated with drought hazards, a recurrent feature of the region's climate. This may be influenced by higher-than-normal global temperatures associated with the phenomenon [4], which are frequently associated with below-average rainfall in southern Africa [5]. As a result, countries in the region experience severe and extensive droughts, and in the case of South Africa, some of the major drought events over the last four decades occurred in 1982/84, 1991/92, 2004/05, 2015/16, and the most recent one was experienced in 2018/20 [6]. The consequences of increased drought events linked to El Niño events include impacts on the availability of water resources, communities' livelihoods, and environmental and human health [7]. Previous studies have described a strong and consistent teleconnection among ENSO and temperature patterns worldwide [2,8,9] while [10] reported this at a regional scale. This can be further justified by the work of Halpert and Ropelewski [11], which was one of the first studies to establish a teleconnection between El Niño events to southeastern Africa temperatures. Teleconnections have the capability to modify regional circulation patterns [12], and as such, an El Niño signal has evidently modulated fluxes of moisture advected onto the southern African plateau [13,14]. Notable El Niño events that have been associated with severe droughts occurred in 1991/92, 2002/03, and 2015/16, with the one occurring in 1997/98 resulting in dry conditions [14]. In 1992, a more critical El Niño event was reported compared to those experienced in 2016, 1998, and 1983 [5]. On a regional scale, the 2014/15 and 2015/16 El Niño events exacerbated agricultural production challenges around the Limpopo Province. This resulted in drought events that affected different water-dependent sectors, i.e., recreational, agricultural, industrial, and domestic. While water security in South Africa is diverse, the least developed and rural areas experience severe and direct risks. This is a result of the high spatial–temporal variability of rainfall and general semi-arid climate characteristics of the country [15]. South Africa's water resources are linked through inter-catchment transfers and managed as large integrated systems. Such a management system is helpful in decreasing the probable risk of collapse through the collective use of resources and harmonising climatic variability over vast geographic areas [16].

The Luvuvhu River Catchment (LRC) in the northernmost part of South Africa, which hosts the Albasini, Nandoni, and Vondo dams, experienced a decrease in mean annual precipitation (MAP) of 14% between 1991 and 1992 [17], resulting in the severe 1991/92 event. Since rainfall is a major contributor to reservoir water resources, a reduction in annual precipitation negatively impacts water supply to different water users. Studies focused on drought such as [18–22] have been conducted in the LRC. However, these aimed to characterise, assess, and predict drought events and not much attention has been directed to evaluate the relationship between El Niño-induced drought and reservoir water resources. There are studies that have examined El Niño-induced drought in various part of the world, e.g., [23,24]; meanwhile, Ref. [25] examined water resource management strategies in times of ENSO over the transboundary Santa Cruz River Basin. The latter studies focused on the impacts of ENSO on agricultural production, water quality, and reservoirs, with less attention on the statistical correlation between reservoir water resources and El Niño-induced drought. From a water management perspective, a study that seeks to investigate drought impacts on reservoir water resources is significant, more so now that the frequency of such events has been predicted to be on an increasing trend due to climate change-related disasters [26]. In the case of Sub-Saharan Africa, drought episodes are often linked to desertification, which may result in serious land degradation of semi-arid and arid environments emanating from climate change, as well as human activities such as unsustainable and inefficient farming practices [27]. The latter further indicate that exacerbate desertification from an increase in drought frequency and severity resulting from climate change could lead to a loss in vegetation cover, a reduction in carbon sinks, and the release of additional greenhouse gases to the environment. The management of droughts resulting in El Niño events is one of the significant challenges of the 21st century. Thus, information on the influence of El Niño on drought occurrence can aid in disaster risk reduction and management. This study aims to explore the critical impact of drought on reservoir water resources in the semi-arid region of South Africa.

2. Materials and Methods

2.1. The Study Area

The Albasini Dam as shown in Figure 1 is located 22 km southeast of the town of Louis Trichardt within the Luvuvhu River Catchment (LRC) in Limpopo Province. The Albasini Dam is found 45 km west (upstream) of the most recent constructed Nandoni Dam in the Luvuvhu River. Constructed on the Luvuvhu River stem in 1952, the Albasini Dam covers a surface area of 350 ha with a maximum carrying capacity of 252×10^5 m^3. The biggest consumer of the Albasini Dam freshwater is the irrigated agriculture in the Levubu Valley [28]. This area is known for its richness in tropical fruits such as bananas, avocados, mangos, and macadamia nuts amongst others. More water from the dam is used by trees that may result in lowered reservoir water levels (RWLs) as the dam is characterised by a high downstream tree density [29]. Average temperatures in summer are in the 30 °C ranges whilst the winter temperatures vary between 20 °C and 25 °C with a mean annual precipitation of 608 mm.

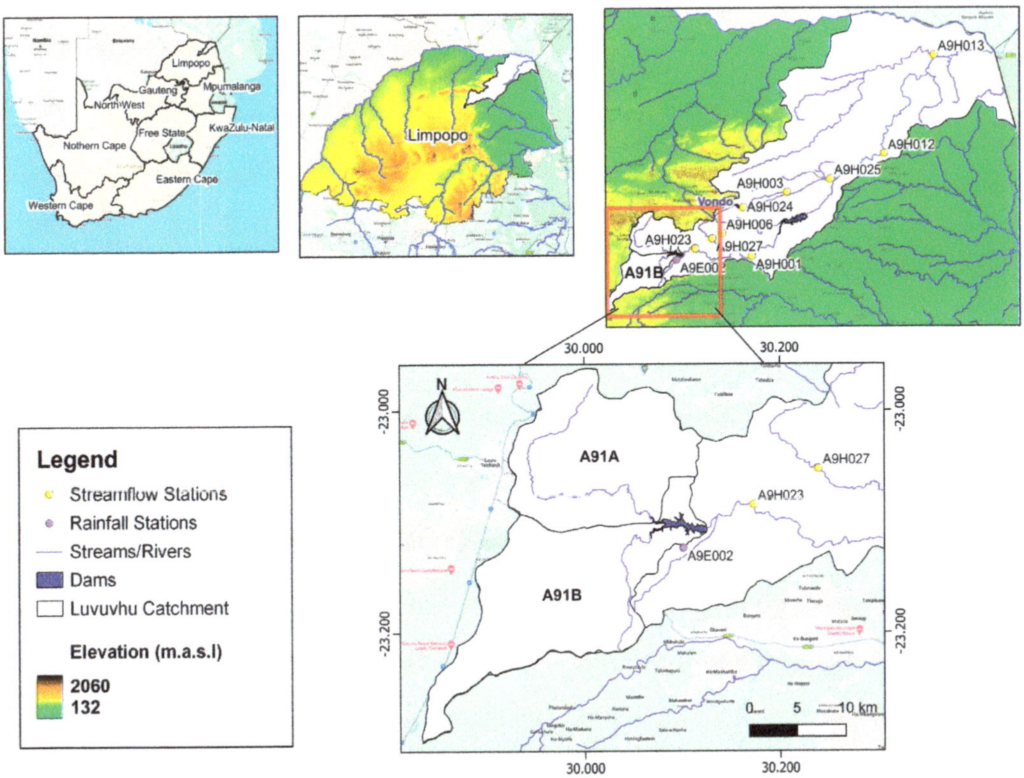

Figure 1. Map of the study area showing the location of Albasini Dam in the context of South Africa, Limpopo Province and the LRC (accession numbers).

2.2. Datasets

To achieve the objective of this study, secondary monthly mean rainfall, streamflow, and reservoir storage data from 1990 to 2020 were obtained from the National Department of Water and Sanitation, South Africa. The Niño 3.4 monthly climate time series was downloaded from National Oceanic and Atmospheric Administration repository (https://psl.noaa.gov/data/timeseries/monthly/NINO34/ accessed on 15 June 2022). Rainfall and streamflow data were used determine the standardised precipitation index (SPI) and standardised streamflow index (SSI), respectively. Meanwhile the reservoir water levels were correlated with the SPI and SSI to determine the influence of drought episodes on reservoir water resources.

2.3. Data Analysis Methods

2.3.1. Standardised Indices

The SPI and SSI were used to describe meteorological and hydrological drought, respectively. These were formulated at the 6- and 12-month timescales. The 6-month SSI and

SPI indicate medium-term changes in streamflow and precipitation time series, respectively, whereas the 12-month SPI and SSI indicate long-term patterns [30]. The SPI was used to evaluate rainfall patterns within respective reservoir catchment areas for 30 years. The SPI [31] was developed for drought monitoring in the USA; however, the index has been extensively adapted in many regions of the world. The authors of [32] reported that the gamma distribution better fitted the precipitation time series than the exponential time series, and as such, this former distribution was adopted for this study. The probability density function that defines the Gamma distribution is given by Equation (1):

$$G(x) = \frac{1}{\beta^{\alpha}\tau^{\alpha}} x^{\alpha-1} e^{-x/b} \tag{1}$$

where $\alpha > 0$, α is a shape factor and $\beta > 0$, β is a scale factor.

$$\acute{\Gamma}(\alpha) = \int_0^{\infty} y^{a-1} e^{-y} \tag{2}$$

where, $\acute{\Gamma}(\alpha)$ is the gamma function. Computation of the SPI involves fitting a gamma probability density function to a given frequency distribution of a precipitation total for a particular station. The cumulative probability is given by Equation (3):

$$G(x) = \int_0^{x} g(x)dx = \frac{1}{\beta^{\dot{\alpha}}\tau^{\dot{\alpha}}} \int_0^{x} x^{\alpha-1} e^{-x/b} \tag{3}$$

By letting $t = x/\beta$, Equation (3) becomes Equation (4):

$$G(x) = \frac{1}{\acute{\Gamma}\dot{\alpha}} \int_0^{x} t^{a-1} e^{-t} dt \tag{4}$$

The gamma function is undefined for x equal to 0 and a precipitation distribution may contain zeros. Therefore, to compute the cumulative probability, Equation (5) is used:

$$H(x) = q + (1-q)Gx \tag{5}$$

where q is the probability of a zero. If m is the number of zeros in a precipitation time series, then q can be approximated by m/n. The cumulative probability, H(x), is then altered to the standard normal arbitrary variable Z with the mean equal to zero and a variance of one, which is the value of the SPI [33]. SSI was computed by following the same steps as applied in computing SPI using streamflow data, and like SPI, this index shows a proper gamma distribution [34]. McKee [31] defined different drought SPI values to a drought category, i.e., −0.99 to 0.99 is classified as near normal or mild conditions, −1 to −1.49 shows moderate category, while −1.50 to −1.99 and −2.0 or greater depict severe and extreme drought categories, respectively.

2.3.2. Drought Characterisation

To characterise historical droughts, the standardised indices time series were used. For this study, drought events will be characterised based on duration, severity, and intensity. Drought duration denoted by M is the months between the beginning and an end of a drought event; the beginning is included in the computations, whereas the end is excluded [35]. To obtain drought duration, the total of the durations of each drought event is divided by the number of drought events [36] as shown in Equation (6):

$$M = \frac{\sum_{i=1}^{n} d_i}{n} \tag{6}$$

where d_i is the duration of the i-th drought event in a particular area and n is the total number of drought events in that area. Drought severity (S_e) is the absolute value of sum of an index value during a drought event and is given by Equation (7):

$$S_{e} \int = |\sum_{j=1}^{m} index_j|_e \quad (7)$$

where e is the drought event, j is the month, and $index_j$ is the index value in month j. Drought intensity is given by the division of drought severity with drought duration. Intensity is directly proportional to severity of drought: when one increases, the other also increases [37]. Intensity is calculated using the following Equation (8):

$$DI_e = \frac{S_e}{M} \quad (8)$$

where M is the drought duration, S_e drought severity, and DI_e the intensity of a drought event.

2.3.3. Trend Analysis

The Break for Additive Seasonal and Trend (BFAST) method as presented in Equation (9) was used to decompose reservoir water levels and drought indices time series.

$$y_a = m + T_a + S_a + R_a \quad (9)$$

In Equation (9), m is the mean, T is trend component, S is seasonal component, and R is the random component of the time series. Mann–Kendall (MK) non-parametric trend test [38,39] was used for drought trend analysis. This trend analysis method has been recommended by the WMO as an approach to computing trends of hydro-meteorological time series. The method has been widely applied in testing trends of climatological time series [40]. The test statistic S is calculated using Equation (13).

$$S = \sum_{k=1}^{n-1} \sum_{j=k+1}^{n} sng\left(y_j - y_k\right) \quad (10)$$

$$Sng\left(y_j - y_k\right) = \begin{Bmatrix} +1 \text{ if } \left(y_j - y_k\right) > 0 \\ 0 \text{ if } \left(y_j - y_k\right) = 0 \\ -1 \text{ if } \left(y_j - y_k\right) < 0 \end{Bmatrix}$$

The average value of S is E[S] = 0. The value of the S statistic is associated with the test statistics τ, and this is computed using Equation (11):

$$\tau = \frac{S}{n(n-1)/2} \quad (11)$$

This study considered a 5% confidence level, where the null hypothesis of no trend was rejected if $|z| > 1.96$. The extended MK trend, known as the Sequential MK (SQMK) [41] is used to detect the turning points of trend in the time series. The SQMK creates two-time series, a progressive $(u(t))$ and a retrograde $(u'(t))$. Equation (12) defines the statistic t_i:

$$t_i = \sum_{j=1}^{i} n_i \quad (12)$$

Equations (13) and (14) are used in calculating the mean and variance t_i:

$$E(t_i) = \frac{i(i-1)}{4} \quad (13)$$

and

$$Var(t_i) = \frac{i(i-1)(2i-5)}{72} \quad (14)$$

Statistic $u(t_i)$ sequential values, which are standardised, are calculated by Equation (15):

$$u(t_i) = \frac{t_i - E(t_i)}{\sqrt{Var(t_i)}} \quad (15)$$

Equation (15) gives a forward sequential progressive statistic. To calculate the backward/retrograde statistic values ($u'(t)$), the same time series ($x_1, x_2, x_3, \ldots, x_n$) is used; however, the tail end of the time series is the starting point in this calculation. By combining the forward and backward sequential statistics, it makes it possible for the recognition of the estimated start of a developing trend [42]. When plotted, if ($u(t)$) and $\left(u'(t)\right)$ cross each other and diverge beyond the ±1.96 (95% confidence interval) threshold, a statistically significant trend exits in a time series. The region where they cross each other is the time where the trend turning point begins.

2.3.4. Wavelet Analysis

The wavelet transform coherence (WTC) is presented by a ratio of the cross-spectrum to the product of the spectrum of each time series and further proposes local correlation between two variables' time series within frequency bands and time intervals [43]. In comparison with other correlation techniques, wavelet coherence analysis can acquire the coherence coefficients at different spatial locations and scales while considering the spatial phase, thereby obtaining the spatial heterogeneity of the association across scales [44]. The WTC technique has frequency and phase variation analysis capabilities across time in a signal at several scales [45,46]. Since climatic variables vary over space and time, the authors of [47] suggested that WTC offers improved analyses of climate time series as these features can be identified through the frequency domain. Therefore, the Morlet Carlo wavelet and coherence analysis was used to quantify the relationship between El Niño-induced drought and water resources for the current study. Additional information and the mathematics on the wavelet analysis applied in this study can be obtained in [48,49].

3. Results

3.1. Drought Assessment

SPI time series were analysed at two timescales (6 and 12 months) in this study, and these were used to analyse historical drought in the Albasini Dam Catchment for a duration of 30 hydrological years (January 1990 to December 2020). Figure 2 shows SPI and SSI results for stations A9E002 and A9H023, respectively. The analysis of all SPI timescales detected all the major drought events between 1990 and 2020 (1991/92, 1995–1996, 2002–2003, 2005–2006, and 2015–2016). The SPI-6 drought classification's percentage occurrence of mild, moderate, severe, and extreme drought conditions ranged between 66.49%, 23.94%, 8.51%, and 1.06%, respectively. The SPI-12 had a percentage occurrence of 66.28% for mild drought, 21.51% for moderate drought, and 12.21% for severe drought conditions. The most extreme drought cases were in 2006 for SPI-6 and between 2002 and 2004 and 2005 and 2006 for SPI-12. The station reported 36, 25, and 24 drought months between 2002 and 2006, 2008 and 2010, and 2015 and 2016, respectively. These three droughts months' durations are the highest in the SPI-12 analysis. The results of the highest duration in 2002–2006 for SPI-12 were similar to findings by Kumar et al. [50]. Catchment A91A at the SPI-12 timescale did not detect any extreme dry drought months, but the SPI-6 timescale detected two extreme dry months. Moderate droughts for SPI timescales had a minimum of −1.49 and SPI-6 had a maximum of −1, whereas SPI-12 had a maximum of −1.04.

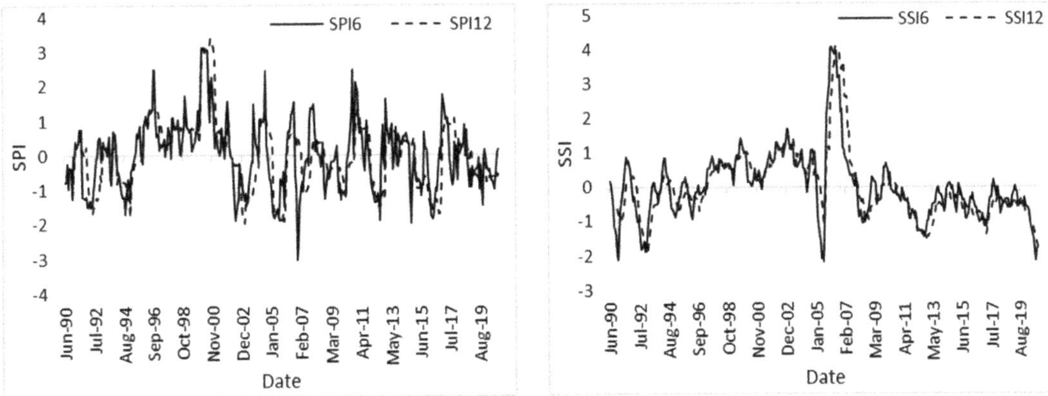

Figure 2. Meteorological (SPI) and hydrological (SSI) drought time at 6- and 12-month timescales for stations A9E002 and A9H023, respectively.

The SSI detected all the major drought events (1991/1992, 2002/2003, and 2015/2016) that have been previously reported in literature over the study period [11,51,52]. SSI reported that the 1991/92 drought had the highest magnitude for both 6- and 12-month timescales, which can be categorised similarly to the extreme drought found by the authors of [51]. The 2004/05 drought is also seen as an extreme drought event, and this agrees with the findings of [52]. The drought period of 2014/15 for this study experienced mild drought and 2016/17 showed severe streamflow drought conditions, particularly from the month of May 2016 to April 2017. The SSI-6 indicates that the ADC experienced mild drought for 160 months and 30 months of moderate drought, with 12 and 5 months of severe and extreme drought, respectively, from 1990 to 2020. SSI-12 has numerous mild droughts than the SSI-6 timescale between 1990 and 2020. Both indices at both timescales detected that mild drought was more dominant throughout the entire study period. This means that the study area experienced mild drought for a majority of the months in the study period. This study found that the increase in timescales increases with drought severity since a majority of the severe droughts were reported at both the 6- and 12-month timescales considered in this study.

3.2. Historical Drought Characterisation

Three drought characteristics (i.e., duration, severity, and intensity) were analysed in this study for both SPI and SSI and the results are presented in Table 1. The table shows the longest duration, strongest severity, and highest intensity of drought events for the indices considered at 6- and 12-month timescales. The longest drought duration between 2010 and 2014 was shown for SSI-6 and between 2010 and 2020 for SSI-12. SPI reported the lowest duration between 2002 and 2003 and 2002 and 2006 at SPI-6 and SPI-12, whereas SSI recorded the highest duration for both timescales.

Table 1. Duration, severity, and intensity of drought events for different periods as depicted by SPI and SSI.

Station	Drought Indicator		Longest		Strongest		Highest	
			Year	Duration	Year	Severity	Year	Intensity
A9E002	SPI	6	2002–2003	19	2004–2005	−18.36	2016	−1.18
		12	2002–2006	36	2002–2006	−39.21	1992–1993	−1.18
A9H023	SSI	6	2010–2014	50	1992–1993	−20.34	2005	−1.35
		12	2010–2020	150	2010–2020	−89.36	2010–2020	−3.57

The SSI showed the strongest severity between 2010 and 2020, which was −89.36, and was then followed by 2002–2006 (−39.21), and the one with the least severity was observed between 1992 and 1993 (−20.34). This therefore shows that the 2014/2016 drought was more severe compared to the 1992 drought. Drought severity was found to increase with increasing timescale. SSI reported the highest intensity of −1.35 for SSI-6 and −3.57 for SSI-12. In the case of drought intensity increasing with an increased timescale, the results were more evident between 2010 and 2020, and 2005. The relationship between duration and severity is important for the drought intensity results and analysis. Intensity is directly proportional to severity, which means high intensity corresponds to high severity.

3.3. Drought Trends and Their Significance

The trends for both SPI and SSI indices were found to be negative. The study found that at both timescales, the decreasing trends were significant. Figure 3 shows that the BFAST extracted non-linear trend fitted with a linear regression line to show the direction of trend. Like the MK trend findings, the BFAST non-linear trend depicts a downward trend which signifies that over the 30 years of study, drought events were on an increase in catchment A91A.

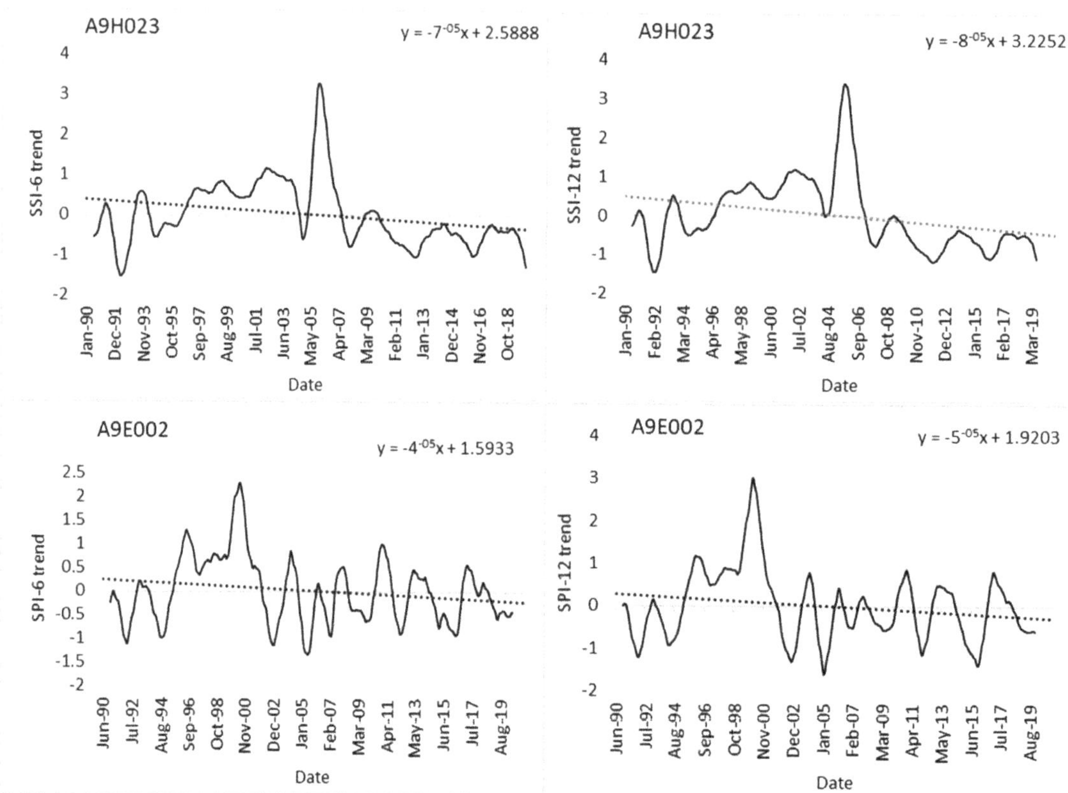

Figure 3. SPI and SSI BFAST trend at 6- and 12-month timescales for stations A9H023 and A9E002.

Sequential Mann–Kendall (SQMK) test statistic for monthly mean standardised indices at the 6th and 12th timescales time series clearly detects the statistically significant change points, and this is shown in Figure 4. Figure 4a,b show a trend increase in 1995; this was followed by a continuous significant decreasing trend after 2005. The SSI at both timescales showed a significant negative trend from the early 2000s. This therefore indicated that streamflow drought is increasing in the study area. These findings are consistent with the MK test and BFAST as well as the drought periods reported in the literature.

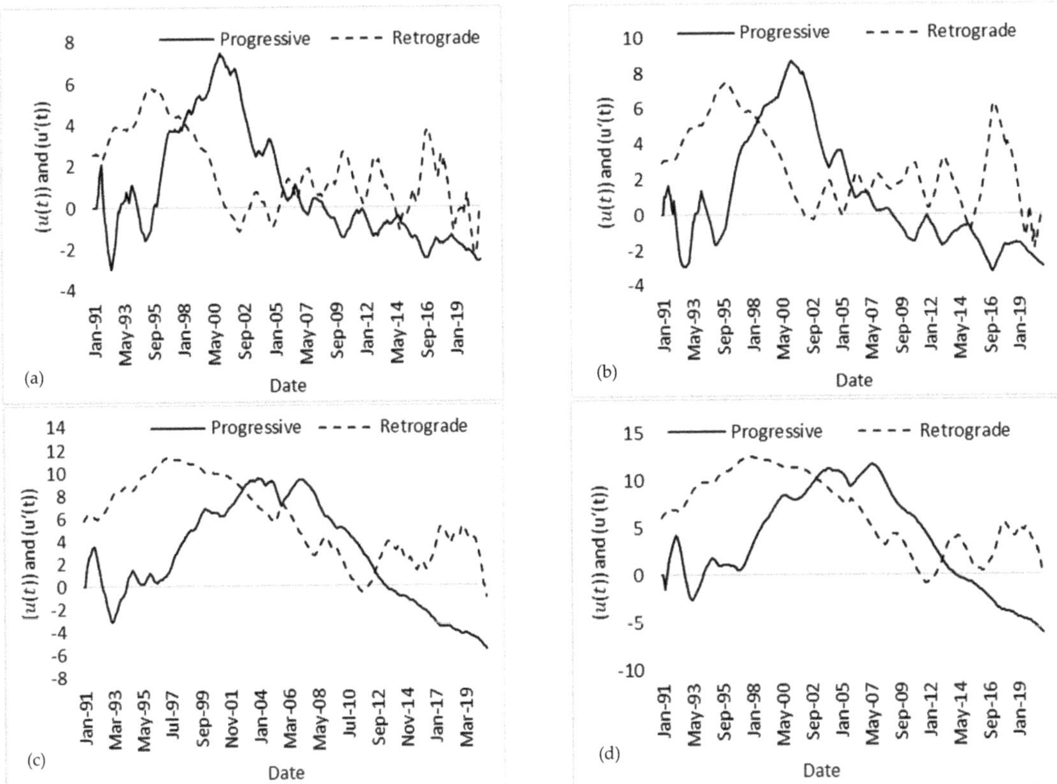

Figure 4. Forward and backward series of SQMK test for (**a**) SPI6, (**b**) SPI12, (**c**) SSI6, and (**d**) SSI12 for sub-catchment A91A.

3.4. Correlation of Drought Events and Reservoir Water Levels

Figure 5 shows a comparison between reservoir water levels and SPI and SSI at 6- and 12-month timescales for major drought events reported over the study area (i.e., 1991/1992, 1995/1995, 2003/2004, 2005/2006, and 2014/2016). For all the drought periods, reservoir water levels decreased as SSI and SPI decreased at both timescales. For the 1995/1996 and 2005/2006 droughts, an increase in reservoir water levels was noted at the end of 1995 and 2005.

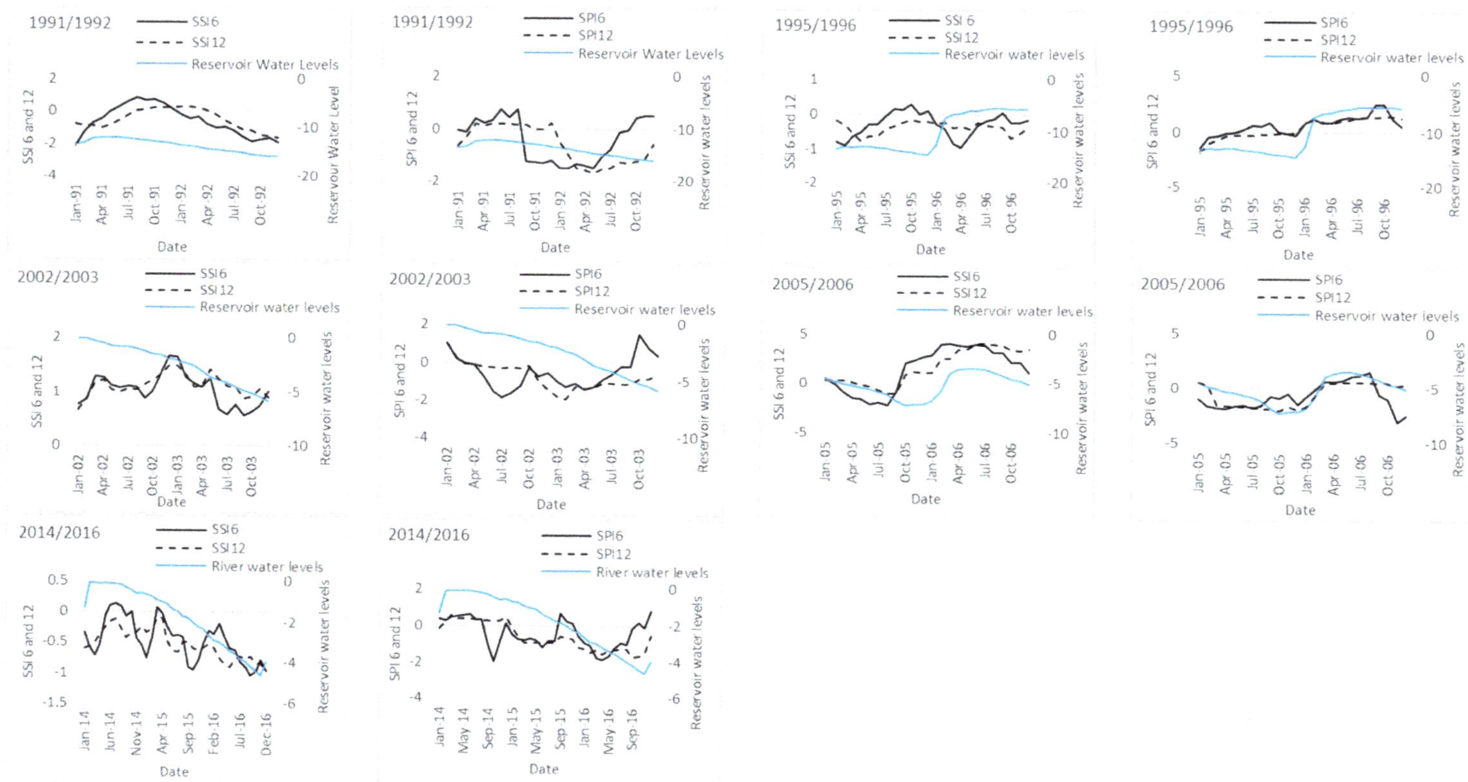

Figure 5. Relationship between SPI and SSI at 6- and 12-month timescales and reservoir water levels for major droughts that have been reported over the study area between 1990 and 2020.

Figure 6 shows the Pearson correlation between the reservoir water levels, Niño 3.4, SSI, and SPI for the study period between 1990 and 2020. The correlation was low at the 6-month timescale compared to the 12-month timescale. Niño time series showed a moderate negative correlation of −0.4 with SPI and a low positive relationship with SSI at both timescales. The correlation between indices and reservoir water levels was found to be 0.36 for SPI-12, while the rest were low and ranged between 0.16 and 0.21.

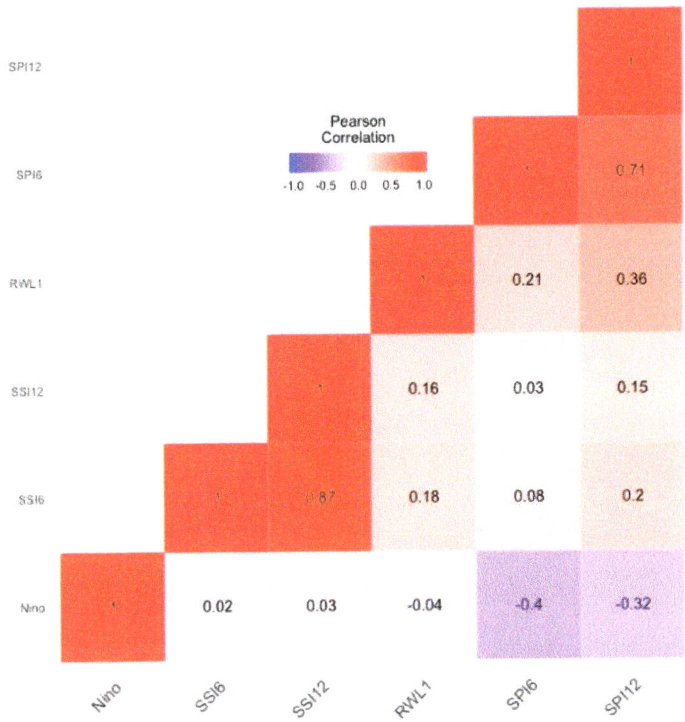

Figure 6. Correlation heatmap between Niño, SPI6, SPI12, SSI6, SSI12 and RWLs.

To further illustrated the relationship between drought and reservoir water levels, the study computed the coefficient of determination as shown in Figure 7. The SPI-12 and water levels possess a strong correlation with an R^2 of 0.7343. The correlation for SSI-6 and SSI-12 timescales are considered to have a moderate correlation with reservoir water levels. The correlation categories based on [53] relates an R^2 of 0.00–0.10 as negligible, 0.10–0.39 as a weak correlation, 0.40–0.69 as moderate correlation, 0.70–0.89 as strong correlation, and 0.90–1.00 as a very strong correlation. This, therefore, shows that there exists a relationship between reservoir water resources and drought over the catchment and this correlation is more pronounced at higher drought indices timescales.

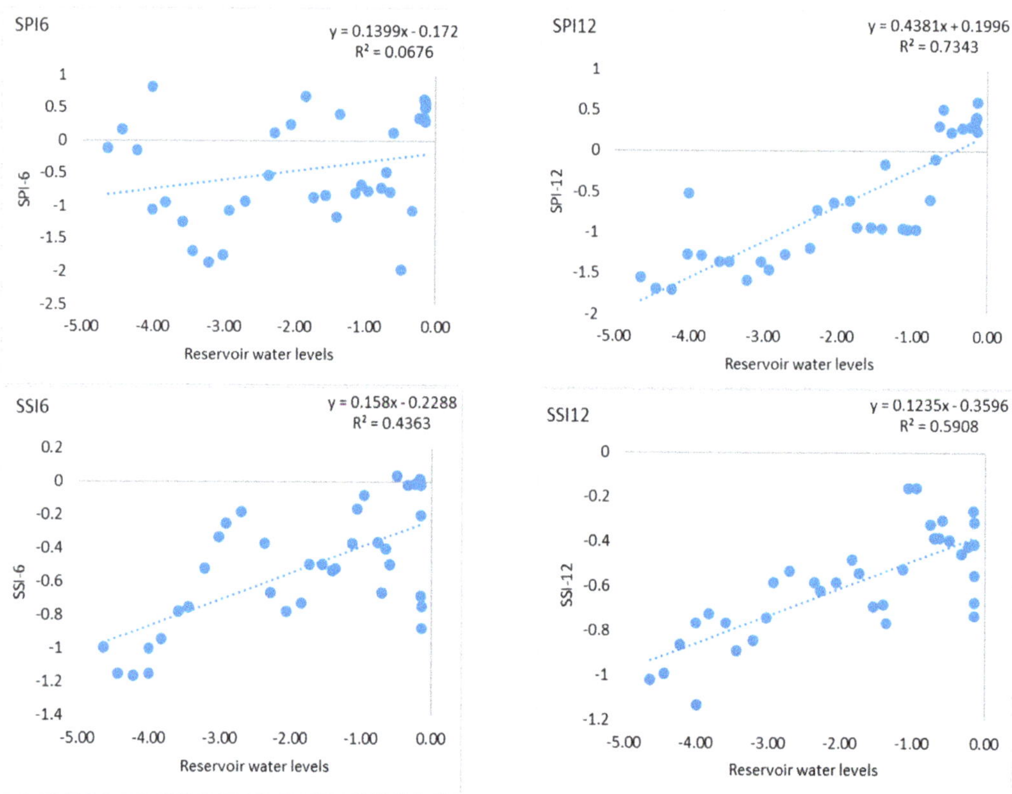

Figure 7. Scatterplot showing SPI and SSI correlation against reservoir water levels.

3.5. Wavelet Coherence Analysis

To investigate the interdependence among SSI, SPI, RWLs, and Niño 3.4, WTC was used. The wavelet coherence analysis between SSI and SPI at the 6th and 12th timescales with Niño 3.4 time series and reservoir water levels is shown in Figure 8a–h. The u-shaped solid line in all the figures represents the cone of influence (COI) [54] and is the only region considered for analysis. The thick solid line within the COI represents the 95% significance regions of the confidence interval [49], and it is within this area where the significance interdependence between Niño 3.4 and the drought indices as well as reservoir water level is studied. Wavelet coherence reveals that the coherence and phase lag between two time series are both a function of time and frequency [55]. SPI showed an anti-phase relationship with the Niño 3.4 time series at both timescales between the 32nd and 64th month. Meanwhile, a high wavelet coherence is observed between the 22nd and 28th month in Figure 8e,f and the upward arrows in the same figures indicate that Niño 3.4 leads the streamflow drought in the study area. SPI showed a strong in-phase relationship with reservoir water level compared to SSI and this was the strongest at the 6th timescale. This in-phase relationship is seen roughly after the 32 months, and this intensified between 2006 and 2016.

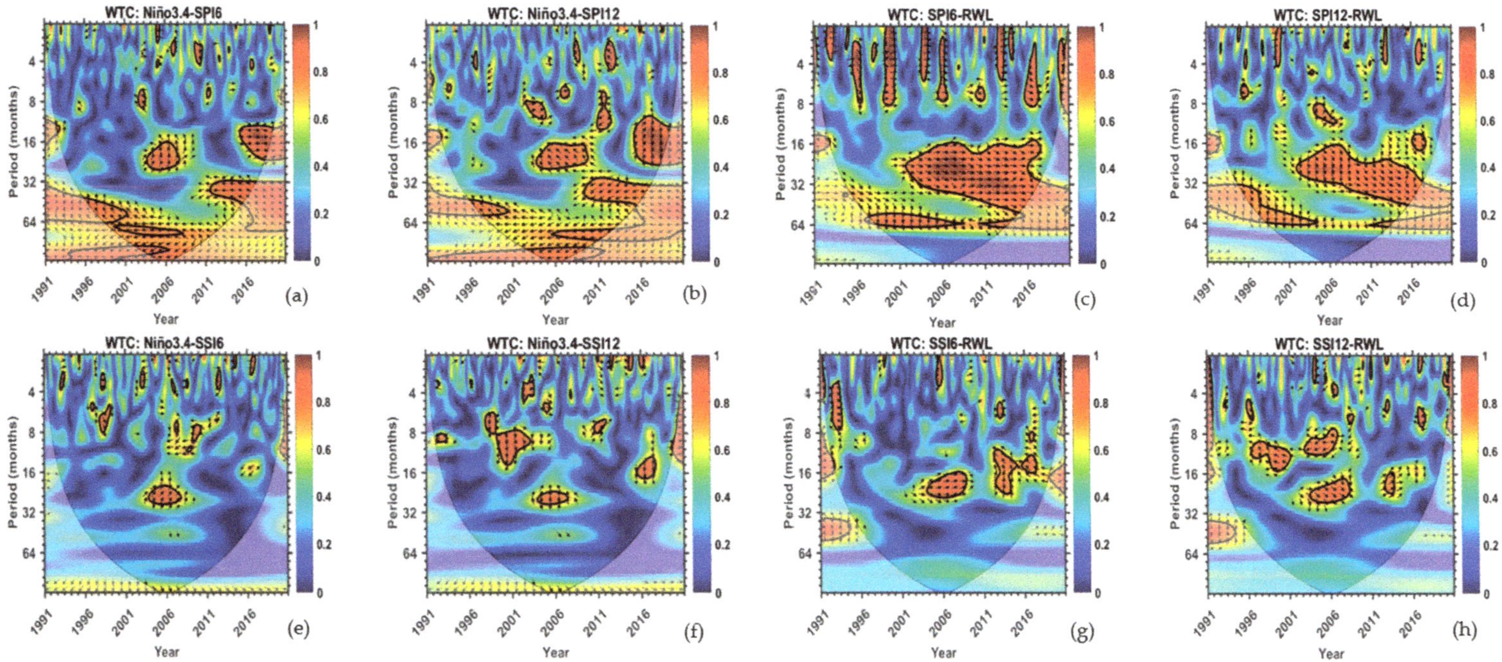

Figure 8. Wavelet coherence between (**a**) Niño 3.4 and SPI12, (**b**) Niño 3.4 and SPI12, (**c**) SPI6 and RWLs, (**d**) SPI12 and RWLs, (**e**) Niño 3.4 and SSI6, (**f**) Niño 3.4 and SSI12, (**g**) SSI6 and RWLs, and (**h**) SSI12 and RWLs. The phase relationship is represented by arrows (in-phase →; anti-phase ←). The black solid line is the cone of influence and indicates a 95% confidence level; contours delimit the region where there is a strong correlation.

4. Discussion

Using SPI and SSI as drought quantifying parameters, this study evaluated the relationship between El Niño-induced drought and reservoir water level. Both indices detected that a majority of drought events were reported to have occurred over the region at the 6th and 12th timescales and these were classified at different categories. From the findings of the study, high streamflow drought duration was observed between 2010 and 2020 and this coincided with one of the strongest El Niño event experiences in the southern hemisphere. Furthermore, as this is the longest duration observed on the streamflow index, this will have convoluted effects on the amount of water making its way to the reservoir. It should be noted that, in addition to natural controls, anthropogenic activities, e.g., reservoir operations and abstractions, can influence streamflow drought event characteristics [56]. However, the authors of [57] suggested different approaches to investigate human influences on streamflow drought. One such approach is the straightforward comparison of meteorological droughts with streamflow droughts as demonstrated in this study. Although studies such as [58,59] critiqued this approach, this study considered both meteorological and hydrological relationship deficits at different timescales. Further statistical analyses were employed to demonstrate the usefulness and effectiveness of the applied approach.

Over the considered study period, trends of both streamflow and rainfall drought show a negative linear trend and this is further supported by the SQMK findings which shows a negative progressive trend at both timescales for SPI and SSI. The general relationship shown in Figure 5 indicates a proportional relationship between the drought indicators and reservoir water level. This relationship is supported by [60] which looks at the "Day zero" phenomenon in Cape Town; the study reported a significant drop in the storage of reservoirs supplying water to 3.7 million people in the metropolitan area. This drop was the result of a three-year rainfall deficit, which resulted in the 2015–2017 drought in southwestern South Africa. The study findings further show a correlation between reservoir water levels and the drought quantifying parameters. With the coefficient of determination showing a higher correlation at the 12th timescale, and the 6th timescale of the SPI showing the least correction. Wavelet coherence analysis showed a significant relationship between reservoir water levels and SPI at both timescales, while the same was not the case for SSI. This strong relationship between SPI and reservoir water levels was observed between 2006 and 2016 in the region of the 20th and 40th months. Moderate-to-strong correlations between hydrological and climatic drought indices point to other factors having an effect on the water levels [61]. Such factors may be attributed to anthropogenic abstraction such as water reservoir abstractions for domestic and agricultural demand. Due to reservoir water levels dropping significantly during the major drought periods, this is an indication of the influence which the phenomenon has on water resource availability. This is evident from the wavelet coherence analysis, which showed that the Niño 3.4 time series leads to streamflow drought, which may affect surface water availability. This is therefore a further indication that during drought events, there is a deficit in rainfall, which affects the amount of water flowing into the reservoirs. This may further be exacerbated by anthropogenic drought. Anthropogenic drought is the compound multidimensional and multiscale phenomenon governed by a combination of factors (i.e., natural water variability, climate change, human decision and activities, and altered micro-climate conditions due to changes in land and water management) [62].

5. Conclusions

This study explored the impacts of drought on reservoir water resources in the Albasini Dam Catchment in the Levubu Valley. The study findings showed a positive relationship between rainfall and streamflow drought with reservoir water levels. A further decreasing trend of streamflow and rainfall was observed. These findings have an implication of surface water availability and sustainability in this area, where there is a critical need for agricultural water demand. In addition to the observed decreasing trends of hydroclimatic variables, together with unpredictable impacts of climate change on rainfall and drought magnitudes and frequencies, the implication on water resources is eminent. To ensure the sustainable management of water resources in this area to circumvent future shortages, management strategies should be improved. These improved management strategies should aim to minimise drought impacts with the development of adaptation plans which consider the multidimensional and multiscale nature of drought.

Author Contributions: Conceptualization, F.I.M. and L.M.; methodology, F.I.M., L.M. and N.M.; software, N.M.; formal analysis, F.I.M., L.M., S.M. and N.M.; data curation, N.M.; writing—original draft preparation, F.I.M.; writing—review and editing F.I.M., L.M., S.M. and N.M.; visualization, N.M. All authors have read and agreed to the published version of the manuscript.

Funding: This research received no external funding.

Institutional Review Board Statement: Not applicable.

Informed Consent Statement: Not applicable.

Data Availability Statement: All data used in the study are freely available from the South Africa Department of Water and Sanitation Hydrological Services, https://www.dws.gov.za/Hydrology/, accessed on 15 July 2021.

Acknowledgments: Authors acknowledge the Department of Water and Sanitation in South Africa for freely providing data used in this study.

Conflicts of Interest: The authors declare no conflicts of interest.

References

1. Govender, R.L.; Grab, S.W. Assessing the impact of El Niño-Southern Oscillation on South African temperatures during Austral summer. *Int. J. Climatol.* **2018**, *39*, 143–156. [CrossRef]
2. World Meteorological Organisation. *El Niño/Southern Oscillation*; WMO: Geneva, Switzerland, 2014; Volume 1145, pp. 2–4. Available online: http://www.wmo.int/pages/prog/wcp/wcasp/documents/JN142122_WMO1145_EN_web.pdf (accessed on 27 June 2016).
3. Bartholomew, H.; Jin, S. ENSO effects on land skin temperature variations; A global study from satellite remote sensing and NCEP/NCAR reanalysis. *Climate* **2013**, *1*, 53–73. [CrossRef]
4. Davey, M.K.; Brookshaw, A.; Ineson, S. The probability of the impact of ENSO on precipitation and near-surface temperature. *Clim. Risk Manag.* **2014**, *1*, 5–24. [CrossRef]
5. Ibebuchi, C.C. Revisiting the 1992 severe drought episode in South Africa: The role of El Niño in the anomalies of atmospheric circulation types in Africa south of the equator. *Theor. Appl. Climatol.* **2021**, *146*, 723–740. [CrossRef]
6. Mahlalela, P.T.; Blamey, R.C.; Hart, N.C.G.; Reason, C.J.C. Drought in the Eastern Cape region of South Africa and trends in rainfall characteristics. *Clim. Dyn.* **2020**, *55*, 2743–2759. [CrossRef] [PubMed]
7. Mukheibir, P.; Sparks, D. *Water Resources Management Strategies in Response to Climate Change in South Africa, Drawing on the Analysis of Coping Strategies Adopted by Vulnerable Communities in the Northern Cape Province of South. Africa in Times of Climate Variability*; WRC Report No. 1500/1/06, 398; Water Research Commission: Pretoria, South Africa, 2006.
8. Banholzersand, D.S. The Influence of different El Niño types on global average temperature. *Geophys. Res. Lett.* **2014**, *41*, 2093–2099. [CrossRef]
9. Cai, W.; Borlace, S.; Lengaigne, M.; van Rensch, P.; Collins, M.; Vecchi, G.; Timmermann, A.; Santoso, A.; McPhaden, M.J.; Wu, L.; et al. Increasing frequency of extreme El Niño events due to greenhouse warming. *Nat. Clim. Chang.* **2014**, *4*, 111–116. [CrossRef]
10. Chowdary, J.S.; John, N.; Gnanaseelan, C. Interannual variability of surface air-temperature over India: Impact of ENSO and Indian Ocean Sea surface temperature. *Int. J. Climatol.* **2014**, *34*, 416–429. [CrossRef]
11. Halpert, M.S.; Ropelewski, C.F. Surface temperature pattern associated with Southern Oscillation. *J. Clim.* **1992**, *5*, 577–593. [CrossRef]
12. Gray, W.M.; Sheaffer, J.D. *El Niño and QBO influences on tropical cyclone activity, In Teleconnections Linking Worldwide Climate Anomalies: Scientific Basis and Social Impacts*; Glantz, M.H., Katz, R.W., Nicholls, N., Eds.; Cambridge University Press: Cambridge, UK, 1991.
13. Pinault, J.L. The anticipation of the ENSO: What resonantly forced baroclinic waves can teach us (Part II). *J. Mar. Sci. Eng.* **2018**, *6*, 63. [CrossRef]
14. Reason, C.J.C.; Jagadheesha, D. A model investigation of recent ENSO impacts over southern Africa. *Meteorol. Atmos. Phys.* **2005**, *89*, 181–205. [CrossRef]

15. Basson, M.S. Water development in South Africa. In Proceedings of the UN-Water International Conference, Zaragoza, Spain, 3–5 October 2011; Available online: http://www.un.org/waterforlifedecade/green_economy_2011/pdf/session_1_economic_instruments_south_africa.pdf (accessed on 20 June 2022).
16. Basson, M.; Van Rooyen, J. Practical application of probabilistic approaches to the management of water resource systems. *J. Hydrol.* **2001**, *241*, 53–61. [CrossRef]
17. International Water Management Institute (IWMI); Agricultural Research Council (ARC). *Limpopo Basin Profile*; CGIAR Challenge Program on Water and Food: Pretoria, South Africa, 2003; p. 132.
18. Mazibuko, S.M.; Mukwada, G.; Moeletsi, M.E. Assessing the frequency of drought/flood severity in the Luvuvhu River catchment, Limpopo Province, South Africa. *Water SA* **2021**, *47*, 172–184. [CrossRef]
19. Mathivha, F.I.; Sigauke, C.; Chikoore, H.; Odiyo, J.O. Short-term and medium-term drought forecasting using generalised additive models. *Sustainability* **2020**, *12*, 6. [CrossRef]
20. Masupha, T.E.; Moeletsi, M.E. Analysis of potential future droughts limiting maize production, in the Luvuvhu River catchment area, South Africa. *Phys. Chem. Earth* **2018**, *105*, 44–51. [CrossRef]
21. Mpandeli, N.S.; Maponya, P.I. Coping with climate variability in Limpopo Province, South Africa. *Peak J. Agric. Sci.* **2013**, *1*, 54–64.
22. Kabanda, T.A. Climatology of Long-Term Drought in the Northern Region of the Limpopo Province of South Africa. Ph.D. Thesis, University of Venda, Thohoyandou, South Africa, 2004.
23. Tilahun, S.; Demeke, K. The influence of El Niño-induced drought on cyanobacterial community structure in a shallow tropical reservoir (Koka Reservoir, Ethiopia). *Aquat. Ecol.* **2019**, *53*, 61–77. [CrossRef]
24. Ainembabazi, J.H. The 2015-16 El Niño-Induced Drought Crisis in Southern Africa: What Do We Learn from Historical Data? In Proceedings of the 2018 Conference, Vancouver, BC, Canada, 28 July–2 August 2018; International Association of Agricultural Economists: Vancouver, BC, Canada, 2018.
25. Sprouse, T.W.; Vaughan, L.F. Water Resource Management in Response to El Niño/Southern Oscillation (ENSO) Droughts and Floods. In *Climate and Water. Advances in Global Change Research*; Diaz, H.F., Morehouse, B.J., Eds.; Springer: Dordrecht, The Netherlands, 2003. [CrossRef]
26. Naumann, G.; Alfieri, L.; Wyser, K.; Mentaschi, L.; Betts, R.A.; Carrao, H.; Spinoni, J.; Vogt, J.; Feyen, L. Global changes in drought conditions under different levels of warming. *Geophys. Res. Lett.* **2018**, *45*, 3285–3296. [CrossRef]
27. Gizaw, M.S.; Gan, T.Y. Impact of climate change and El Niño episodes on droughts in sub-Saharan Africa. *Clim. Dyn.* **2017**, *49*, 665–682. [CrossRef]
28. Dagada, K. Influence of Climate Change on Flood and Drought Cycles and Implications on Rainy Season Characteristics in Luvuvhu River Catchment, South Africa. Master's Thesis, University of Venda, Thohoyandou, South Africa, 2016.
29. Mokgoebo, J.M.; Kabanda, T.A.; Gumbo, J.R. Assessment of the Raparian Vegetation Changes Downstream of Selected Dams in Vhembe District, Limpopo Province Base on Historical Aerial Photography. In *Environmental Risks*; Mihai, F., Grozavu, A., Eds.; IntechOpen: Rijeka, Croatia, 2018. [CrossRef]
30. WMO. *Standardized Precipitation Index: User Guide*; WMO: Geneva, Switzerland, 2012.
31. Mckee, T.B.; Doesken, N.J.; Kleist, J. The relationship of drought frequency and duration to time scales. In Proceedings of the 8th Conference on Applied Climatology, Anaheim, CA, USA, 17–22 January 1993; pp. 17–22.
32. Aksoy, H. Use of gamma distribution in hydrological analysis. *Turk. J. Eng. Environ. Sci.* **2000**, *24*, 419–428.
33. Tsakiris, G.; Loukas, A.; Pangalou, D.; Vangelis, H.; Tigkas, D.; Rossi, G.; Cancelliere, A. Drought characterization [Part 1. Components of drought planning. 1.3. Methodological component]. In *Drought Management Guidelines Technical Annex*; CIHEAM: Zaragoza, Spain, 2007; pp. 85–102.
34. Shamshirbad, S.; Hashemi, S.; Salimi, H.; Samadianfard, S.; Asadi, E.; Shadkani, S.; Kargar, K.; Mosavi, A.; Nabipour, N.; Chau, K.-W. Predicting Standardised Streamflow Index for Hydrological Drought using machine learning Models. *Eng. Appl. Comput. Fluid Mech.* **2020**, *14*, 342–343. [CrossRef]
35. Tan, C.; Yang, J.; Li, M. Temporal-Spatial variation of drought indicated by SPI and SPEI in Ningxia Hui Autonomous Region, China. *Atmosphere* **2015**, *6*, 1399–1421. [CrossRef]
36. Shah, R.; Bharadiya, N.; Manekar, V. Drought Index Computation Using Standardised Precipitation Index (SPI) method for Surat District, Gujarat. *Aquat. Procedia* **2015**, *4*, 1243–1249. [CrossRef]
37. Guttman, N.B. Accepting the Standardized Precipitation Index: A Calculation Algorithm. *J. Am. Water Resour. Assoc.* **1999**, *35*, 311–322. [CrossRef]
38. Mann, H.B. Non-parametric tests against trend. *Econometrica* **1945**, *13*, 245–259. [CrossRef]
39. Kendall, M.A.; Stuart, A. *The Advanced Theory of Statistics*, 2nd ed.; Charles Griffin: London, UK, 1967.
40. Zhang, X.; Lucie, A.V.; Hogg, W.D.; Niitsoo, A. Temperature and precipitation trends in Canada during the 20th century. *Atmosphere* **2000**, *38*, 395–429. [CrossRef]
41. Sneyers, S. *On the Statistical Analysis of Series of Observations*; Technical note no. 143, WMO No. 725 415; Secretariat of the World Meteorological Organization: Geneva, Switzerland, 1990; p. 192.
42. Jain, V.K.; Rivera, L.; Zaman, K.; Espos, R.A.; Sirivichayakul, C.; Quiambao, B.P.; Rivera-Medina, D.M.; Kerdpanich, P.; Ceyhan, M.; Ener, C.; et al. Vaccine for prevention of mild and moderate-to-severe influenza in children. *N. Engl. J. Med.* **2013**, *369*, 2481–2491. [CrossRef]

43. Aladwani, J. Wavelet Coherence and Continuous Wavelet Transform-Implementation and Application to the Relationship between Exchange Rate and Oil Price for Importing and Exporting Countries. *Int. J. Energy Econ. Policy* **2023**, *13*, 531–541. [CrossRef]
44. Peng, J.; Qiao, R.; Liu, Y.; Blaschke, T.; Li, S.; Wu, J.; Xu, Z.; Liu, Q. A wavelet coherence approach to prioritizing influencing factors of land surface temperature and associated research scales. *Remote Sens. Environ.* **2020**, *246*, 111866. [CrossRef]
45. Torrence, C.; Compo, G.P. A Practical Guide to Wavelet Analysis: With Significance and Confidence Testing. University of Colorado at Boulder, Program in Atmospheric and Oceanic Sciences. 2011. Available online: https://atoc.colorado.edu/research/wavelets/ (accessed on 29 September 2022).
46. Restrepo, J.C.; Higgins, A.; Escobar, J.; Ospino, S.; Hoyos, N. Contribution of low-frequency climatic–oceanic oscillations to streamflow variability in small, coastal rivers of the Sierra Nevada de Santa Marta (Colombia). *Hydrol. Earth Syst. Sci.* **2019**, *23*, 2379–2400. [CrossRef]
47. Baghanam, A.H.; Nourani, V.; Norouzi, E.; Vakili, A.T.; Gökçekuş, H. Application of Wavelet Transform for Bias Correction and Predictor Screening of Climate Data. *Sustainability* **2023**, *15*, 15209. [CrossRef]
48. Grinsted, A.; Moore, J.C.; Jevrejeva, S. Application of the cross wavelet transform and wavelet coherence to geophysical time series. *Nonlin. Processes Geophys.* **2004**, *11*, 561–566. [CrossRef]
49. Torrence, C.; Compo, G.P. A practical guide to wavelet analysis. *Bull. Am. Meteorol. Soc.* **1998**, *79*, 61–78. [CrossRef]
50. Kumar, R.; Musuuza, J.L.; Van Loon, A.F.; Teuling, A.J.; Barthel, R.; Ten Broek, J.; Mai, J.; Samaniego, L.; Attinger, S. Multiscale evaluation of the standardized precipitation index as a groundwater drought indicator. *Hydrol. Earth Syst. Sci.* **2015**, *12*, 7405–7436. [CrossRef]
51. Archer, E.R.M.; Landman, W.A.; Tadross, M.A.; Marumbwa, F.M. Understanding the evolution of 2014–2016 summer rainfall seasons in Southern Africa. *Clim. Risk Manag.* **2017**, *16*, 22–28. [CrossRef]
52. Rojas, O.; Li, Y.; Cumani, R. *Understanding the Drought Impact of El Niño on Global Agricultural Areas: An Assessment Using FAO's Agricultural Stress Index (ASI)*; FAO: Rome, Italy, 2014; pp. 9–23.
53. Schober, P.; Boer, C. Correlation Coefficients: Appropriate Use and Interpretation. *Anesth. Analg.* **2018**, *126*, 1764–1765. [CrossRef]
54. Loua, R.T.; Bencherif, H.; Mbatha, N.; Bègue, N.; Hauchecorne, A.; Bamba, Z.; Sivakumar, V. Study on Temporal Variations of Surface Temperature and Rainfall at Conakry Airport, Guinea: 1960–2016. *Climate* **2019**, *7*, 93. [CrossRef]
55. Chang, C.; Glover, G.H. Time-frequency dynamics of resting-state brain connectivity measured with fMRI. *NeuroImage* **2010**, *50*, 81–98. [CrossRef]
56. Tijdeman, E.; Hannaford, J.; Stahl, K. Human influences on streamflow drought characteristics in England and Wales. *Hydrol. Earth Syst. Sci.* **2018**, *22*, 1051–1064. [CrossRef]
57. Van Loon, A.F.; Stahl, K.; Di Baldassarre, G.; Clark, J.; Rangecroft, S.; Wanders, N.; Gleeson, T.; Van Dijk, A.I.J.M.; Tallaksen, L.M.; Hannaford, J.; et al. Drought in a humanmodified world: Reframing drought definitions, understanding, and analysis approaches. *Hydrol. Earth Syst. Sci.* **2016**, *20*, 3631–3650. [CrossRef]
58. Haslinger, K.; Koffler, D.; Schöner, W.; Laaha, G. Exploring the link between meteorological drought and streamflow: Effects of climate-catchment interaction. *Water Resour. Res.* **2014**, *50*, 2468–2487. [CrossRef]
59. Barker, L.J.; Hannaford, J.; Chiverton, A.; Svensson, C. From meteorological to hydrological drought using standardised indicators, Hydrol. *Earth Syst. Sci.* **2016**, *20*, 2483–2505. [CrossRef]
60. Pascale, S.; Kapnick, S.B.; Delworth, T.L.; Cooke, W.F. Increasing risk of another Cape Town "Day Zero" drought in the 21st century. *Proc. Natl. Acad. Sci. USA* **2020**, *117*, 29495–29503. [CrossRef]
61. Saase, R.; Schütt, B.; Bebermeier, W. Analyzing the Dependence of Major Tanks in the Headwaters of the Aruvi Aru Catchment on Precipitation. Applying Drought Indices to Meteorological and Hydrological Data. *Water* **2020**, *12*, 2941. [CrossRef]
62. AghaKouchak, A.; Mirchi, A.; Madani, K.; Di Baldassarre, G.; Nazemi, A.; Alborzi, A.; Anjileli, H.; Azarderakhsh, M.; Chiang, F.; Hassanzadeh, E.; et al. Anthropogenic Drought: Definition, Challenges and Opportunities. *Rev. Geophys.* **2021**, *59*, e2019RG000683. [CrossRef]

atmosphere

MDPI

Article

Research on the Terrain Characteristics of Changbai Mountain and Their Impact on Precipitation and Wind Distribution

Li Liang [1], Wanxiu Ai [2,*], Xiaodan Yang [1] and Luqiang Zhao [1]

[1] Public Meteorological Service Centre, China Meteorological Administration, Beijing 100081, China; liangli@cma.gov.cn (L.L.); yangxd@cma.gov.cn (X.Y.); zhaolq@cma.gov.cn (L.Z.)
[2] National Climate Centre, China Meteorological Administration, Beijing 100081, China
* Correspondence: aiwx@cma.gov.cn

Abstract: The terrain of Changbai Mountain has great influence on the distribution of atmospheric flows and the occurrence and development of precipitation. However, quantitative studies on the real terrain characteristics and the terrain effect on precipitation distribution in this region are scant at present. This study quantitatively analyzes the regional characteristic of topographic perturbations and the relationship between terrain, wind, and precipitation in Changbai Mountain region by using a spectral analysis of the two-dimensional discrete cosine transform. Three domains with relatively heavy summer precipitation are selected as the study region. The results indicate that the overall terrain of the Changbai Mountain region exhibits anisotropic characteristics. The terrain spectra of domain B are less than those of domains A and C across the whole wavelength (λ) bands, indicating that the large-scale topographic perturbations of domain B are relatively weak. The largest topographic spectral peak of domain C shows the most pronounced undulation of terrain among the three domains. The dominant wavelengths of terrain height variance for domains A and C, both close to the respective maximum wavelengths, indicate more prominent large-scale topographic perturbations. For domain A, the variation of the precipitation spectra is consistent with that of the wind spectra at the wavelength bands of λ < 390 km, showing a high correlation between wind field and the occurrence of rainfall. The inverse relationship at larger wavelengths indicates that multiple factors contribute to the occurrence of rainfall. For domain B, there is consistency in the fluctuations of terrain spectra, precipitation spectra, and wind spectra at the wavelength bands of λ < 278.3 km, implying that the smaller-scale terrain has an important effect on the occurrence of summer precipitation. For domain C, the variations of terrain spectra, precipitation spectra, and wind spectra are almost consistent across the whole wavelength bands, indicating that the large-scale terrain and minor terrain both play a crucial role in atmospheric uplift and the occurrence and development of summer rainfall.

Keywords: Changbai Mountain; two-dimensional discrete cosine transform; terrain spectra; wind spectra; precipitation spectra; additive synthesis; the dominant wavelength

1. Introduction

Rainstorms are common disastrous weather events in China, which have caused serious losses of people's lives and property. In recent years, extreme rainstorm events have gradually increased in the context of global warming [1–3]. With the development of observational equipment and numerical models, scholars have made significant progress in understanding the formation mechanisms of rainstorms. Forecast accuracy of rainstorms, especially regional organized systematic rainstorms, has also been improved to a certain extent [4–8]. However, it is still a universal problem to predict the occurrence and intensity of rainstorms over complex terrain, which plays a very important role in triggering precipitation and affecting its spatial and temporal distribution. Major mountain barriers can significantly modulate atmospheric flows and precipitation over mountainous areas through dynamic processes, such as upslope ascent, leeside descent, and associated gravity wave activities [9–17]. A rough terrain can lead to increased turbulence near the ground due to the strong disturbance of the terrain to the air flows and the complex flow pattern generated by the terrain. The wind speed and its direction can also be changed within a rough terrain because of the topographic obstruction and relief. In some cases, rough terrain can cause air currents to separate, creating vortices and backflows. These flow features can cause localized climate and weather phenomena.

As a topographic rainstorm usually shows characteristics of suddenness, small coverage, and short duration, it is difficult to give an accurate early warning and prediction. Therefore, it often leads to mountain torrents, debris flows, landslides, and other meteorological secondary disasters. Megacities built near mountains are vulnerable to natural disasters. Rainstorms and other disastrous weather could well cause serious losses of local people's lives, property, and economy. For example, an extreme weather event occurred in northeast China on 13 July 2017, with the peak precipitation reaching 296 mm. The extensive rainfall accompanied by severe convection weather that included hailstones, thunderstorms, and strong winds caused flooding damage and other geological disasters, such as debris flows and landslides, in some areas. The rainstorm affected nearly one million people and resulted in the direct economic loss of several hundred million yuan [18,19].

Most of the rainstorm events in northeast China are closely related to complex terrain [20–23]. The Changbai Mountain region stretches for thousands of kilometers from southwest to northeast, across the east of Jilin, Liaoning, and Heilongjiang Provinces, with its chief part located in the southeast of Jilin Province. It is the common birthplace of Tumen River, Yalu River, and Songhua River. With unique geographical location, it is a weather- and climate-sensitive area in northeast China, of which the central part is located in the transitional windward zone between the Changbai Mountain region and the Songliao Plain. The semi-mountainous terrain is an area where meso- and micro-scale systems occur frequently. The local severe weather is the result of the combination of terrain and system development. Therefore, a systematic study on the topography of the Changbai Mountain region is required for accurate forecast of rainstorms over the complex terrain.

Pielke and Kennedy (1980) [24] suggested that terrain characteristics can be truly represented in spectrum space, and terrain spectra can quantitatively reflect the effect of topographic dynamic forcing on the atmosphere. Young and Pielke (1983) [25] and Young et al. (1984) [26] found that the spectral analysis of terrain cross-sections can be used to investigate the terrain spectra. Since then, the one-dimensional operation along several adjacent terrain sections has been extensively used for high computational efficiency [25–28]. Steyn and Ayotte (1985) [29] proposed that if the terrain of a given area is directional, the one-dimensional spectral analysis method may not be accurate enough to deal with the terrain. Salvador et al. (1999) [30] determined a grid size of 2 km for meso-scale models of the east coast of Spain based on a two-dimensional spectral analysis. However, it is computationally time-consuming to calculate the two-dimensional terrain spectra for a larger study region, in spite of the rapid development of computational resources [31]. Wang and Wang (2004) [32] revealed the physical mechanism of the topographic effect on the westerly perturbation over central and eastern China using the one-dimensional weighted-average spatial spectral method, which was also used to study the mechanism of terrain forcing on precipitation distribution in the Tibet Plateau by Shu et al. (2006) [33].

However, the one-dimensional or two-dimensional spectral analysis methods mentioned above all used traditional discrete Fourier transform (DFT). Denis et al. (2002) [34] proposed that spectral analysis of atmospheric fields on limited-area grids using Fourier transforms can lead to the aliasing or projection of the large-scale trend on the high-wavenumber components, though the DFT is well suited for global atmosphere. The two-dimensional discrete cosine transform (2D-DCT) is suitable for spectral analysis of data over a limited area. Since then, the DCT has been widely applied in spectral analysis of atmospheric fields on limited-area grids [35–39]. For example, Huang and Cui (2016) [39] used 2D-DCT to decompose the terrain height field and torrential rainfall distribution field in Sichuan, China.

Although the terrain of Changbai Mountain plays an important role in atmospheric flows and occurrence and development of rainfall, there is currently a lack of quantitative research on the real terrain characteristics and the terrain effect on precipitation distribution in this region. The aim of this study was to quantitatively explore the effect of topographic dynamic forcing on precipitation distribution in the Changbai Mountain region by using 2D-DCT. The regional characteristic of topographic perturbations can be accurately depicted based on the terrain characteristics represented in spectrum space. The relationships among terrain dynamic forcing, wind and precipitation in spectrum space, and the possible reasons for the high incidence of local rainstorms were also investigated by analyzing the distributions of terrain spectra, wind spectra, and precipitation spectra. In this paper, the data and the spectral analysis method are presented in Section 2. The precipitation distribution characteristics, the terrain spectra of the Changbai Mountain region, and spectral analysis of terrain, precipitation, and wind field are investigated in Section 3. The discussion and conclusions are summarized in Section 4.

2. Data and Methods

2.1. Data

The latest Shuttle Radar Topographic Mission (SRTM) 90 m digital elevation database, originally produced by National Aeronautics and Space Administration (NASA), was used for the algorithm of terrain spectra. The observed precipitation was from the merged hourly precipitation product, which had a horizontal resolution of $0.1^\circ \times 0.1^\circ$. It is based on the optimum interpolation technique through the combination of the hourly precipitation observed by automatic weather stations and retrieved from Climate Prediction Center Morphing (CMORPH) technique satellite data [40]. The wind data at 850 hPa were from the European Centre for Medium-Range Weather Forecasts (ECMWF) global data, which had a horizontal resolution of $0.125^\circ \times 0.125^\circ$. Data of wind and precipitation from June to September of each year from 2008 to 2022 were used in this study.

2.2. Methods

Spectral decomposition of terrain height field, wind field at 850 hPa, and precipitation field in the Changbai Mountain domain was determined by the algorithm of 2D-DCT. The DCT was first applied to atmospheric spectral analysis by Denis et al. (2002) [34]. In the spectral analysis of atmospheric fields in limited-area domains, fields on the grids are generally aperiodic and dominated by large-scale factors whose wavelengths are greater than the domain in most cases. Trying to use a standard Fourier transform on limited-area domains can lead to the aliasing of the large-scale trend on the high-wavenumber components, thus destroying all usefulness of spectra at high wavenumbers. The DCT is particularly well adapted for spectral analysis of atmospheric fields on a limited area, comparing favorably with the periodic Fourier transform. It does not need any trend removal or any other prior modification before applying the DCT and can overcome the problem of the aliasing of large-scale variance into shorter scales [39].

An algorithm of spectral computation has been coded, based on Denis et al.'s (2002) study, which used the 2D-DCT. For a two-dimensional physical field $h(i,j)$ of N_i by N_j grid points, the direct and inverse DCT are respectively defined as

$$F(m,n) = \beta(m)\beta(n)\sum_{i=0}^{N_i-1}\sum_{j=0}^{N_j-1} h(i,j)\times\cos(\pi m\frac{i+1/2}{N_i})\cos(\pi n\frac{j+1/2}{N_j}) \tag{1}$$

and

$$h(i,j) = \sum_{m=0}^{N_i-1}\sum_{n=0}^{N_j-1} \beta(m)\beta(n)F(m,n)\times\cos(\pi m\frac{i+1/2}{N_i})\cos(\pi n\frac{j+1/2}{N_j}) \tag{2}$$

with $\beta(m) = \begin{cases} \sqrt{1/N_i} & m=0 \\ \sqrt{2/N_i} & m=1,2,\ldots,N_i-1 \end{cases}$ and $\beta(n) = \begin{cases} \sqrt{1/N_j} & n=0 \\ \sqrt{2/N_j} & n=1,2,\ldots,N_j-1 \end{cases}$.

Here, $F(m,n)$ is the spectral coefficient corresponding to the (m,n) adimensional wavenumbers. The spectral variance can be computed from the spectral coefficients:

$$\sigma^2(m,n) = F^2(m,n)/(N_i N_j) \tag{3}$$

In order to analyze the relationship between terrain height spectral variance, total precipitation spectral variance, and wind spectral variance, the two-dimensional spectral coefficients undergo additive synthesis to obtain the relationship between variance and total wavelength. Instead of adding the contribution $\sigma^2(m, n)$ only to wavenumber k as in Denis et al. (2002) [34], an alternative method is to assign this contribution between the two wavenumbers k and k +1 using the distance weight coefficients [38]. Different from the allocation method of variance in Ricard et al. (2013) [38], the allocation is adjusted so that the closer the distance, the stronger the weight. The weight coefficients are distributed into wavenumbers k and $k + 1$ as follows:

$$\begin{cases} a_{m,n} = \frac{\alpha'(k+1)-\alpha(m,n)}{\alpha'(k+1)-\alpha'(k)} \\ b_{m,n} = \frac{\alpha(m,n)-\alpha'(k)}{\alpha'(k+1)-\alpha'(k)} \end{cases} \tag{4}$$

In Equation (4), $\alpha(m, n)$ is the standardized two-dimensional wavenumber, and the specific meaning of and $\alpha'(k + 1)$ is $\alpha'(k)$, shown in Ricard et al. (2013) [38].

3. Results

3.1. Precipitation Distribution in Changbai Mountain Region and Selection of Study Region

Before spectral decomposition of terrain height field, wind field, and precipitation field, the precipitation in the Changbai Mountain region is analyzed first. Figure 1 shows the distribution of monthly average precipitation in the Changbai Mountain region from June to September during the period of 2008 to 2022. The precipitation in this region is concentrated from June to August, especially in August (Figure 1c), which is mainly distributed in the southwest of the Changbai Mountain region. For a better understanding of the relationship between the spatial distribution of precipitation and topography, the terrain height of the Changbai Mountain region and the total precipitation from June to August averaged over 2008–2022 in this area are represented in Figure 2. As shown in Figure 2b, the rainband presents a northeast–southwest distribution and the precipitation is mainly concentrated in the areas between 40° N and 42° N and between 47° N and 48° N with three main centers at 129° E, 47° N, 125° E, 41° N, and 127° E, 41.2° N, which have values of 450, 750, and 550 mm, respectively. The three precipitation centers, distributed in steep terrain areas, are located in the Xiaoxing'an Mountain area in the northwest of the Changbai Mountain region, the southwest part of the main Changbai Mountain range, and the eastern part of Jilin Province, respectively. Based on the distribution characteristics of precipitation, three regions, namely domain A with limits of 126° E–130° E, 45.5° N–49° N (rectangle A in Figure 2), domain B with limits of 122.4° E–126.2° E, 40° N–42.5° N (rectangle B in Figure 2), and domain C with limits of 126.2° E–129.4° E, 41.2° N–45° N (rectangle C in Figure 2), are selected for the spectral analysis.

The terrain height fields of the three study areas are shown in Figure 3. It can be seen that the western terrain of all three areas is relatively low, whereas the remaining areas show the towering and complex terrain with crisscrossing gullies. In Figure 3a, though domain A is relatively low-lying compared to the other two domains, there are still many high mountains and deep valleys and ravines, mainly distributed in the northeast of this region, which is conducive to the accumulation and uplift of water vapor, thus forming rainfall in this region. In Figure 3b, the mountain range presents a northeast–southwest orientation in domain B, of which the southern part shows a trumpet-shaped topography, which plays an essential role in the lifting of the low-level southwest air flow, thus resulting in rainfall mainly distributed in the trumpet-shaped area. In Figure 3c, the high mountains are mainly located in the southeast of domain C, where the northwest terrain is relatively flat. The terrain height increases gradually from the northwest to the southeast, and the precipitation is mainly distributed in the southern region of the complex terrain.

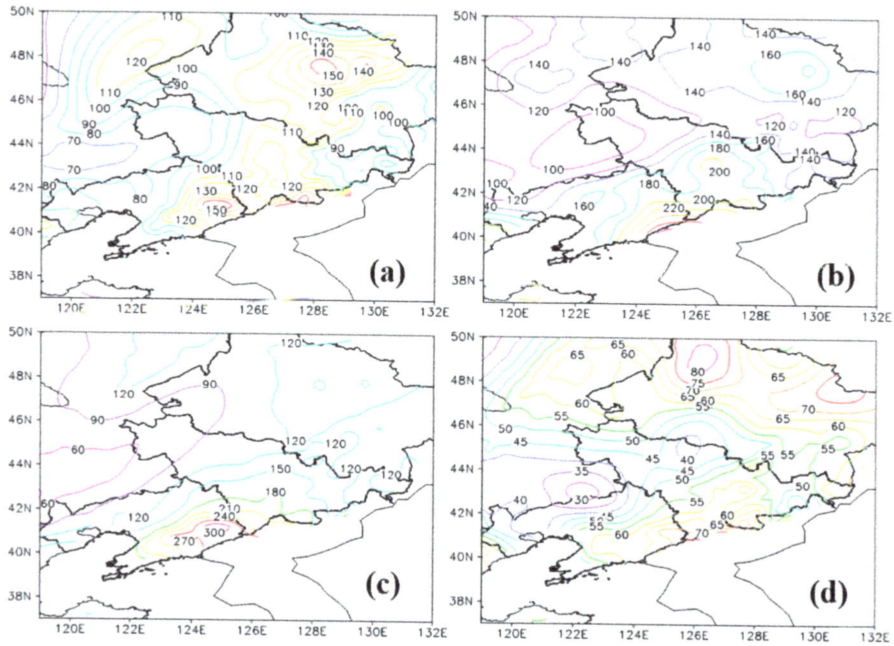

Figure 1. Distribution of monthly average precipitation in Changbai Mountain region from 2008 to 2022 (contour, unit: mm) in (**a**) June, (**b**) July, (**c**) August, (**d**) September.

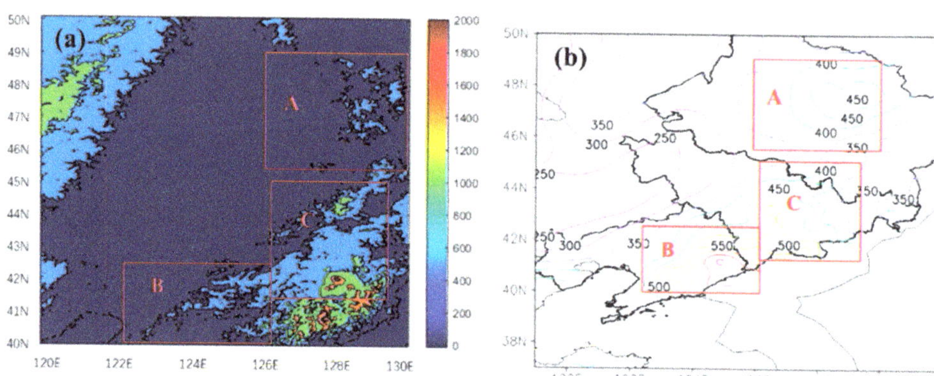

Figure 2. (**a**) Topographic height of Changbai Mountain region (shade, unit: m) and (**b**) Distribution of the total precipitation from June to August averaged over 2008–2022 in Changbai Mountain region (contour, unit: mm). The red rectangles indicate the three study areas (A, B, and C) in this study.

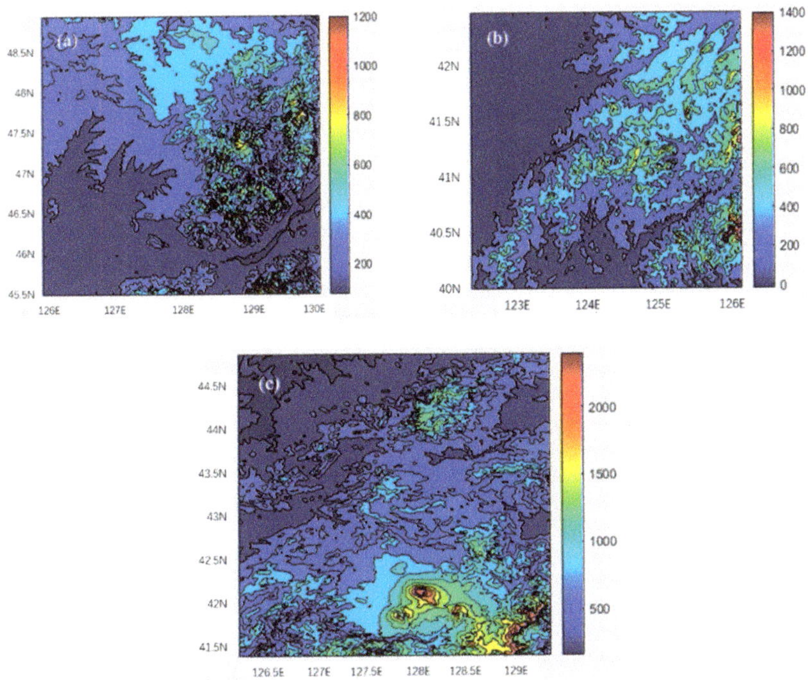

Figure 3. Terrain height fields in the study areas A (**a**), B (**b**), and C (**c**). The study area locations are plotted in Figure 2.

3.2. The Terrain Spectra of Changbai Mountain Region

The 2D-DCT routine is applied over the height values of the three study areas in the Changbai Mountain region, and the resulting spectra depict the distribution of terrain height variances in either wavenumber (k) or wavelength ($\lambda = 1/k$) (Figure 4). Terrain height variance spectra reveal the effect of topographic dynamic forcing on the atmosphere, and the greater the spectral energy then the stronger the topographic dynamic forcing [41]. As shown in Figure 4, the terrain height variance in each domain declines with increasing k, implying that the topographic dynamic forcing weakens with the gentleness of the topographic relief. In order to investigate whether there are isotropic features in each study area, the distributions of terrain height variance versus wavelength (km) with the wavenumber (0,0) as the starting point, and the angle (θ) between the rays (as shown in Figure 4) and the x-axes, are represented in Figures 5–7, respectively. Because of the discreteness of terrain height variance and the integer variables of the zonal and meridional wavenumbers, the angles meeting with tan θ= $1/4$, $1/2$, 1, 2, and 4 are used for analyzing. It can be seen that, along these angles in each study area, the longer wavelength presents a larger terrain height variance, and the relatively short wavelength corresponds to the smaller terrain height variance. The variance of the whole wavelength bands has an uptrend with increasing λ, in spite of several departures. Also, the majority of discrete points of terrain height variance are scattered at shorter wavelengths (λ < 1 km), under which the values of the variance are basically less than 0.001 m^2.

A power law relationship between the terrain height variance (σ^2) and wavelength λ in the form of $\sigma^2 = a\lambda^b$ is fitted using the least squares method, and the relevant parameters a, b and the correlation coefficient r can be obtained by the fitting algorithm. Using the F-criterion with a significance level of 0.05, parameters a and b are significant. The coefficient a, representing the terrain height variance when λ = 1 km, reflects the intensity of topographic forcing. The exponent b, which is the spectral slope of the fitting curve in the log–log space, indicates the terrain smoothness. A smaller b shows a greater terrain height variance in shorter wavelength bands and, videlicet, a more pronounced small-scale perturbation. The values of a, b, and r at the wavelength bands of $\lambda \geq 1$ km for the five sections of the three study areas are given in Table 1. It can be found that the values of a and b have considerable differences among these sections in each domain. The average fitting slopes b of the three study areas are 2.98, 2.70, and 2.91, respectively, indicating a pronounced perturbation of large-scale topography in domains A and C. The minimum b of 2.70 implies that the perturbation of small-scale topography is more significant in domain B. The exponent b varies from 2.70 to 2.98, indicating that the terrain of the Changbai Mountain region is complex and there is a difference in the terrain features of different areas, though domains B and C are adjacent.

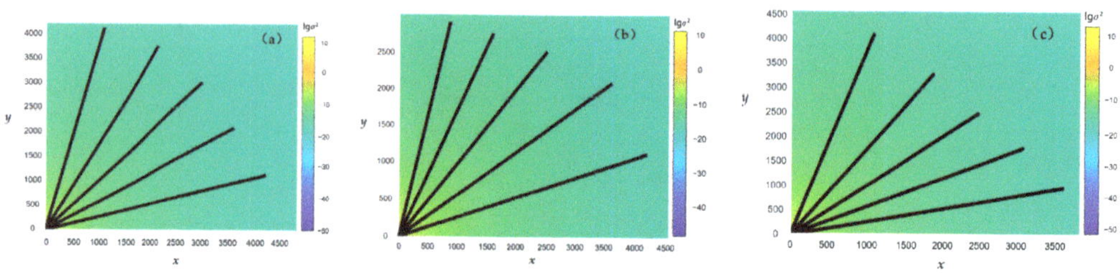

Figure 4. The logarithm distribution (lgσ^2) of two-dimensional terrain height spectral variance (σ^2) in domains A (**a**), B (**b**), and C (**c**). x and y represent the zonal and meridional wavenumber axes, respectively.

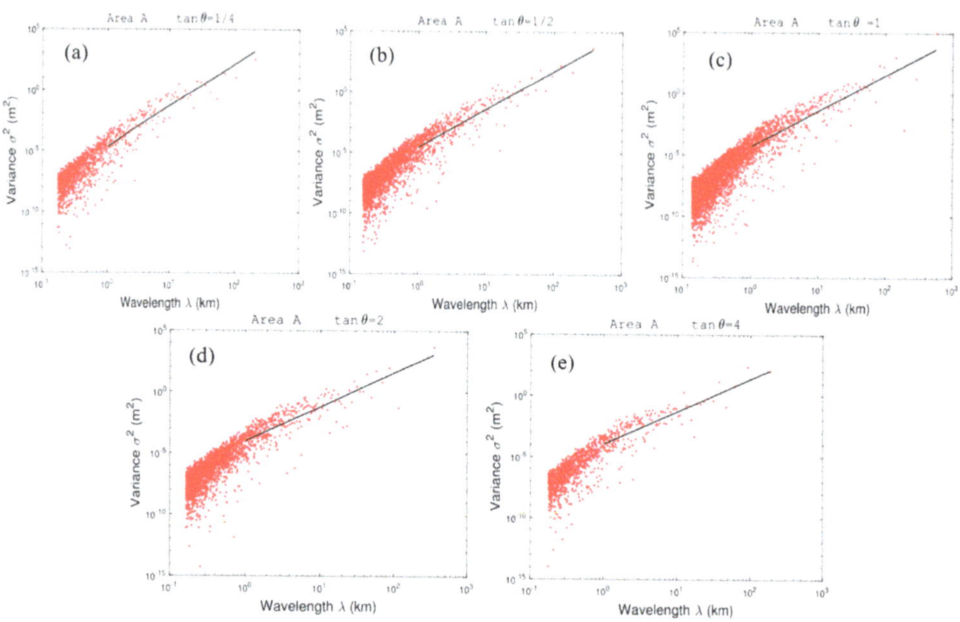

Figure 5. The distribution of terrain height variance (σ^2) versus wavelength (λ) with the wavenumber (0,0) as the starting point and the angle (θ) between the rays (as shown in Figure 4a) and the x-axes meeting with tanθ = 1/4 (**a**), 1/2 (**b**), 1 (**c**), 2 (**d**), and 4 (**e**), respectively, for domain A.

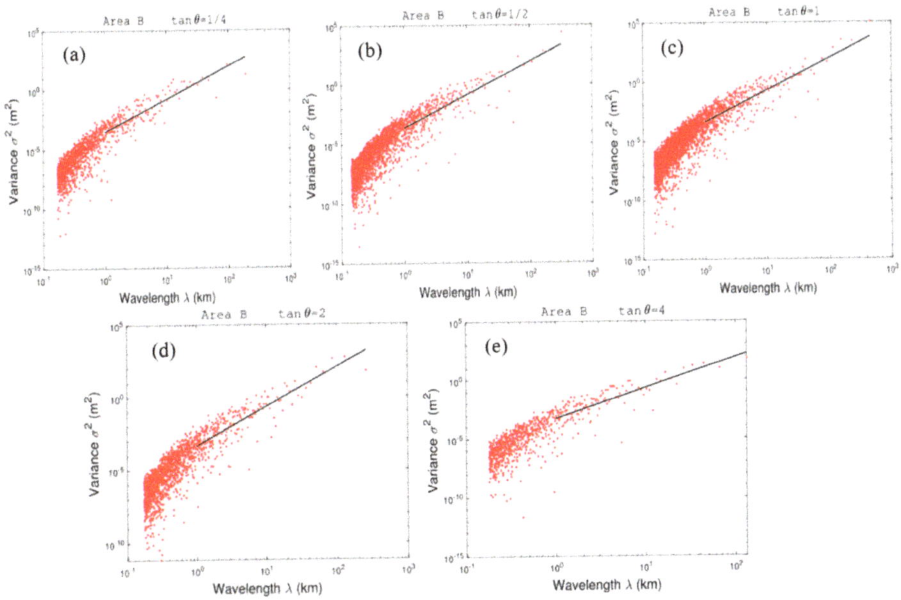

Figure 6. The distribution of terrain height variance (σ^2) versus λ with the wavenumber (0,0) as the starting point and the angle (θ) between the rays (as shown in Figure 4b) and the x-axes meeting with $\tan\theta = 1/4$ (**a**), $1/2$ (**b**), 1 (**c**), 2 (**d**), and 4 (**e**), respectively, for domain B.

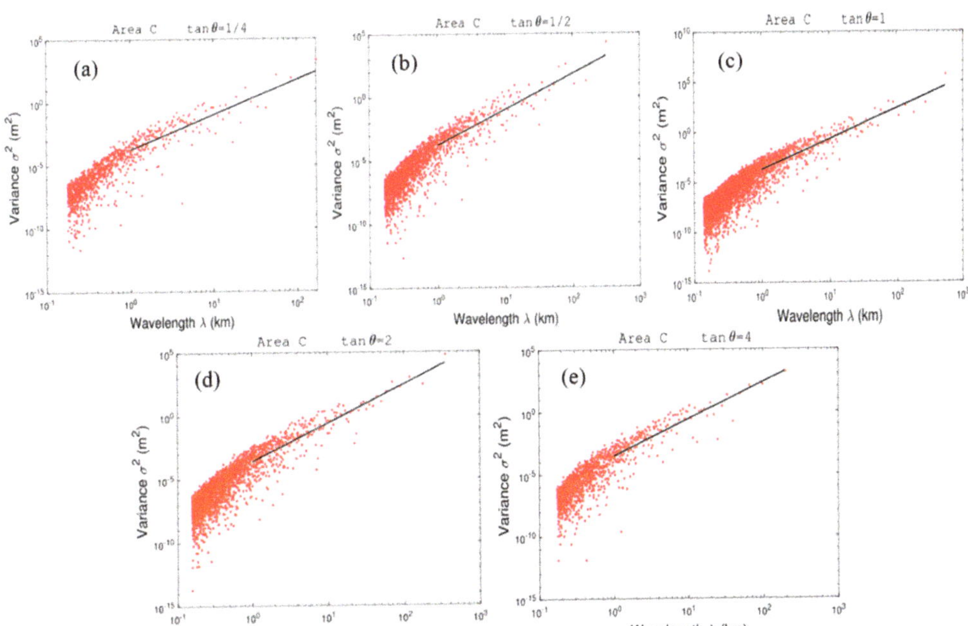

Figure 7. The distribution of terrain height variance (σ^2) versus wavelength λ with the wavenumber (0,0) as the starting point and the angle (θ) between the rays (as shown in Figure 4c) and the x-axes meeting with $\tan\theta = 1/4$ (**a**), $1/2$ (**b**), 1 (**c**), 2 (**d**), and 4 (**e**), respectively, for domain C.

Table 1. Parameters of the least square fitting relation $\sigma^2 = a\lambda^b$ ($\lambda \geq 1$ km) for the five sections in each domain.

Domain	tan θ	1/4	1/2	1	2	4	Average
A	$A(10^{-4})$	0.21	0.27	0.50	0.94	1.07	0.60
	b	3.40	3.14	2.90	2.79	2.66	2.98
	r	0.94	0.93	0.93	0.93	0.92	
B	$A(10^{-4})$	2.90	2.37	3.80	4.50	5.41	3.80
	b	2.75	2.78	2.70	2.70	2.59	2.70
	r	0.93	0.94	0.91	0.92	0.91	
C	$A(10^{-4})$	1.90	1.70	1.87	2.38	3.03	2.18
	b	2.74	2.77	3.03	3.04	2.95	2.91
	r	0.90	0.90	0.91	0.92	0.91	

The values of maximum and minimum b in each domain can be obtained from Table 1. The calculated ratios of the range of b (the difference between the maximum and minimum) to the average value of b for domains A, B, and C are 24.83%, 2.96%, and 10.31%, respectively. Similarly, the ratios of the range of a to the average a for domains A, B, and C are 143.33%, 80%, and 61.01%, respectively. Obviously, the ratios of 24.83% and 143.33%, corresponding to domain A, are the largest in the three study areas, which shows the most obvious anisotropy in the three domains. The minimum ratio of 2.96%, corresponding to domain B, indicates that the anisotropy in this region is the least significant. However, the coefficient a varies considerably from 2.37 to 5.41, and the ratio of the range of a to the average a reaches 80%, implying differing geographic coverage among these sections of domain B. It shows that the overall terrain of the Changbai Mountain region is complex and the anisotropy characteristics are relatively obvious, indicating that simply using one-dimensional terrain spectra may not better present the overall terrain characteristics of this region.

3.3. *The Spectral Analysis of Terrain, Precipitation, and Wind Field*

In order to investigate the correlation between terrain spectra, total precipitation spectral variance, and average wind spectral variance, the relationship between variance and total wavelength is obtained by performing additive synthesis on the two-dimensional spectral coefficients. Figure 8 shows the distributions of synthesized terrain height variance, total precipitation spectral variance, and average wind spectral variance from June to August averaged over 2008–2022 with respect to λ in the three study areas. It can be seen that the majority of terrain height variance (solid line) is distributed at larger wavelengths ($\lambda > 100$ km), under which the values of the variance are basically greater than 1000 m^2. The dominant wavelengths (DWs) corresponding to the topographic spectral peak represent the most evident undulation of terrain and, videlicet, the maximum topographic dynamic forcing. When the wavelength is shorter than the DW, the forcing mainly decreases with decreasing λ, in spite of several departures. In Figure 8b, the DW is 278.3 km, corresponding to the maximum spectral energy of 4200 m^2 for domain B, less than the values for domains A and C. It indicates that the large-scale topographic perturbations of domain B are relatively weak and the topographic relief of domain B is gentler than that of the other two areas. However, the fluctuation of terrain height variance is more pronounced when $\lambda <$ DW, indicating significant smaller-scale topographic perturbations of domain B, which is consistent with the result of smaller b for domain B obtained by the previous analysis. Moreover, the maximum topographic spectral energy of 2.6×10^4 m^2 for domain C is larger than the values for the other two domains, indicating that the undulation of terrain in domain C is the most pronounced. Also, as shown in Figure 8a and c, the overall fluctuations of terrain height variance are relatively similar, and the DWs are both close to the respective maximum wavelengths for domains A and C, indicating more prominent large-scale topographic perturbations of domains A and C compared to that of domain B.

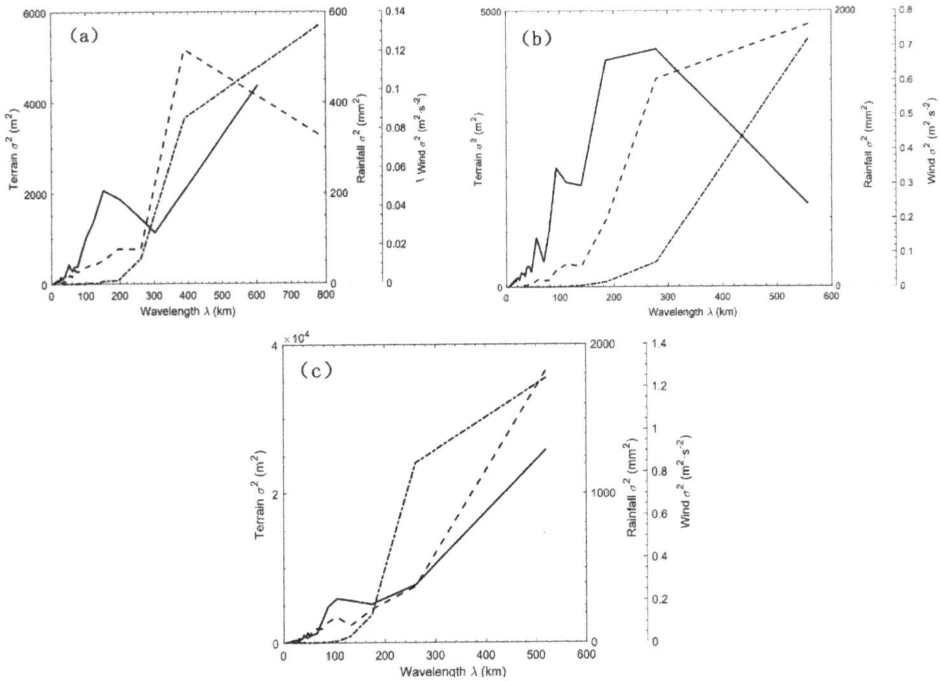

Figure 8. The distributions of terrain height variance (solid line), total precipitation variance (dashed line), and average wind spectral variance (dotted line) from June to August averaged over 2008–2022 versus wavelength λ for domains A (**a**), B (**b**), and C (**c**).

Further insight into the total precipitation spectral variance (dashed line in Figure 8) and wind spectral variance (dotted line in Figure 8) is gained. It can be seen that the precipitation variances of the three study areas are not exactly the same. The DWs correspond to the spectral peak of precipitation in precipitation spectrum space. As shown in Figure 8a, the DW of precipitation is 390 km, corresponding to the maximum spectral energy of 520 mm^2 for domain A. When $\lambda < 390$ km, the precipitation spectral variance decreases with decreasing λ, whereas the trend of spectral variance is downward with increasing λ when $\lambda > 390$ km. The DWs of precipitation in Figure 8b,c are both close to the respective maximum wavelengths for domains B and C, indicating that the precipitation spectral variance increases with increasing λ in the whole wavelength bands. Similarly, the wind spectral variance of all three domains has an upward spectral energy trend with increasing λ, corresponding to the DWs of wind at the maximum wavelengths. However, differences of the slopes at different wavelength bands exist objectively. A large number of observational facts confirm that the spectral energy of wind displays spectral segments of a -3 slope at the synoptic scale and a $-5/3$ slope at the meso-scale over wavelengths of <500 km [42–47]. Moreover, the wind spectra show a transition characteristic, from dependence at a large-scale range to dependence at a meso-scale range [43,47–50]. For example, Yu et al. [47] suggested that the average scale of spectral transition is around 300 km when investigating the characteristics of the horizontal kinetic energy spectra of an eastward-moving southwest vortex. Sun et al. [50] found a distinct transition of the wind spectra at a scale of ~400 km. As shown in Figure 8a, the transition scale is at a wavelength of 390 km for domain A, showing the slope transition characteristic of the wind spectra from a large scale to meso-scale. Similarly, the transition scales for domain B and domain C are at a wavelength of 280 km and 260 km, respectively.

A further analysis is undertaken on the relation between the terrain height spectra, wind spectra, and precipitation spectra. Obviously, there is a strong correlation between the variances of the terrain height, total precipitation, and wind field in respective study regions.

For domain A, the variation of the precipitation spectral variance is consistent with that of the wind spectral variance at the wavelength bands of $\lambda < 390$ km (the DW of precipitation), implying a high correlation between wind field and the occurrence of rainfall. The inverse relationship between terrain spectra and precipitation spectra appears at larger wavelengths of $\lambda > 390$ km, indicating that the precipitation may be dominated by non-topographic factors, and the effect of topographic uplift is not obvious in this region. The occurrence of rainfall is probably the result of the combination of topographic action and the development of weather systems, and the frequency of topographic forcing is not the same as the frequency of the systems. The weather systems that lead to precipitation in this region mainly include the eastward-moving upper trough, the northward-moving north China cyclone, the anomalous cold vortex activity, the high- and low-level jets, frontal systems, and gravity waves. The mountain terrain has a precipitation enhancement effect under some circumstances.

In Figure 8b, it can be seen that there is consistency in the fluctuations of terrain spectra, precipitation spectra, and wind spectra at the wavelength bands of $\lambda < 278.3$ km (the DW of terrain), especially at the range of 50–100 km. The consistency shows that the small-scale terrain in this region plays an important role in the uplift of atmosphere and therefore the occurrence and development of summer precipitation. When $\lambda > 278.3$ km, there is a reverse relationship of the variation between the terrain height variance and the precipitation spectral variance, whereas the fluctuations of precipitation spectra and wind spectra are relatively consistent. It indicates that the large-scale topographic perturbations of domain B are relatively weak, and precipitation is mainly dominated by certain precipitation weather systems.

For domain C, the topographic spectral energy mainly increases with increasing λ, though there are sporadic departures (100 km $< \lambda <$ 140 km) from this trend for the actual complex terrain. The precipitation spectral variance also has a spectral energy upward trend with increasing λ, in spite of several departures. The fluctuations of terrain spectra, precipitation spectra, and wind spectra show a significant positive correlation across the whole wavelength bands. The occurrence of resonance between terrain, wind field, and precipitation implies that the large-scale terrain and minor terrain both play a crucial role in atmospheric uplift and the occurrence and development of summer rainfall in domain C. The precipitation systems are impacted by continuous periodic topographic forcing. When the frequency of topographic forcing equals or almost equals the vibration frequency of the precipitation systems or the background fields, the precipitation systems resonate with topographic forcing, thus resulting in the formation of precipitation.

4. Discussion and Conclusions

The terrain of Changbai Mountain has great influence on the distribution of meso-scale atmospheric flows and the occurrence and development of precipitation through dynamic processes. However, quantitative studies on the real terrain characteristics and the terrain effect on precipitation distribution are scant at present, and even fewer studies reference the Changbai Mountain region. This study quantitatively analyzes the distributions of terrain spectra, wind spectra, and precipitation spectra in this region by using a spectral analysis of 2D-DCT. The regional characteristic of topographic perturbations, the distribution characteristics of summer precipitation, and the relationship between terrain, wind field, and precipitation in spectrum space are discussed. The main conclusions are presented below.

The summer precipitation in the Changbai Mountain region is mainly concentrated in August, which is mostly distributed in the southwest of the Changbai Mountain region. The rainband of the summer precipitation presents a northeast–southwest distribution and the precipitation is mainly concentrated in the areas between 40° N and 42° N and between 47° N and 48 °N, of which the terrain is steep. Based on the distribution characteristics of precipitation, three domains with relatively heavy precipitation are selected as the study region for the spectral analysis. The results of the two-dimensional spectral decomposition of the terrain height field in each domain indicate that the overall terrain of the Changbai Mountain region shows anisotropic characteristics. The anisotropy of the terrain in domain A is the most obvious among the three domains, while the terrain in the southwest of the main Changbai Mountain range shows the least significant anisotropy. The complex terrain implies that only using one-dimensional terrain spectra may not better present the overall terrain characteristics of this region, whereas the terrain spectra performing the 2D-DCT routine can successfully demonstrate the anisotropic characteristics.

Through the comparison among the three spectral curves of synthesized terrain height variance with respect to λ in the three study areas, the values of the variance for domain B are basically less than the values for domains A and C in the whole wavelength bands, indicating that the large-scale topographic perturbations of domain B are relatively weak, and the topographic relief is gentler than that of the other two domains. However, the smaller-scale topographic perturbations of domain B are prominent because of undulation of terrain in the shorter wavelength bands. The dominating peak value of terrain height variance for domain C is larger than the values for the other two domains, indicating that the undulation of terrain in domain C is the most pronounced. Also, the dominant wavelengths of terrain height variance for domains A and C are both close to the respective maximum wavelengths and the overall fluctuation of terrain height variance for domain A is similar to that for domain C, indicating more prominent large-scale topographic perturbations of domains A and C compared to that of domain B. The distributions of the total precipitation spectral variance of the three domains are not exactly the same. The DW corresponding to the spectral peak of precipitation for domain A is 390 km, whereas the DWs for domains B and C are close to the maximum wavelengths, respectively. It indicates that the precipitation spectra of domains B and C both increase with increasing λ in the whole wavelength bands. Similarly, the wind spectral variance of all the three domains has an upward trend with increasing λ and shows the slope transition characteristic from a large scale to meso-scale.

Based on the analysis of the relation between the terrain height spectra, wind spectra, and precipitation spectra, the variation of the precipitation spectral variance is consistent with that of the wind spectral variance in the wavelength bands of $\lambda < 390$ km for domain A, showing a high correlation between wind field and the occurrence of rainfall. The inverse relationship between terrain spectra and precipitation spectra at larger wavelengths indicates that both terrain and non-topographic factors contribute to the occurrence of rainfall and the terrain effect is weakened. For domain B, there is consistency in the fluctuations of terrain spectra, precipitation spectra, and wind spectra in the wavelength bands of $\lambda < 278.3$ km, implying that the dynamic forcing of the smaller-scale terrain has an important effect on the occurrence of summer precipitation. When $\lambda > 278.3$ km, a reverse fluctuation exists between the terrain spectra and the precipitation spectra, indicating that the large-scale topographic perturbations are relatively weak, and precipitation is mainly dominated by certain precipitation weather systems. For domain C, the variations of terrain spectra, precipitation spectra, and wind spectra are almost consistent across the whole wavelength bands. The occurrence of resonance between terrain, wind field, and precipitation implies that the large-scale and smaller-scale terrain both play a crucial role in atmospheric uplift and the occurrence and development of summer rainfall in domain C.

The study accurately characterizes the regional characteristics of topographic disturbance over the area of focus by using 2D-DCT. The improved method of additive synthesis is first proposed to calculate the two-dimensional spectral coefficients in order to analyze the relationship between terrain height spectral variance, total precipitation spectral variance, and wind spectral variance. Representing the terrain, wind field, and precipitation in spectrum space simultaneously can more truly reflect the distribution characteristics of the actual terrain, wind field, and precipitation and also indicate the specific response relationship between them, which is conducive to in-depth understanding of the impact of complex terrain on the summer precipitation at different scales. Moreover, terrain height variance plays a key role in the simulation of meso-scale atmospheric flows [51]. The ability to simulate local circulation and rainfall accurately relies heavily on resolving the important terrain features over the study area. Research on the terrain characteristics and its impact on precipitation and wind distribution can help to focus on the dominant terrain that modulates atmospheric flows and precipitation and determine the required grid spacing for meso-scale models of the focus region.

Author Contributions: Author Contributions: Conceptualization, L.L.; Methodology, L.L. and W.A.; Formal analysis, L.L. and W.A.; Investigation, L.L.; Data curation, W.A. and X.Y.; Writing—original draft, L.L.; Writing—review & editing, L.L., W.A., X.Y. and L.Z.; Visualization, L.L.; Supervision, W.A., X.Y. and L.Z.; Funding acquisition, X.Y. and L.Z. All authors have read and agreed to the published version of the manuscript.

Funding: This study was funded by the National Key Research and Development Plan (Grant No. 2022YFC3003000, No. 2022YFC3003004 and No. 2023YFC3006800).

Institutional Review Board Statement: Not applicable.

Informed Consent Statement: Not applicable.

Data Availability Statement: The SRTM digital elevation data used in this work are available from the National Aeronautics and Space Administration (NASA) (https://srtm.csi.cgiar.org/ (accessed on 16 January 2024)). The observed precipitation data are available from the corresponding author on request. The wind data are available from the European Centre for Medium-Range Weather Forecasts (ECMWF) (https://cds.climate.copernicus.eu/cdsapp#!/dataset/reanalysis-era5-pressure-levels?tab=form (accessed on 16 January 2024)).

Acknowledgments: The authors thank the anonymous reviewers for their suggestions, which helped to improve the manuscript.

Conflicts of Interest: The authors declare no conflict of interest.

References

1. Sun, W.Y.; Mu, X.M.; Song, X.Y.; Wu, D.; Cheng, A.F.; Qiu, B. Changes in extreme temperature and precipitation events in the Loess Plateau (China) during 1960–2013 under global warming. *Atmosphere* **2016**, *168*, 33–48. [CrossRef]
2. Zhao, Y.F.; Zou, X.Q.; Cao, L.G.; Xu, X. Changes in precipitation extremes over the Pearl River Basin, southern China, during 1960–2012. *Quat. Int.* **2014**, *333*, 26–39. [CrossRef]
3. Domroes, M.; Schaefer, D. Recent climate change affecting rainstorm occurrences: A case study in East China. *Clim. Past* **2008**, *4*, 303–309. [CrossRef]
4. Chen, S.J.; Wang, W.; Lau, K.H.; Zhang, Q.H.; Chung, Y.S. Mesoscale convective systems along the Meiyu front in a numerical model. *Meteorol. Atmos. Phys.* **2000**, *75*, 149–160. [CrossRef]
5. Hu, B.W.; Cui, C.G.; Fang, C.H. Causes of a two-day successively heavy rain along the Changjiang Valley in the eastern Hubei Province during 21–22 July 1998. *Chin. J. Atmos. Sci.* **2001**, *25*, 479–491. (In Chinese)
6. Ding, Z.Y.; Zhang, X.Q.; Shou, S.W. Analysis of relation of South Asia High and rainstorm caused by northwesterly upper-level jet. *J. Appl. Meteorol. Sci.* **2002**, *13*, 671–679. (In Chinese)
7. Chen, J.; Liu, L. Characteristics of urban rainstorm and its disaster cause over Beijing in flood season of 2011. *Torrential Rain Disasters* **2011**, *30*, 282–287. (In Chinese)
8. Wang, C.X.; Gao, S.T.; Liang, L.; Ding, D.F.; Gong, H.N. Multi-scale characteristics of moisture transport during a rainstorm process in North China. *Atmos. Res.* **2014**, *145–146*, 189–204. [CrossRef]
9. Scorer, R.S. Theory of waves in the lee of mountains. *Q. J. R. Meteorol. Soc.* **1949**, *75*, 41–56. [CrossRef]
10. Mclntyre, M.E. On Long's hypothesis of no upstream influence in uniformly stratified or rotating flow. *J. Fluid Mech.* **1972**, *52*, 209–243. [CrossRef]
11. Klemp, J.B.; Lilly, D.K. Numerical simulation of hydrostatic mountain waves. *J. Atmos. Sci.* **1978**, *35*, 78–107. [CrossRef]
12. Tucker, D.F.; Reiter, E.R. Modeling heavy precipitation in complex terrain. *Meteorol. Atmos. Phys.* **1988**, *39*, 119–131. [CrossRef]
13. Johnson, G.L.; Hanson, C.L. Topographic and atmospheric influences on precipitation variability over a mountainous watershed. *J. Appl. Meteorol.* **1995**, *34*, 68–87. [CrossRef]
14. Aebischer, U.; Schar, C. Low-level potential vorticity and cyclogenesis to the lee of the Alps. *J. Atmos. Sci.* **1998**, *55*, 186–207. [CrossRef]
15. Jiang, Q.F. Precipitation over concave terrain. *J. Atmos. Sci.* **2006**, *63*, 2269–2288. [CrossRef]
16. Jiang, Q.F. Precipitation over multiscale terrain. *Tellus* **2007**, *59*, 321–335. [CrossRef]

17. Steeneveld, G.J.; Holtslag, A.A.M.; Nappo, C.J.; van de Wiel, B.J.H.; Mahrt, L. Exploring the possible role of small-scale terrain drag on stable boundary layers over land. *J. Appl. Meteorol. Climatol.* **2008**, *47*, 2518–2530. [CrossRef]
18. Jia, B.S.; Jia, F.N.; Zhang, W.T.; Liu, G.Y.; Cui, Z.Q. Analysis of heavy rainstorm weather process in Jilin province from July 13 to 14, 2017. *Mod. Agric. Sci. Technol.* **2021**, *4*, 176–180. (In Chinese)
19. Yu, J.H.; Gong, Y.F.; Mao, W.S. Comparative analysis on two extreme severe precipitation events in Yongji county, Jinlin province in July 2017. *J. Chengdu Univ. Inf. Technol.* **2019**, *3*, 287–296. (In Chinese)
20. Zhang, L.; Li, Z.C. The reason of a heavy rain event in Nenjiang valley in August, 1998. *Meteorol. Mon.* **2003**, *8*, 7–12. (In Chinese)
21. Zheng, X.Y.; Zhang, Y.Z.; Bai, R.H. *Rainstorm in Northeast China*; Meteorology Press: Beijing, China, 1992; pp. 142–145.
22. He, B.H.; Sun, J.Q.; Yu, E.T.; Wang, H.J.; Zhang, M.Q.; Hua, W. Simulation study on the Influence of the great Khingan Strip and Changbai Mountain on summer rainfall in Northeast China. *Clim. Environ. Res.* **2020**, *3*, 268–280. (In Chinese)
23. Liu, C.H.; Wang, Y.; Yan, Q.; Tan, Z.H.; Liu, S.; Jiao, H.R.; Jin, Y. Impact of flow over and flow around caused by super low-level jet on a sudden rainstorm over the Changbai Mountains. *Torrential Rain Disaster* **2023**, *3*, 273–282. (In Chinese)
24. Pielke, R.A.; Kennedy, E. *Mesoscale Terrain Features. Report UVA-ENV SCI-MESO-1980-1*; University of Virginia: Charlottesville, VA, USA, 1980; p. 19.
25. Young, G.S.; Pielke, R.A. Application of terrain height variance spectra to mesoscale modeling. *J. Atmos. Sci.* **1983**, *40*, 2555–2560. [CrossRef]
26. Young, G.S.; Pielke, R.A.; Kessler, R.C. A comparison of the terrain height variance spectra of the Front Range with that of a hypothetical mountain. *J. Atmos. Sci.* **1984**, *41*, 1249–1252. [CrossRef]
27. Srinivasan, K.; Ramanathan, N. Terrain variance spectra for Indian Western Ghats. *Proc.-Indian Natl. Sci. Acad. Part A* **1994**, *60*, 133–138.
28. Ramanathan, N.; Srinivasan, K. An estimation of optimum grid size for Kashmir Valley by spectral method. *J. Appl. Meteorol.* **1995**, *34*, 2783–2786. [CrossRef]
29. Steyn, D.G.; Ayotte, K.W. Application of two-dimensional terrain height spectra to mesoscale modeling. *J. Atmos. Sci.* **1985**, *42*, 2884–2887. [CrossRef]
30. Salvador, R.; Calbo, J.; Millan, M.M. Horizontal grid size selection and its influence on mesoscale model simulations. *J. Appl. Meteorol.* **1999**, *38*, 1311–1329. [CrossRef]
31. Wang, C.X.; Liang, L.; Zhang, W.C.; Gao, S.T.; Yang, S. The Impact of Improved Topographic Resolution on the Distribution of Terrain Spectra and Grid-Size Selection for Mesoscale Models. *Atmosphere* **2022**, *13*, 708. [CrossRef]
32. Wang, W.T.; Wang, Y. A spectral analysis of satellite topographic profile: A coincident pattern between latitudinal topographic and westerly perturbation on the lee side of Qinghai-Tibet Plateau. *J. Nanjing Univ.* **2004**, *40*, 304–317. (In Chinese)
33. Shu, S.J.; Wang, Y.; Li, Y. Effect of topographic perturbation on the precipitation distribution in Tibetan Plateau. *Adv. Water Sci.* **2006**, *17*, 585–591. (In Chinese)
34. Denis, B.; Cote, J.; Laprise, R. Spectral decomposition of two-dimensional atmospheric fields on limited-area domains using the discrete cosine transform (DCT). *Mon. Weather Rev.* **2002**, *130*, 1812–1829. [CrossRef]
35. Zheng, Y.J.; Jin, Z.Y.; Chen, D.H. Kinetic energy spectrum analysis in a semi-implicit semi-Lagrangian dynamical frame work. *Acta Meteorol. Sin.* **2008**, *70*, 371–386. (In Chinese)
36. Jiang, C.; Shen, X.S. Assessment of the simulative performance of the GRAPES model on the convective boundary layer based on the large eddy simulations. *Acta Meteorol. Sin.* **2013**, *71*, 879–890. (In Chinese)
37. Prein, A.F.; Holland, G.J.; Rasmussen, R.M. Importance of regional climate model grid spacing for the simulation of heavy precipitation in the Colorado Headwaters. *J. Clim.* **2013**, *26*, 4848–4857. [CrossRef]
38. Ricard, D.; Lac, C.; Riette, S.; Legrand, R.; Mary, A. Kinetic energy spectra characteristics of two convection-permitting limited-area models AROME and Meso-NH. *Q. J. R. Meteorol. Soc.* **2013**, *139*, 1327–1341. [CrossRef]
39. Huang, Y.J.; Cui, X.P. Spectral characteristics of terrain in the Sichuan basin and the horizontal grid size selection for a mesoscale model. *Acta Meteorol. Sin.* **2016**, *1*, 114–126. (In Chinese)
40. Pan, Y.; Shen, Y.; Yu, J.J.; Zhao, P. Analysis of the combined gauge-satellite hourly precipitation over China based on the OI technique. *Acta Meteorol. Sin.* **2012**, *6*, 1381–1389. (In Chinese)
41. Pielke, R.A. *Mesoscale Meteorological Modeling*; Academic Press: SanDiego, CA, USA, 1984; p. 599.
42. Boer, G.J.; Shepherd, T.G. Large-scale two-dimensional turbulence in the atmosphere. *J. Atmos. Sci.* **1983**, *40*, 164–184. [CrossRef]
43. Nastrom, G.D.; Gage, K.S. A climatology of atmospheric wavenumber spectra of wind and temperature observed by commercial aircraft. *J. Atmos. Sci.* **1985**, *42*, 950–960. [CrossRef]
44. Cho, J.Y.N.; Newell, R.E.; Barrick, J.D. Horizontal wavenumber spectra of winds, temperature, and trace gases during the Pacific Exploratory Missions: 1. Climatology. *J. Geophys. Res.* **1999**, *104*, 5697–5716. [CrossRef]
45. Cho, J.Y.N.; Lindborg, E. Horizontal velocity structure functions in the upper troposphere and lower stratosphere: 1. Observations. *J. Geophys. Res. Atmos.* **2001**, *106*, 10223–10232. [CrossRef]
46. Shikhovtsev, A.Y.; Kovadlo, P.G.; Lezhenin, A.A.; Korobov, O.A.; Kiselev, A.V.; Russkikh, I.V.; Kolobov, D.Y.; Shikhovtsev, M.Y. Influence of Atmospheric Flow Structure on Optical Turbulence Characteristics. *Appl. Sci.* **2023**, *13*, 1282. [CrossRef]
47. Yu, S.W.; Zhang, L.F.; Wang, Y.; Peng, J. Mesoscale Horizontal Kinetic Energy Spectra of an Eastward-Moving Southwest Vortex. *Atmosphere* **2022**, *13*, 653. [CrossRef]
48. Waite, M.L.; Snyder, C. Mesoscale energy spectra of moist baroclinic waves. *J. Atmos. Sci.* **2013**, *70*, 1242–1256. [CrossRef]

49. Peng, J.; Zhang, L.; Luo, Y.; Zhang, Y. Mesoscale Energy Spectra of the Mei-Yu Front System. Part I: Kinetic Energy Spectra. *J. Atmos. Sci.* **2014**, *71*, 37–55. [CrossRef]
50. Sun, Y.; Zhang, F. Intrinsic versus practical limits of atmospheric predictability and the significance of the butterfly effect. *J. Atmos. Sci.* **2016**, *73*, 1419–1438. [CrossRef]
51. Wang, C.X.; Gao, S.T.; Ran, L.K.; Liang, L. Proof of the monotonicity of grid size and its application in grid-size selection for mesoscale models. *Adv. Atmos. Sci.* **2015**, *32*, 1005–1015. [CrossRef]

atmosphere MDPI

Communication

Vegetation and Evapotranspiration Responses to Increased Atmospheric Vapor Pressure Deficit across the Global Forest

Rihong Wen [1,2,†], Meiou Qin [3,†], Peng Jiang [1,2,4,*], Feiyun Yang [5], Bin Liu [4], Mengyuan Zhu [6], Yuan Fang [1,4], Yichen Tian [1,4] and Bo Shang [7,*]

1 Institute of Atmospheric Environment, China Meteorological Administration, Shenyang 110166, China; fangyuan@iaesy.cn (Y.F.)
2 Panjin National Climate Observatory, Panjin 124000, China
3 Regional Climate Center of Shenyang, Shenyang 110166, China
4 Liaoning Ecological Meteorology and Satellite Remote Sensing Center, Shenyang 110166, China
5 China Meteorological Administration Training Center, Beijing 100081, China
6 College of Agronomy, Shenyang Agricultural University, Shenyang 110161, China
7 Changchun Meteorological Service, Changchun 130051, China
* Correspondence: jiangpenglnqx@iaesy.cn (P.J.); shangbobosea@163.com (B.S.)
† These authors contributed equally to this work.

Abstract: A forest is vulnerable to drought and plays important roles in the regulation of carbon and water cycling in a terrestrial ecosystem. Atmospheric vapor pressure deficit (VPD) has been identified as an increasingly major factor in plant functioning and has been established as a main contributor to recent drought-induced plant mortality, independent of other drivers associated with climate change. However, most previous studies have focused on the effects of climate warming and CO_2 enrichment on vegetation growth, without considering the effects of an increased VPD on vegetation growth and evapotranspiration (ET) in forest ecosystems. This could lead to a large uncertainty in estimating the variability in forest carbon sinks. Based on the long-term satellite data, we investigated the response of the leaf area index (LAI) and ET to the VPD via a partial correlation analysis in this study. We also examined the temporal variability in the partial coefficients within a ten-year moving window. The results showed that over 50% of the region displayed a negative partial correlation between the LAI, ET, and VPD, and those pixels were mainly concentrated in North America and the plains of Eastern Europe. Regions with a negative trend of partial correlation in both the LAI and ET are mostly located in the plains of Eastern Europe and the Siberian Plain of western Russia, while the positive trend is mainly in South America. The plains of Eastern Europe are becoming drier, which was proved by the interannual trend of the Standardized Precipitation Evapotranspiration Index (SPEI) and soil water content (SWC). Additionally, the LAI and ET in those areas exhibited a significant positive correlation with the SWC based on the moving window average. This study suggests that the role of the VPD on vegetation will become increasingly prominent in the context of future climate change for the forest ecosystem.

Keywords: vapor pressure deficit; LAI; evapotranspiration; water availability; global forests

1. Introduction

The terrestrial biosphere is believed to have provided a net sink for ~20% of the carbon dioxide emitted by fossil fuel burning and industry over the last three decades [1], with the majority estimated to occur in forests [2]. The growth of forests is thus believed to retard anthropogenic climate change by slowing the rate of carbon dioxide (CO_2) accumulation in the atmosphere [3]. Previous studies have mainly investigated the response of forest productivity to climate warming, changes in precipitation patterns, CO_2 fertilization effects, and human activity. With climate change, an atmospheric VPD has been recorded in a sharp rise since the late 1990s [4] and played an important role in driving global vegetation dynamics. However, the response of forest productivity to increasing VPD remains poorly characterized, limiting both our understanding of why it occurs and our ability to predict its continued future existence.

An increasing VPD may exert a strong impact on the carbon and water cycle in global forests. A rising VPD can significantly reduce stomatal conductance and limit the actual photosynthetic rate [5–8]. Thus, a widespread constraint of the rising VPD on global vegetation growth [4,9] or carbon sinks [10,11] has been reported in recent studies. Extended periods of high VPD have been recognized as an important determinant of large-scale tree mortality [12] and as a driver of wildfire [13]. The negative effects of an increasing VPD on vegetation growth and ET have been widely observed in forest ecosystems in water-limited regions [14]. In contrast, recent work has reported that in most of the wettest parts of subtropical and tropical forests, vegetation photosynthesis (or productivity) and ET tend to increase with a rising VPD [7,15,16]. These discrepancies suggest a large uncertainty in the impacts of an increasing VPD on the carbon and water cycle across the global forests and thus call for more investigations on this.

The carbon and water cycle responses to the increasing VPD may be regulated by soil water availability across the global forests. When the soil water availability is limited, the xylems or roots of plants sense the low water supply via an abscisic acid signal and subsequently stomatal closure in response to an increasing VPD [17]. Studies examining the dependence of the VPD impacts on soil water availability in forest ecosystems have primarily been conducted on the site scale, such as a study on a forest in south central Indiana, USA [5]. Despite that, the relationships between water availability and spatial VPD effects may be disturbed by other factors. For one, although the spatial distribution of global forests is mainly located in water-rich regions, the seasonal dryness always exerts assignable impacts on photosynthesis (or productivity) and ET [7,18,19]. Also, changes in water availability could also affect VPD effects [4], for example, the significant negative VPD effects observed in Western Europe could be due to the apparent trend toward aridification [20]. Therefore, the relationships between water availability and VPD impacts across the global forests need to be further quantified on a large spatial scale.

In sum, there is growing evidence that an increasing VPD plays an important role in driving global vegetation dynamics [4,9,11]. However, very few studies have investigated the VPD effects across global forests. Here, we used long-term satellite observations of the LAI and ET to parse the VPD impacts across the global forests. We first analyzed the spatial distribution of the response of the ET and LAI to the VPD across the global forests; subsequently, we examined whether the impacts of the VPD on ET and the LAI can be regulated by the SWC; then, we investigated the temporal changes in the VPD impacts and then analyzed the correlation of these changes with changes in the water conditions.

2. Materials and Methods

The Global Land Surface Satellite LAI (GLASS LAI) was used in this study. This product was generated based on Advanced Very High Resolution Radiometer (AVHRR) and Moderate Resolution Imaging Spectroradiometer (MODIS) surface reflectance data via general regression neural networks [21,22]. In particular, preprocessing should be conducted on the surface reflectance so as to mitigate the effects caused by clouds, aerosols, and sensor noise. The GLASS LAI product has been widely used in climate change, ecological environment monitoring, and other fields due to its unique features, including a long-term span (from 1982 to present), fine resolution (0.05 degree and 8-day interval), high quality, and accuracy [23,24]. The monthly actual ET and root-zone SWC data came from the Global Land Evaporation Amsterdam Model (GLEAM, v3.5a). This dataset spans over 40 years, from 1980 to 2021, at a spatial resolution of 0.25 degree [25]. The monthly air temperature (*Ta*), precipitation, and actual vapor pressure (AVP) were derived from CRUST 4.04 (0.5 degree), and the monthly solar radiation (SR) was provided by the ERA5 reanalysis products (0.1 degree). Then, the VPD could be calculated according to the saturated vapor pressure (SVP) and AVP following Equations (1) and (2) [4]. The

Standardized Precipitation Evapotranspiration Index (SPEI) was used to quantify the dry/wet conditions from 1982 to 2015 and the three-month SPEI was derived from the SPEI base v2.5 at a 0.5 resolution [26,27]. For the subsequent analysis, all the data were resampled to 0.5 degrees to align to the same resolution.

$$\mathrm{SVP} = 0.611 \times exp\left(\frac{17.27 \times Ta}{273.3 + Ta}\right) \quad (1)$$

$$\mathrm{VPD} = \mathrm{SVP} - \mathrm{AVP} \quad (2)$$

We implemented a partial correlation analysis to explore the responses of the LAI and ET to the VPD during the growing season from 1982 to 2015 after controlling the effect of the Ta, precipitation, and SR. Furthermore, we detected the temporal variability in the partial coefficients within a ten-year moving window via the Mann–Kendall test [28,29]. To access the temporal dynamics of the water availability from 1982 to 2015, we also conducted the Mann–Kendall test on the SWC and SPEI. Finally, we used the Pearson correlation analysis to estimate the relationship between the SWC and the effects of the VPD on the LAI and ET based on a ten-year moving mean value. All the statistical analyses were carried out in R (version 4.0.5) software.

3. Results

Overall, the negative partial correlation coefficients occupied a large proportion for both the LAI (Figure 1a) and ET (Figure 1b), with the negative partial correlations accounting for 68.6% (39.0% of pixels were significant) and 56.6% (34.9% of pixels were significant) of the total study area, respectively. The negatively correlated areas were concentrated in the Northern Hemisphere, mainly distributed in regions including North America and the plains of Eastern Europe. In contrast, the positive correlation pixels of the LAI and ET to the VPD were mainly observed in South America, Siberia, and southern China, and only a few correlations were significant.

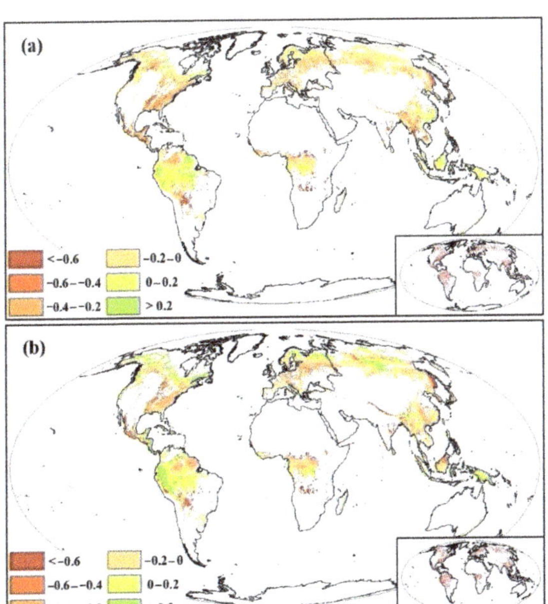

Figure 1. The spatial patterns of the partial correlation between LAI (**a**), ET (**b**), and VPD across the global forest ecosystem. The insets map in lower right represents the significance at the level of 0.05.

As shown in Figure 2, the proportions of the negative trend in the moving window partial correlation between the LAI and VPD were 62.1% (40.0% of the pixels were significant), and mainly distributed in the northern and eastern parts of North America, the plains of Eastern Europe, and the Siberian Plain region of western Russia. The regions that exhibited a positive trend were mainly located in South America, accounting for 53.9% of the area. As for ET, it displayed an increased trend on the whole, with 54.0% (34.4% of the pixels were significant) being positive. Similar to the LAI, the negative values were also found mainly in the northern and eastern parts of North America, the plains of Eastern Europe, and the Siberian Plain region of western Russia. Additionally, about 62.3% of the regions in Africa displayed a negative trend in ET, with 47.7% of the pixels experiencing a significant decrease.

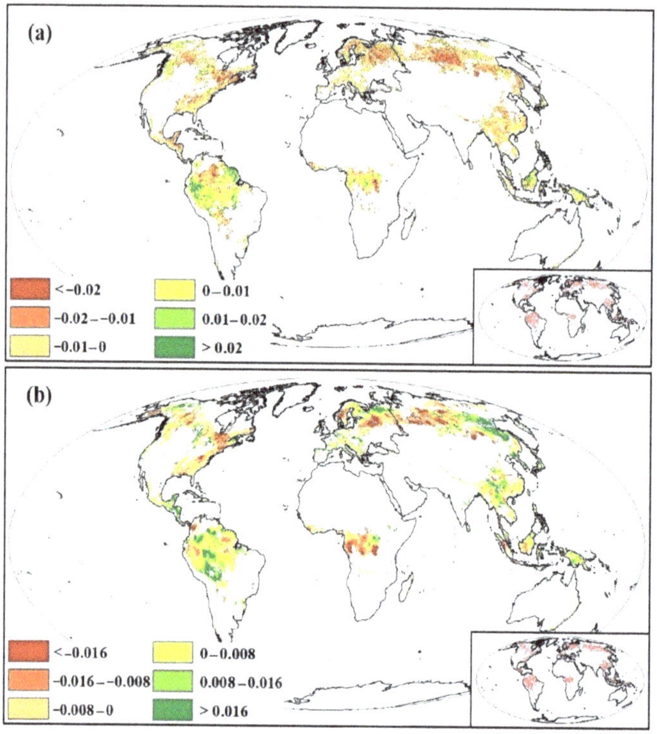

Figure 2. The trend of moving window in partial correlation between LAI (**a**), ET (**b**), and VPD across the global forest ecosystem in spatial pattern. The inset maps in lower right represent the significance at the level of 0.05.

The temporal trend of the growing season SPEI and SWC are shown in Figure 3. In general, the SPEI and SWC appeared as predominantly positive and negative trends, respectively, with the former accounting for 63.2% (9.3% of the pixels were significant) and the latter for 56.4% (23.4% of the pixels were significant). The pixels with increased SPEI values were concentrated in eastern North America, the Siberian plains of western Russia, and western South Africa, while the decreasing pixels were located in the plains of Eastern Europe, southern Siberia, the subtropical regions of China, and the central regions of South America (Figure 3a). As for the SWC (Figure 3b), the areas with increasing pixel values were mainly in North America and parts of Russia, while other areas such as the plains of Eastern Europe, southern Siberia, and central Africa were where the SWC decreases.

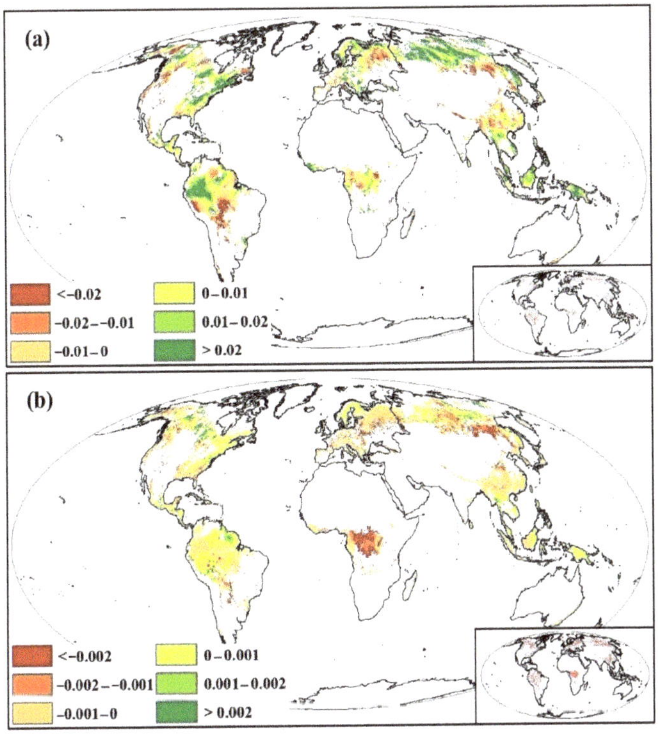

Figure 3. The spatial patterns of temporal trend for SPEI (**a**) and SWC (**b**) across the global forest ecosystem. The insets map in lower right represents the significance at the level of 0.05.

To further clarify the relationship between the LAI, ET, and SWC, we performed the correlation analysis based on the ten-year moving window averages. As a whole, the positive correlation pixels for the LAI and SWC (Figure 4a) accounted for 53.8% (29.5% of the pixels were significant), mainly distributed in the plains of Eastern Europe, southern Siberia, the subtropics of China, and parts of the Amazon. In comparison, the negatively correlated pixels mainly occurred in the Siberian plains of western Russia, western North America, central Africa, and parts of South America. The proportion of significantly negatively correlated pixels was relatively small, only occupying 23.6% of the study area. We also observed that about 50.1% (27.5% of the pixels were significant) of the pixels displayed a positive correlation between the ET and SWC (Figure 4b), most of them distributed in the plains of Eastern Europe, eastern Russia, central Africa, and parts of the Amazon. For those negatively correlated pixels, they were mainly located in the Siberian Plain in western Russia, eastern and southwestern China, and parts of the Amazon.

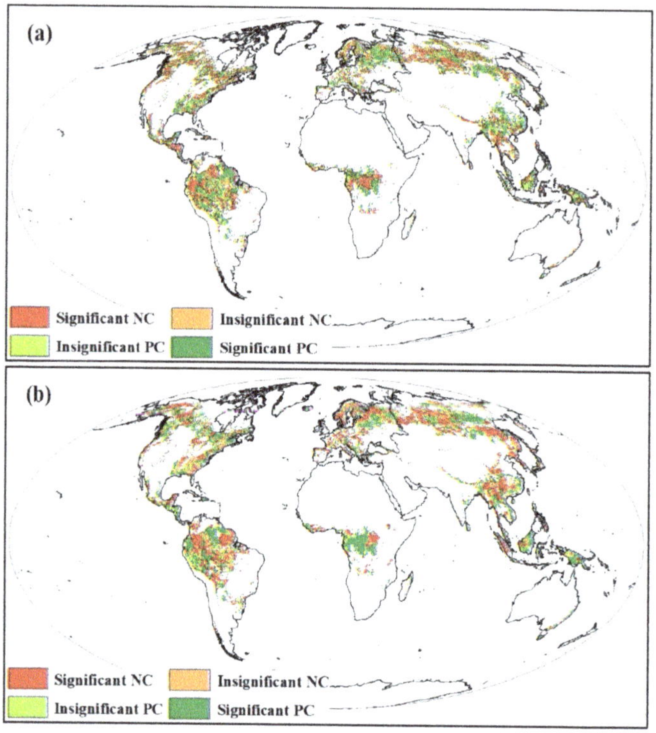

Figure 4. The spatial patterns of correlation between the effects of LAI (**a**), ET (**b**), and SWC based on ten-year moving window average across the global forest ecosystem. Note: NC and PC represent negative correlation and positive correlation, respectively.

4. Discussion

Overall, the response of the forest LAI and ET to the VPD varied with the spatial water gradients. The VPD may also indicate atmospheric dryness. Increased atmospheric water demand caused by an increasing VPD reduces plant photosynthesis and transpiration by controlling the stomatal activity and xylem conductance [14,30]. A typical plant response to an increasing VPD is to close the stomata to minimize the water loss and avoid excessive water tension in the xylem at the expense of reducing or stopping photosynthesis [5,31]. This negative response of plant photosynthesis and transpiration to an increasing VPD has always been observed in water-limited regions [4,5]. This is because, in these regions, xeric species tend to adopt a more conservative sequence to prevent severe tissue damage through tighter stomatal regulation and higher resistance to embolism [32]. In contrast, the LAI and ET can resist the atmospheric dryness caused by an increasing VPD across the mostly forest regions, especially for the water-rich regions. Most forests are distributed in water-rich regions. Although annual precipitation in most boreal forest regions is always less than 500 mm, they are not water-limited due to the high soil water-holding capacity of high soil organic matter [33] and low water loss under thermal constraint. Mesic species adopt a riskier sequence through looser stomatal regulation to maximize carbon uptake at the expense of hydraulic safety [32]. Plant root systems could partly determine how a certain soil water availability is translated into leaf water potential, which is strongly correlated with stomatal activity [34].Forests can extract deeper soil water through vigorous root systems and allow them to maintain higher leaf water potential and stomatal openness under dryness stress (such as under a VPD) [7,19]. In addition, the subtropical and tropical forests always have light limitations and thus plants could tend to maintain their stomata openness at the cost of water loss in response to an increasing VPD [7,35]. Furthermore, new leaves have a higher photosynthetic capacity than the leaves they replaced in the dry season, compensating for the negative stomatal response to increased dryness [15]. In line with our study, a recent global study also has shown that the VPD has weakly negative or even positive impacts on vegetation productivity in most of boreal southern China, India, and tropical rainforests [16].

In contrast, there is a significant negative response of ET and the LAI to the VPD in the seasonally dry regions of forests. Although these forest regions are always classified as water-rich regions, the seasonal dryness could lead to the negative response of the LAI and ET to the VPD. The majority of seasonally dry ecosystems become more isohydric during the sunnier and drier seasons, as they are most at risk of hydraulic failure [18,19]. Like other forests in the water-rich regions, forests could also utilize soil water via deep roots to resist the atmospheric water deficits [36], but forests may restrain stomatal activity to avoid hydraulic failure as the SWC continues to decrease during the dry seasons [7]. Similar to seasonal drought regions, tropical forests also experience seasonal dryness, whereas their co-variation patterns are different. Except for the plant water use strategy and soil water conditions, these differences could also be attributed to plant phenology developments. Certainly, we expect that future research combining multiple factors (water use strategy, phenology developments, water availability, etc.) is urgently needed to help shed light on these differences.

Climate scenarios and our observations present an enhancement of the water deficit and heatwaves over the past few decades in Eastern Europe [37,38]. In particular, frequent heatwaves, evidenced by marked decreases in the ESI, co-occur with high temperatures and low SWC, amplifying the negative impacts of a VPD on vegetation growth and ET by pushing atmospheric dryness up to the peak [8,39]. Thus, the negative response of the LAI and ET is exacerbated by a drop in the water supply in Eastern Europe on a temporal scale.

With an improving water supply, the positive regulatory effects disappeared in the moisture-rich regions of Central Siberia, suggesting that the positive co-variation was withering on a temporal scale. This is in reasonable agreement with recent work in humid regions, where the positive sensitivity of the vegetation growth to the VPD in wet gradients is less than in other gradients [7,40]. Given the increasing likelihood for wet and cloudy conditions to coincide over the past 30 years, as evidenced by decreases in SR, the decreases seem not to be responsible for the VPD because an adequate water supply makes it less possible for atmospheric dryness to occur. It is possible that a water surplus can limit vegetation growth due to waterlogging or the fact that temperatures and solar radiation limit productivity or ET in some wetter years [40,41].

In this study, we mainly used the remote sensing datasets combined with a statistical correlation analysis to analyze the response of the LAI and ET to the VPD. This may increase uncertainties in the VPD impacts. Many indicators (e.g., leaf hydraulic traits) related to the plants' water use strategy were not measured in this study, which may influence the ecological response mechanisms of the VPD effects [17,42]. In view of the above, future field experiments on extreme drought should comprehensively measure indicators of the plant water use strategy. In addition, we expect future earth system models to fully consider the effects of the VPD on the forest ecosystem structure and functions.

5. Conclusions

This study utilized long-term remote sensing data to evaluate the response of the LAI and ET to the VPD via multiple statistical analyses across the global forest ecosystem. The results showed that negative partial correlations were found in over 50% of the regions between the LAI, ET, and VPD, indicating the suppression effect of vegetation growth by the VPD. Moreover, this suppression effect is getting stronger, especially in the plains of Eastern Europe and the Siberian Plain of western Russia. We also observed a decreasing trend for both the SPEI and SWC in the plains of Eastern Europe, which shows the region is becoming increasingly drier. Furthermore, the regions of the LAI and ET in the above area exhibited a significant positive correlation with the SWC based on the moving window average. This study would deepen our understanding of vegetation and ET responses to an increased atmospheric VPD across the global forest in the context of climate change. Our research potentially improves our knowledge and understanding of the negative impacts of the VPD on forests and the predictions of forest ecosystem functioning and bio-diversity in response to climate change.

Author Contributions: Conceptualization, R.W., M.Q., P.J. and B.S.; methodology, P.J. and R.W.; software, R.W. and M.Q.; data curation, M.Z. and Y.T.; writing—original draft preparation, R.W. and M.Q., writing—review and editing, P.J., B.S., F.Y., B.L., M.Z., Y.F. and Y.T.; visualization, F.Y., M.Q. and B.L.; supervision, P.J. and B.S. All authors have read and agreed to the published version of the manuscript.

Funding: This work was supported by the National Key R&D Program of China (Grant No. 2022YFF0801301), the Key Project of Liaoning Provincial Meteorological Bureau (Grant No. LNCP202205), the Annual Fund Project of Liaoning Provincial Meteorological Bureau for 2024, and the Key Laboratory Project of Agricultural Meteorological Disasters in Liaoning Province in 2023 (Grant No. 2023SYIAEKFMS30).

Institutional Review Board Statement: Not applicable.

Informed Consent Statement: Not applicable.

Data Availability Statement: The codes used in this study are available upon request from the authors. The data are not publicly available due to privacy.

Acknowledgments: We acknowledge the support of all co-authors for their constructive and helpful comments and the organization of this study.

Conflicts of Interest: The authors declare that this research was conducted in the absence of any commercial or financial relationships that could be construed as potential conflicts of interest.

References

1. Le Quéré, C.; Andrew, R.M.; Friedlingstein, P.; Sitch, S.; Pongratz, J.; Manning, A.C.; Korsbakken, J.I.; Peters, G.P.; Canadell, J.G.; Jackson, R.B.; et al. Global carbon budget 2017. *Earth Syst. Sci. Data* **2018**, *10*, 405–448. [CrossRef]
2. Pan, Y.; Birdsey, R.A.; Fang, J.; Houghton, R.; Kauppi, P.E.; Kurz, W.A.; Phillips, O.L.; Shvidenko, A.; Lewis, S.L.; Canadell, J.G.; et al. A large and persistent carbon sink in the world's forests. *Science* **2011**, *333*, 988–993. [CrossRef]
3. Pugh, T.A.M.; Lindeskog, M.; Smith, B.; Poulter, B.; Arneth, A.; Haverd, V.; Calle, L. Role of forest regrowth in global carbon sink dynamics. *Proc. Natl. Acad. Sci. USA* **2019**, *116*, 4382–4387. [CrossRef]
4. Yuan, W.; Zheng, Y.; Piao, S.; Ciais, P.; Lombardozzi, D.; Wang, Y.; Ryu, Y.; Chen, G.; Dong, W.; Hu, Z.; et al. Increased atmospheric vapor pressure deficit reduces global vegetation growth. *Sci. Adv.* **2019**, *5*, eaax1396. [CrossRef]
5. Sulman, B.N.; Roman, D.T.; Yi, K.; Wang, L.; Phillips, R.P.; Novick, K.A. High atmospheric demand for water can limit forest carbon uptake and transpiration as severely as dry soil. *Geophys. Res. Lett.* **2016**, *43*, 9686–9695. [CrossRef]
6. Ding, J.; Yang, T.; Zhao, Y.; Liu, D.; Wang, X.; Yao, Y.; Peng, S.; Wang, T.; Piao, S. Increasingly important role of atmospheric aridity on Tibetan alpine grasslands. *Geophys. Res. Lett.* **2018**, *45*, 2852–2859. [CrossRef]
7. Chen, N.; Song, C.; Xu, X.; Wang, X.; Cong, N.; Jiang, P.; Zu, J.; Sun, L.; Song, Y.; Zuo, Y.; et al. Divergent impacts of atmospheric water demand on gross primary productivity in three typical ecosystems in China. *Agric. For. Meteorol.* **2021**, *307*, 108527. [CrossRef]
8. Fu, Z.; Ciais, P.; Prentice, I.C.; Gentine, P.; Makowski, D.; Bastos, A.; Luo, X.; Green, J.K.; Stoy, P.C.; Yang, H.; et al. Atmospheric dryness reduces photosynthesis along a large range of soil water deficits. *Nat. Commun.* **2022**, *13*, 989. [CrossRef] [PubMed]
9. Liu, L.; Gudmundsson, L.; Hauser, M.; Qin, D.; Li, S.; Seneviratne, S.I. Soil moisture dominates dryness stress on ecosystem production globally. *Nat. Commun.* **2020**, *11*, 4892. [CrossRef]
10. Green, J.K.; Seneviratne, S.I.; Berg, A.M.; Findell, K.L.; Hagemann, S.; Lawrence, D.M.; Gentine, P. Large influence of soil moisture on long-term terrestrial carbon uptake. *Nature* **2019**, *565*, 476–479. [CrossRef]
11. He, B.; Chen, C.; Lin, S.; Yuan, W.; Chen, H.W.; Chen, D.; Zhang, Y.; Guo, L.; Zhao, X.; Liu, X. Worldwide impacts of atmospheric vapor pressure deficit on the interannual variability of terrestrial carbon sinks. *Natl. Sci. Rev.* **2022**, *9*, nwab150. [CrossRef] [PubMed]
12. Williams, A.P.; Allen, C.D.; Macalady, A.K.; Griffin, D.; Woodhouse, C.A.; Meko, D.M.; Swetnam, T.W.; Rauscher, S.A.; Seager, R.; Grissinomayer, H.D. Temperature as a potent driver of regional forest drought stress and tree mortality. *Nat. Clim. Chang.* **2013**, *3*, 292–297. [CrossRef]

13. Seager, R.; Hooks, A.; Williams, A.P.; Cook, B.; Nakamura, J.; Henderson, N. Climatology, variability, and trends in the US vapor pressure deficit, an important fire-related meteorological quantity. *J. Appl. Meteorol. Climatol.* **2015**, *54*, 1121–1141. [CrossRef]
14. Novick, K.A.; Ficklin, D.L.; Stoy, P.C.; Williams, C.A.; Bohrer, G.; Oishi, A.C.; Papuga, S.A.; Blanken, P.D.; Noormets, A.; Sulman, B.N.; et al. The increasing importance of atmospheric demand for ecosystem water and carbon fluxes. *Nat. Clim. Chang.* **2016**, *6*, 1023–1027. [CrossRef]
15. Green, J.; Berry, J.; Ciais, P.; Zhang, Y.; Gentine, P. Amazon rainforest photosynthesis increases in response to atmospheric dryness. *Sci. Adv.* **2020**, *6*, eabb7232. [CrossRef]
16. Song, Y.; Jiao, W.; Wang, J.; Wang, L. Increased Global Vegetation Productivity Despite Rising Atmospheric Dryness over the Last Two Decades. *Earth's Future* **2022**, *10*, e2021EF002634. [CrossRef]
17. Rogiers, S.Y.; Greer, D.H.; Hatfield, J.M.; Hutton, R.J.; Clarke, S.J.; Hutchinson, P.A.; Somers, A. Stomatal response of an anisohydric grapevine cultivar to evaporative demand, available soil moisture and abscisic acid. *Tree Physiol.* **2012**, *32*, 249–261. [CrossRef]
18. Konings, A.G.; Gentine, P. Global variations in ecosystem-scale isohydricity. *Glob. Chang. Biol.* **2017**, *23*, 891–905. [CrossRef]
19. Jiang, P.; Meinzer, F.C.; Wang, H.; Kou, L.; Dai, X.; Fu, X. Belowground determinants and ecological implications of shrub species' degree of isohydry in subtropical pine plantations. *New Phytol.* **2020**, *226*, 1656–1666. [CrossRef]
20. Stevenson, S.; Coats, S.; Touma, D.; Cole, J.; Lehner, F.; Fasullo, J.; Otto-Bliesner, B. Twenty-first century hydroclimate: A continually changing baseline, with more frequent extremes. *Proc. Natl. Acad. Sci. USA* **2022**, *119*, e2108124119. [CrossRef]
21. Xiao, Z.; Liang, S.; Wang, J.D.; Chen, P.; Yin, X.J.; Zhang, L.Q.; Song, J.L. Use of general regression neural networks for generating the GLASS leaf area index product from time-series MODIS surface reflectance. *IEEE Trans. Geosci. Remote Sens.* **2014**, *52*, 209–223. [CrossRef]
22. Liang, S.; Cheng, J.; Jia, K.; Jiang, B.; Liu, Q.; Xiao, Z.; Yao, Y.; Yuan, W.; Zhang, X.; Zhao, X.; et al. The Global Land Surface Satellite (GLASS) Product Suite. *Bull. Am. Meteorol. Soc.* **2021**, *102*, E323–E337. [CrossRef]
23. Li, X.; Qu, Y. Evaluation of Vegetation Responses to Climatic Factors and Global Vegetation Trends using GLASS LAI from 1982 to 2010. *Can. J. Remote Sens.* **2019**, *44*, 357–372. [CrossRef]
24. Bai, Y.; Li, S.; Liu, M.; Guo, Q. Assessment of vegetation change on the Mongolian Plateau over three decades using different remote sensing products. *J. Environ. Manag.* **2022**, *317*, 115509. [CrossRef]
25. Martens, B.; Miralles, D.G.; Lievens, H.; van der Schalie, R.; de Jeu, R.A.M.; Fernández-Prieto, D.; Beck, H.E.; Dorigo, W.A.; Verhoest, N.E.C. GLEAM v3: Satellite-based land evaporation and root-zone soil moisture. *Geosci. Model Dev.* **2017**, *10*, 1903–1925. [CrossRef]
26. Beguería, S.; Vicente-Serrano, S.M.; Angulo-Martínez, M. A Multiscalar Global Drought Dataset: The SPEIbase: A New Gridded Product for the Analysis of Drought Variability and Impacts. *Bull. Am. Meteorol. Soc.* **2010**, *91*, 1351–1356. [CrossRef]
27. Vicente-Serrano, S.M.; Beguería, S.; López-Moreno, J.I.; Angulo, M.; El Kenawy, A. A New Global 0.5° Gridded Dataset (1901–2006) of a Multiscalar Drought Index: Comparison with Current Drought Index Datasets Based on the Palmer Drought Severity Index. *J. Hydrometeorol.* **2010**, *11*, 1033–1043. [CrossRef]
28. Piao, S.; Nan, H.; Huntingford, C.; Ciais, P.; Friedlingstein, P.; Sitch, S.; Peng, S.; Ahlstrom, A.; Canadell, J.G.; Cong, N.; et al. Evidence for a weakening relationship between interannual temperature variability and northern vegetation activity. *Nat. Commun.* **2014**, *5*, 5018. [CrossRef]
29. Zu, J.; Zhang, Y.; Huang, K.; Liu, Y.; Chen, N.; Cong, N. Biological and climate factors co-regulated spatial-temporal dynamics of vegetation autumn phenology on the Tibetan Plateau. *Int. J. Appl. Earth Obs. Geoinf.* **2018**, *69*, 198–205. [CrossRef]
30. Grossiord, C.; Buckley, T.N.; Cernusak, L.A.; Novick, K.A.; Poulter, B.; Siegwolf, R.T.W.; Sperry, J.S.; McDowell, N.G. Plant responses to rising vapor pressure deficit. *New Phytol.* **2020**, *226*, 1550–1566. [CrossRef] [PubMed]
31. Running, S.W. Environmental control of leaf water conductance in conifers. *Can. J. For. Res.* **1976**, *6*, 104–112. [CrossRef]
32. Jin, Y.; Hao, G.; Hammond, W.M.; Yu, K.; Liu, X.; Ye, Q.; Zhou, Z.; Wang, C. Aridity-dependent sequence of water potentials for stomatal closure and hydraulic dysfunctions in woody plants. *Glob. Chang. Biol.* **2023**, *29*, 2030–2040. [CrossRef] [PubMed]
33. Stack, S.; Jones, C.; Bockstette, J.; Jacobs, D.F.; Landhäusser, S.M.J.E.E. Surface and subsurface material selections influence the early outcomes of boreal upland forest restoration. *Ecol. Eng.* **2020**, *144*, 105705. [CrossRef]
34. Meinzer, F.C.; Woodruff, D.R.; Marias, D.E.; Smith, D.D.; McCulloh, K.A.; Howard, A.R.; Magedman, A.L. Mapping 'hydroscapes' along the iso- to anisohydric continuum of stomatal regulation of plant water status. *Ecol. Lett.* **2016**, *19*, 1343–1352. [CrossRef] [PubMed]
35. Yu, G.; Zhang, L.; Sun, X.; Fu, Y.; Wen, X.; Wang, Q.; Li, S.; Ren, C.; Song, X.; Liu, Y. Environmental controls over carbon exchange of three forest ecosystems in eastern China. *Glob. Chang. Biol.* **2008**, *14*, 2555–2571. [CrossRef]
36. Jiang, P.; Wang, H.; Meinzer, F.C.; Kou, L.; Dai, X.; Fu, X. Linking reliance on deep soil water to resource economy strategies and abundance among coexisting understorey shrub species in subtropical pine plantations. *New Phytol.* **2020**, *225*, 222–233. [CrossRef]
37. Seneviratne, S.I.; Corti, T.; Davin, E.L.; Hirschi, M.; Jaeger, E.B.; Lehner, I.; Orlowsky, B.; Teuling, A.J. Investigating soil moisture–climate interactions in a changing climate: A review. *Earth-Sci. Rev.* **2010**, *99*, 125–161. [CrossRef]
38. Liu, X.; He, B.; Guo, L.; Huang, L.; Yuan, W.; Chen, X.; Hao, X.; Xie, X.; Zhang, Y.; Zhong, Z.; et al. European Carbon Uptake has Not Benefited from Vegetation Greening. *Geophys. Res. Lett.* **2021**, *48*, e2021GL094870. [CrossRef]

39. Wang, S.; Zhang, Y.; Ju, W.; Porcar-Castell, A.; Ye, S.; Zhang, Z.; Brümmer, C.; Urbaniak, M.; Mammarella, I.; Juszczak, R.; et al. Warmer spring alleviated the impacts of 2018 European summer heatwave and drought on vegetation photosynthesis. *Agric. For. Meteorol.* **2020**, *295*, 108195. [CrossRef]
40. Rigden, A.J.; Mueller, N.D.; Holbrook, N.M.; Pillai, N.; Huybers, P. Combined influence of soil moisture and atmospheric evaporative demand is important for accurately predicting US maize yields. *Nat. Food* **2020**, *1*, 127–133. [CrossRef]
41. Jiao, W.; Wang, L.; Smith, W.K.; Chang, Q.; Wang, H.; D'Odorico, P. Observed increasing water constraint on vegetation growth over the last three decades. *Nat. Commun.* **2021**, *12*, 3777. [CrossRef] [PubMed]
42. Novick, K.A.; Konings, A.G.; Gentine, P. Beyond soil water potential: An expanded view on isohydricity including land–atmosphere interactions and phenology. *Plant Cell Environ.* **2019**, *42*, 1802–1815. [CrossRef] [PubMed]

atmosphere

Article

Estimating Concurrent Probabilities of Compound Extremes: An Analysis of Temperature and Rainfall Events in the Limpopo Lowveld Region of South Africa

Caston Sigauke *,† and Thakhani Ravele †

Department of Mathematical and Computational Sciences, University of Venda Private Bag X5050, Thohoyandou 0950, South Africa; thakhani.ravele@mvulaunivenac.onmicrosoft.com
* Correspondence: caston.sigauke@univen.ac.za; Tel.: +27-15-962-8135
† These authors contributed equally to this work.

Abstract: In recent years, there has been increasing interest in the joint modelling of compound extreme events such as high temperatures and low rainfall. The increase in the frequency of occurrence of these events in many regions has necessitated the development of models for estimating the concurrent probabilities of such compound extreme events. The current study discusses an application of copula models in predicting the concurrent probabilities of compound low rainfall and high-temperature events using data from the Lowveld region of the Limpopo province in South Africa. The second stage discussed two indicators for monitoring compound high temperature and low rainfall events. Empirical results from the study show that elevations ranging from 100–350 m, 350–700 m and 700–1200 m exhibit varying probabilities of experiencing drought, with mild droughts having approximately 64%, 66%, and 65% chances, moderate droughts around 36%, 39%, and 38%, and severe droughts at approximately 16%, 19%, and 18%, respectively. Furthermore, the logistic regression models incorporating the southern oscillation index as a covariate yielded comparable results of copula-based models. The methodology discussed in this paper is robust and can be applied to similar datasets in any regional setting globally. These findings could be useful to disaster management decision makers, helping them formulate effective mitigation strategies and emergency response plans.

Keywords: bivariate extremes; copulas; drought; joint extreme events; rainfall deficit; temperature

1. Introduction

1.1. Overview

The current study discusses applying copula models to predict the concurrent probabilities of compound low-rainfall and high-temperature events using data from the Lowveld region of the Limpopo province in South Africa. In order to develop effective climate adaptation and risk-management strategies, it is therefore important to accurately predict the joint behaviour of high temperature and low rainfall. There is a strong correlation between the frequency and impact of extreme events, such as heat waves and droughts. However, extreme events can have a greater impact when they occur in a particular order or sequence. Drought and heat waves occurring simultaneously have a greater effect than univariate counterparts [1]. A simultaneous occurrence of extreme events can significantly impact the ecosystem and society. Climate change and variability can be mitigated by predicting changes in concurrent climate extremes [2].

A few studies have examined concurrent climatic extremes, but most analysed changes in a single climate variable. Climate extremes, such as extreme temperature and rainfall, are critical in determining drought severity and risk. In order to manage and mitigate natural hazards, it is imperative to understand rainfall and temperature trends, their probabilistic characteristics, and how they relate to future climate changes [2].

1.2. A Survey of the Related Literature

Due to the increased threat to human society and ecosystems caused by extreme weather events, there has been a growing interest in the joint modelling of high temperatures and low rainfall. Several approaches are used in modelling such events, but the copula modelling framework has gained much interest.

An overview of copula modelling is discussed in detail by [3]. The authors investigated the dependence between two random variables using copulas. Although the paper emphasised inference and testing procedures, the authors also presented an application of the proposed methodology to modelling Harricana River data.

Bivariate extreme value-copula models are powerful in modelling the joint distribution of extreme compound events such as temperature and rainfall extremes [4]. The bivariate extreme value-copula model has several advantages. It can capture the tail dependence between compound extreme events such as temperature and rainfall. In addition, it allows for a more flexible joint distribution modelling, including nonlinear relationships between extreme compound events [4]. In support of bivariate extreme-value copulas, ref. [5] argues that extreme-value copulas are among the most commonly used copula families since they can capture asymmetry well and are also known to be very flexible.

A recent study in modelling drought risk using bivariate spatial extreme is that of [6]. The authors used temperature and rainfall data to model meteorological drought. Max-stable processes were used in the study to capture the spatio-temporal dependencies of temperature and rainfall data from the Limpopo Lowveld region of South Africa. Results from this study showed that the Schlather model with various covariance functions was a good fit for both data sets compared to the Smith model based on the Gaussian covariance function. However, in this study, the authors did not estimate concurrent probabilities.

In another study, ref. [7], the author used the multivariate frequency analysis to quantify drought risk in the contiguous United States (CONUS). This was carried out by analysing the temperature and rainfall data of CONUS. Results from this study showed that the dependence between low rainfall and high temperature could be positive, negative, or insignificant and that there were no major changes in the last three years. Serinaldi [7] argues that the probability of occurrence of the compound event depends largely on the variables selected and how they are combined.

Furthermore, ref. [8] used Indian data to investigate the concurrence of meteorological droughts and heatwaves. Both variables' extremes are modelled using the peaks over threshold method. Empirical results from this study suggest that there could be an increase in the frequency of concurrent meteorological droughts and heatwaves in India. Zscheischler and Seneviratne [9] investigated how the dependence structure between meteorological variables affects the frequency of occurrence of multivariate extremes. They argue that to fully understand the changes in climate extremes, including their impacts and the designing of adaptation strategies, it is important to use the multivariate modelling framework.

A review of the different approaches used in the characterisation and modelling compound extremes in hydroclimatology is given by [10]. The approaches discussed include the indicator approach, empirical approach, multivariate distribution, quantile regression, and Markov chain model. The authors highlight the limitations of the data available for modelling extremes and the challenges of modelling asymmetric tail dependencies of multiple events. In another study, ref. [11] conducted a comparative analysis of traditional empirical methods and copula models to estimate the probability of compound climate extremes, i.e., hot, dry and windy events, using data from the central United States of America. In a separate study, ref. [12] used copula models to establish the characteristics and the probability of the occurrence of different combinations of water discharge and several water quality indicators. Empirical results from this study showed that the Gaussian copula is the best function for describing the joint distribution of water discharge and water quality.

McKee et al. [13] utilised the standardised rainfall index (SPI) to classify droughts into four primary categories. Specifically, they defined mild droughts when SPI falls within the range of 0 to −0.99, moderate droughts for SPI between −1 and −1.49, severe droughts for SPI in the range of −1.5 to −1.99, and extreme droughts for SPI less than or equal to −2. The authors argued that for SPI values of −2, −1, 0, 1, and 2, there are associated probabilities that the SPI will be less than or equal to the values above, namely, 0.02, 0.16, 0.5, 0.84, and 0.92, respectively.

Drought is recognised as a complex phenomenon. Esit and Yuce [14] in their study argue that a comprehensive analysis of drought necessitates modelling it with multiple variables. The authors used the SPI to characterise drought and utilised various bivariate copula functions in their study, considering different elevation levels. Carrillo et al. [15] support the modelling of drought considering different elevation levels and claim that considering different elevation levels is important. They argue that in regions characterised by complex topography, including elevation gradients can significantly contribute to an improved understanding of drought modelling.

Using the SPI values for two sub-seasons of the rain season, October to December and January to March, ref. [16] assessed the impact of elevation on the severity of drought and frequency of occurrence using South African data from the Free State province over the period 1960–2013. Empirical results showed that highland areas had the highest frequency of droughts. However, the authors noted that extreme droughts occurred in the low-lying areas. It also stated that variations in altitude have notable impacts on the severity of drought at the onset of the summer compared to the late summer season. In a related study using two drought indices, ref. [17] assessed meteorological drought and wet conditions using data from the KwaZulu-Natal province in South Africa. This study showed increased drought frequency and severity with the most extreme dry periods experienced between the 1992–1993 and 2015–2016 summer seasons.

1.3. Research Highlights

Based on the literature survey in Section 1.2, the highlights and contributions of this study are as follows:

- Elevations ranging from 100–350 m, 350–700 m, and 700–1200 m show varying probabilities of experiencing drought, with mild droughts having approximately 64%, 66%, and 65% chances, moderate droughts around 36%, 39%, and 38%, and severe droughts at approximately 16%, 19%, and 18% respectively.
- Specific elevation clusters exhibit distinct frequency probabilities for mild and moderate drought occurrences, such as 0.43 and 0.03 for 100–350 m, 0.32 and 0.02 for 350–700 m, and 0.02 for mild drought at 700–1200 m elevations.
- Logistic regression models incorporating the Southern Oscillation Index (SOI) as a covariate yielded comparable results to copula-based models, demonstrating strong predictive performance for compound low rainfall and high temperatures during the 2015/2016 season.
- The monitoring system captured the major drought years in Southern Africa between 1970 and 2020, which are the 1982/1983, 1991/1992, 2002/2003, 2015/2016, and 2019/2020 seasons.

The remainder of the paper is structured as follows: Section 2 elaborates on the methodology. Section 3 presents the empirical results, followed by the discussion in Section 4, and the conclusion in Section 5.

2. Methods

It is well known that rainfall is dependent on temperature. As temperature increases, rainfall also increases. However, there is a decrease in rainfall for high temperatures, resulting in meteorological drought. Due to the dependence between these two meteorological variables, their joint distribution is best described by copula functions. The present study is an extension of the work done by [6].

2.1. Bivariate Copulas

A copula C is a joint distribution function of standard uniform random variables and is given as [18].

$$C(u_1, \ldots, u_n) = P(U_1 \leq u_1, \ldots, U_d \leq u_n), \tag{1}$$

where $U_i = U(0,1) \forall i = 1, \ldots, n$. Copulas capture the dependence structure of random variables very well, separately from the marginal distributions. Let Y denote rainfall and X temperature. The distribution function of X and Y is given as ([18])

$$G(y, x) = C(F_Y(y), F_X(x)) = C(v, u), \tag{2}$$

where $U = F_X(x)$ and $V = F_Y(y)$ are standard uniform variables. There are three main classes of copulas: the elliptical, Archimedean, and extreme value copulas. In this study, the elliptical and Archimedean copulas were used. These copulas are known to be flexible in modelling the joint distribution of variables with different marginal distributions.

Drought risk is typically associated with inadequate rainfall. Low rainfall decreases soil moisture, impacting vegetation, agriculture, and water supplies. Temperature, on the other hand, affects evaporation rates, potentially intensifying drought conditions. Considering this, $P(V < v, U > u)$ would provide a more comprehensive understanding of the joint impact of rain and temperature on drought risk. From Figure 1, region 4 is our region of interest in which we need to calculate $P(V < v, U > u)$.

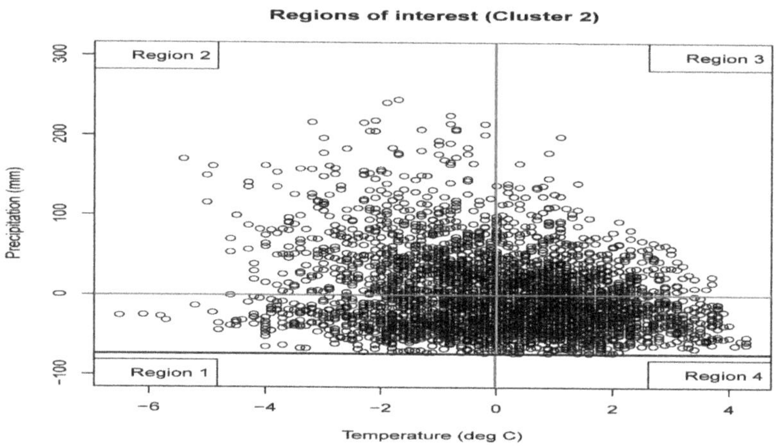

Figure 1. Scatter plot of temperature and rainfall showing the regions of interest for cluster 2. Rainfall and temperature anomalies were used. The blue horizontal line refers to a the reference rainfall while the red line is the reference temperature both used for comparison.

The joint probability $P(V < v, U > u)$ quantifies the probability of experiencing low rainfall with high temperatures above a certain threshold. This probability captures the joint impact of rain and temperature on drought risk by considering situations where both contribute to drought conditions. It helps assess the likelihood of drought risk with insufficient rainfall and high temperatures.

On the other hand, the conditional probability $P(V < v, U > u)$ focuses on rainfall alone, given a specific temperature threshold. While it provides information on the likelihood of low rainfall during high-temperature periods, it does not explicitly capture the joint impact of both variables. Therefore, $P(V < v, U > u)$ is better suited for describing the joint impact of rain and temperature on drought risk as it considers the combined effects of these variables, providing a more holistic understanding of the conditions leading to drought.

2.1.1. Elliptical Copulas

Elliptical copulas are a class of copulas used in multivariate statistics to model the dependence structure between random variables. These copulas are based on elliptical distributions, which include well-known distributions like the multivariate normal (Gaussian) distribution, the Student's t-distribution, and the multivariate Laplace distribution. Elliptical copulas are useful for modelling symmetric, linear, and well-behaved dependencies. Two common models within the elliptical copulas are the Gaussian and the t copulas ([19]).

The choice between Gaussian and t copulas depends on the specific characteristics of the data and the modelling objectives. Gaussian copulas are simpler but may not capture extreme events well, while t copulas offer more flexibility but come with greater complexity and estimation challenges ([19]).

The Gaussian Copula

The Gaussian copula is defined as ([19])

$$C_P^{Gauss} = \Phi(\Phi^{-1}(u_1), \ldots, \Phi(\Phi^{-1}(u_d)), \tag{3}$$

where $\Phi(.)$ is the standard univariate normal distribution function (DF) and $\Phi_P(.)$ represents the joint CDF of $\mathbf{X}$.

The t Copula

The t copula is defined as ([19])

$$C_{v,P}^{t}(u) = t_{v,P}(t_v^{-1}(u_1), \ldots, t_{v,P}^{-1}(u_d)), \tag{4}$$

where P is a correlation matrix, $t_{v,P}$ is the joint DF of $\mathbf{X} \sim t_d(v, 0, P)$ and t_v is the standard univariate DF of a $t-$ distribution with v degrees of freedom.

2.1.2. Archimedean Copulas

Archimedean copulas have been widely used because they are convenient and easy to use ([20]), comprise different families, and possess several nice properties ([21]). The Archimedean copula produces a much better dependency model due to its more tractable mathematical properties. Archimedean copulas contain sufficient dependence models for modelling upper and lower tail dependences. Naifar [18] defines the bivariate Archimedean copula equation as given in Equation (5).

$$C_{arch}(u, v) = \phi^{-1}(\phi(u) + \phi(v)), \tag{5}$$

where ϕ denotes a generator function of the copula, if for all $0 \leq u_1, u_2 \leq 1$. Within the Archimedean copula, there are a variety of different dependency structures. These structures simplify the construction of bivariate distributions in many families ([18]). Two Archimedean copula functions are considered in this study: Frank and Gumbel.

The Frank Copula

Frank copula was first introduced by [22] and is defined by,

$$C_\theta^{Frank}(u, v) = -\frac{1}{\theta} \ln\left(1 + \frac{(e^{-\theta u} - 1)(e^{-\theta v} - 1)}{(e^{-\theta} - 1)}\right), \tag{6}$$

where θ denotes the dependence parameter. The upper and lower Fréchet–Hoeffding bounds can be determined using the θ. Modelling data with weak tail dependence is suitable for this model since it is not tail-dependent ($\lambda_u = \lambda_L = 0$).

The Gumbel Copula

Gumbel copula was first introduced by [23] and is defined by Equation (7).

$$C_\theta^{Gumbel}(u,v) = \exp\left(-\left[(-\ln u)^\theta + (-\ln v)^\theta\right]^{\frac{1}{\theta}} \right); \qquad 0 \leq u,\, v \leq 1, \tag{7}$$

where the parameter $\theta \in [1,\infty)$ controls the degree of dependence between u and v. If $\theta = 1$, the bivariate Gumbel copula converges to complete independence, and if $\theta \to 0$, perfect independence is achieved. The bivariate Gumbel parameter (θ) and Kendall's tau (τ) are integrated by the following formula:

$$\tau_k = 1 - \theta^{-1}. \tag{8}$$

Bivariate Gumbel copula upper (λ^U) and lower (λ^L) tail dependence estimation is carried out by the following functions:

$$\lambda^U = 2 - 2^{\frac{1}{\theta}} \text{ and } \lambda^L = 0.$$

Estimation of the parameters of the copula models will be completed using the maximum likelihood method.

Table 1 presents the relationship between bivariate elliptical and Archimedean copula parameters (θ), tail-dependence coefficients: Kendall's tau (τ) and upper and lower tail dependence, λ^U and λ^L, respectively.

Table 1. Bivariate Archimedean copulas.

Family	θ	Kendal's Tau (τ)	Upper Tail (λ^U)	Lower Tail (λ^L)
Gaussian				
t				
Frank	$-\infty < \theta < \infty$	$1 - \frac{4}{\theta}[D_j(\theta)]$	0	0
Gumbel	$\theta \geq 1$	$\frac{\theta-1}{\theta}$	$2 - 2^{\frac{1}{\theta}}$	0

2.2. *Joint and Conditional Distributions*

2.2.1. Joint Probability Distribution

The joint probability distribution of low rainfall and high temperature is given in Equation (9).

$$P(V < v, U > u) \quad = \quad P(V \leq v) - P(V \leq v, U \leq u) \tag{9}$$

2.2.2. Conditional Distribution

The conditional distribution of compound high temperature and low rainfall is given in Equation (10).

$$\begin{aligned}
P(V < v|U > u) &= \tfrac{P(V<v,U>u)}{P(U>u)} \\
\text{But } P(V < v, U > u) &= P(V \leq v) - P(V \leq v, U \leq u) \text{ Joint Probability} \\
\Longrightarrow P(V < v|U > u) &= \tfrac{P(V\leq v)-P(V\leq v,U\leq u)}{PU>u} \\
&= \tfrac{P(V\leq v)-P(V\leq v,U\leq u)}{1-P(U\leq u)} \\
&= \tfrac{C(v,1)-C(v,u)}{1-C(1,u)} \\
&= \tfrac{v-C(v,u)}{1-u}
\end{aligned} \tag{10}$$

2.3. Monitoring Compound Dry and Hot Events

We use two indicators to monitor compound low rainfall and high temperature events. The first is the standardised compound event indicator (SCEI) discussed in [24]. This indicator is based on the joint probability of the two weather variables, rainfall and temperature, in this study.

$$\text{SCEI} = \Phi^{-1}(F(P(Y \leq y, X > x))) \tag{11}$$

According to [24], lower SCEI values indicate more severe occurrences of compound dry and hot events.

The second indicator assesses the concurrence of low rainfall and high temperature for specific regions. For given thresholds, p_o and t_o of rainfall and temperature, respectively, the occurrence of a compound low rainfall and high temperature event can be defined as given in Equation (12) ([24]).

$$I = \begin{cases} 1, & \text{if } P \leq p_o, T \geq t_o, \\ 0, & \text{otherwise,} \end{cases} \tag{12}$$

Table 2 summarises the drought classification discussed by [24]. We present only part of the table given in [24].

Table 2. SPI Drought classification ([24]).

SPI Values	Drought Category	Time in Category
0 to −0.99	mild drought	24%
1.00 to −1.49	moderate drought	9.2%
1.50 to −1.99	severe drought	4.4%
≤−2.00	extreme drought	2.3%

2.4. Prediction of The Occurrences of Compound Events

To predict the occurrence of compound events, i.e., when $Z = 1$, using the Southern Oscillation Index (SOI) as a covariate, we applied the logistic regression model outlined by [24] and represented by Equation (13).

$$\ln\left[\frac{\theta}{1-\theta}\right] = \alpha + \beta x, \tag{13}$$

where the probability of occurrence $P(Z = 1|x)$ is denoted by θ, α represents the constant, β is the regression coefficient, and x denotes a covariate, which, in this context, is the SOI. Consequently, the one-month-ahead forecast of the probability of compound low rainfall and high temperature events occurring (i.e., $Z = 1$) can be expressed as shown in Equation (14).

$$P(Z_{t+1} = 1|x) = \frac{1}{1 + \exp[-(\alpha + \beta x_t)]} \tag{14}$$

The aim is to evaluate the influence of SOI on the prediction of compound event occurrences, specifically when $Z = 1$.

3. Empirical Results

3.1. Data and Study Area

The study area for this research is found in the Lowveld region of the Limpopo Province in South Africa, situated between latitude −22° and −24° and longitude 30° and 33°. This area is known to experience warm temperatures throughout the year and has an average annual rainfall of 500 mm from October to March ([25]). In this study, we use monthly rainfall and temperature data. We consider only summer months, i.e., October to March.

The top and bottom panels of Figure 2 show maps illustrating the average annual precipitation and the average maximum temperature across the study area, respectively. These maps also identify clusters corresponding to annual average precipitation and maximum temperature within the study region.

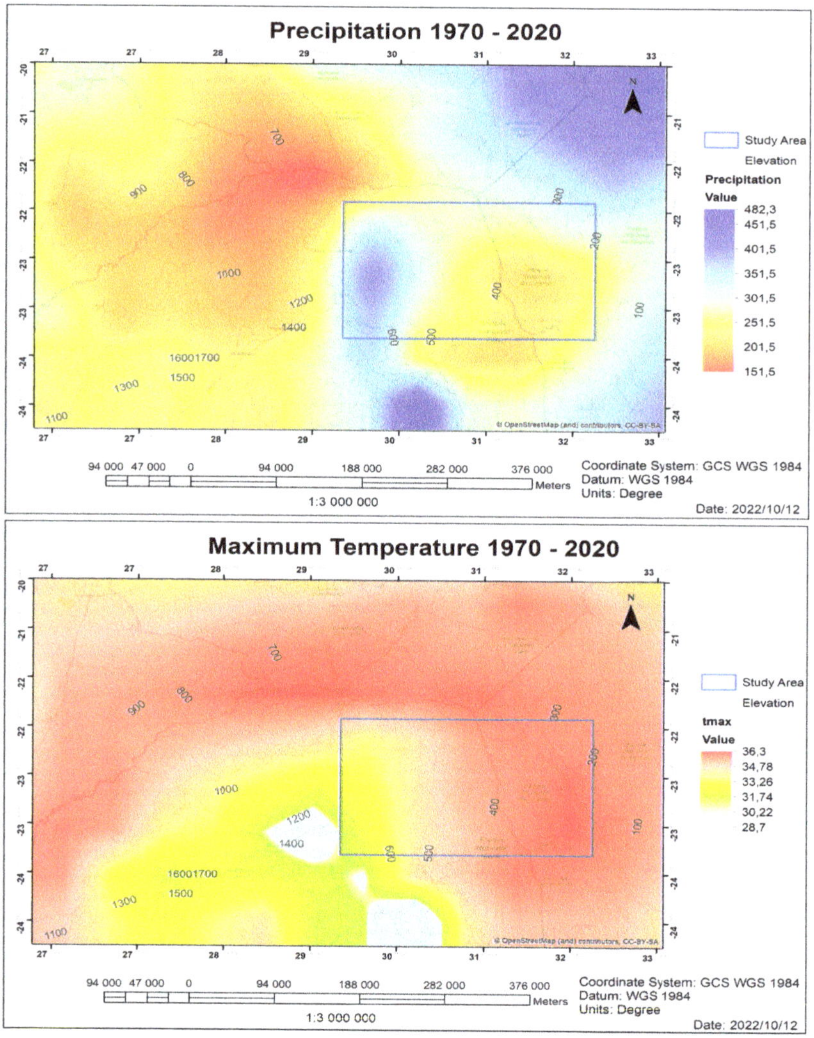

Figure 2. Top panel: Annual average rainfall. **Bottom panel:** Annual maximum temperature. Source: [6].

The study area is split into three subregions based on the following elevation (altitude) intervals: subregion one 100–350 m, subregion two 350–700 m, and subregion three 700–1200 m, respectively. The average temperatures in these three subregions are: 32.9°, 31.9° and 29.4°, respectively. The average rainfall in these three subregions 1–3 during the rainy season, October to March of each year, are 73.4 mm, 73.9 mm, and 110.1 mm, respectively.

This study uses anomalies of both temperature and rainfall data. Using anomalies can help remove long-term trends, seasonality, and other confounding factors, making it easier to analyse and model the joint distribution. The metadata, a summary of the grid points, is given in Table A1. The three groups based on elevation are as follows:

- Cluster 1 (elevation 100–350 m) has the following grid points: r4c6, r2c6, r3c6, r1c6, r4c5, r3c5, r2c5, r2c4, r1c4, r1c5 and r4c4 with $n_2 = 3366$ observations.
- Cluster 2 (elevation 350–700 m) has the following grid points: r2c3, r3c4, r4c3, r3c3, r1c3, r2c2, r3c2,r4c2, r1c2, r1c1 and r2c1 with $n_2 = 3366$ observations.
- Cluster 3 (elevation 700–1200 m) has the following grid points: r4c1 and r3c1 with $n_2 = 612$ observations.

We fitted theoretical distributions to precipitation and temperature data. A summary of the results is given in Table 3.

Table 3. Distribution fitting to the data.

	Elevation 100–350 m		Elevation 350–700 m		Elevation 700–1200 m	
Distribution	Temp	Rain	Temp	Rain	Temp	Rain
Gamma	shape = 2561.8 (13.7) rate = 17.1 (0.42)	shape = 2.46 (0.056) rate = 0.03 (0.0008)	shape = 335.68 (8.18) rate = 10.51 (0.26)	Shape = 2.155 (0.0489) rate = 0.029 (0.0007)	shape = 346.85 (19.82) rate = 11.81 (0.68)	
Weibull						shape = 1.67 (0.02) scale = 123.35

3.2. Exploratory Data Analysis

Figure 3 shows histograms (diagonal) superimposed with kernel densities, pairwise scatter plots (bottom left), and pairwise Kendall's rank correlation coefficient (top right) of temperature and rainfall data for cluster 2 (elevation 350–700 m). The figures for clusters 1 and 3 are given in Figrues A1 and A2, respectively.

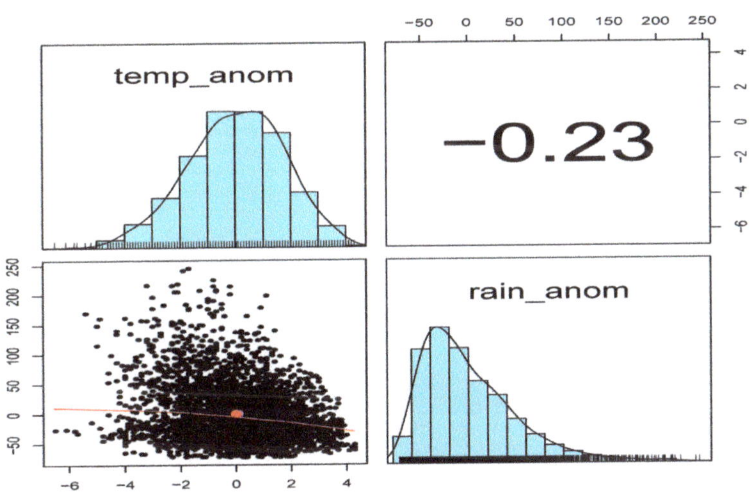

Figure 3. Cluster 2: Histograms (diagonal), a scatter plot of the rainfall and temperature (**bottom left**). The red dot represents the reference point (0,0). Kendall's rank correlation coefficient (**top right**) of temperature and rainfall data for Cluster 2 (elevation 350–700 m).

Based on Kendall's rank correlation coefficient values for Cluster 1 (elevation 100–350 m), Cluster 2 (elevation 350–700 m), and Cluster 3 (elevation 700–1200 m), rainfall and temperature tend to be more negatively correlated for cluster 2 and a very weak correlation in cluster 1.

Figure 4 shows plots of rainfall, SOI, and temperature, accompanied by box plots displaying their distributions for Cluster 1, corresponding to elevations ranging from 100 to 350 m. High rainfall is observed during the December to February period.

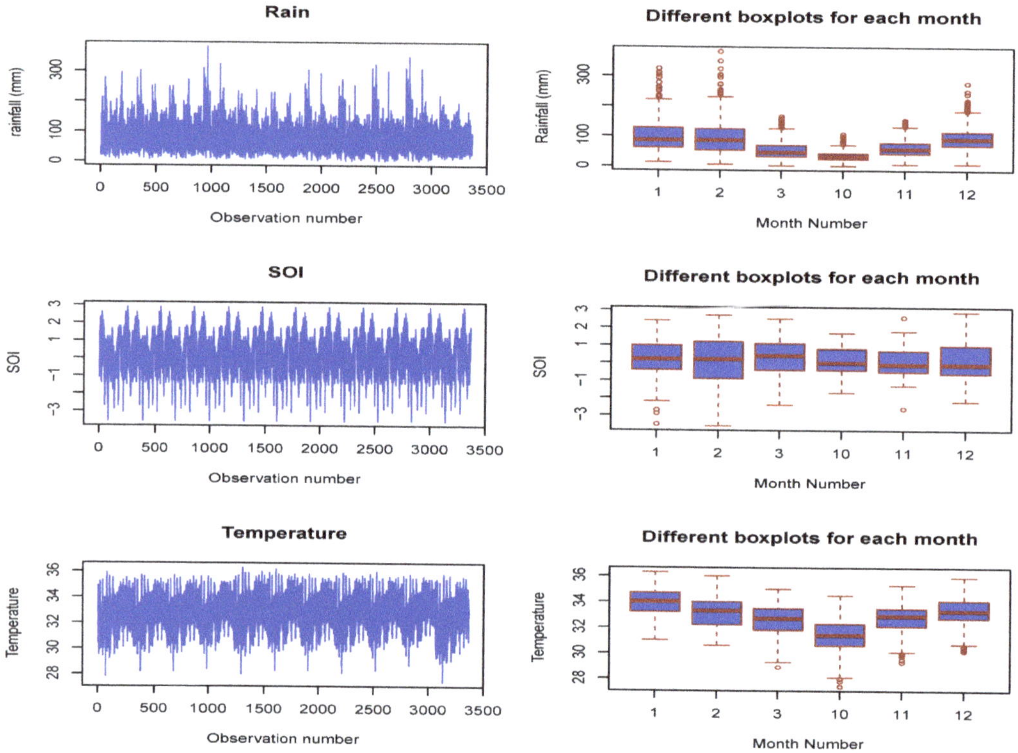

Figure 4. Plots for Cluster 1 (elevation 100–350 m).

3.3. Results

A formal goodness-of-fit test was carried out. The choice of the significance level (α) depends on the trade-off between type I and type II errors. A lower α, such as 0.01 compared to 0.05, reduces the risk of type I error, i.e., false positives. In this study, $\alpha = 0.01$ will be used. The parameter estimates for the best-fitting copula function for all three clusters are given in Table 4.

Table 4. (**a**) Parameter estimation for the copula functions: Cluster 1 (Elevation 100–350 m); (**b**) Parameter estimation for the copula functions Cluster 2 (Elevation 350–700 m); (**c**) Parameter estimation for the copula functions Cluster 3 (Elevation 700–1200 m).

Family	Copula	Estimate (ρ)	ℓ	AIC	BIC	λ^U	λ^L	τ
				(a)				
Arhimedean	Frank	0.1710	1.384	−0.7680	5.353	0	0	0.0190
				(b)				
Arhimedean	Frank	−1.1137	57.45	−112.89	−106.77	0	0	−0.1224
				(c)				
Arhimedean	Frank	−0.8171	5.61	−9.220	−4.803	0	0	−0.0902

Estimating Concurrent Probabilities

Approximate temperature ranges for the different levels of drought severity in the Limpopo Lowveld region of South Africa might be as follows:

- Mild Drought: Slightly above-average temperatures, typically 1–3 °C above the long-term average.
- Moderate Drought: Elevated temperatures, often in the 3–5 °C range above the long-term average.
- Severe Drought: High temperatures, commonly exceeding 5 °C above the long-term average.
- Extreme Drought: High temperatures, potentially reaching 7 °C or more above the long-term average.

Average rainfall and temperature in the study area during the rainy season, October to March, are 74 mm and 32 °C, respectively. The drought characterisation based on the rainfall and temperature ranges of values is given in Table 5. These ranges were based on the information in Table 2 third column, i.e., time in the category given as a percentage.

As discussed in Section 2.1, the joint probability considers the simultaneous occurrence of low rainfall and high temperature, key factors influencing drought conditions. It provides insights into the combined effect of these variables, indicating when both conditions align to contribute to drought risk.

Table 5. Drought characterisation based on rainfall and temperature ranges of values.

Drought Characterisation	Rainfall (mm)	Temperature (Degrees Celsius)
Mild	$59 < Y \leq 74$	$32 < X \leq 35$
Moderate	$52 < Y \leq 59$	$35 < X \leq 37$
Severe	$44 < Y \leq 52$	$37 < X \leq 40$
Extreme	$Y \leq 44$	$X > 40$

Tables 6–8 show the results from the computation of the joint and conditional probabilities for the three elevation groups: 100–350 m, 350–700 m, and 700–1200 m, respectively. The conditional probabilities are significantly higher compared to the joint probabilities. This result is expected since there is some dependency between rainfall and temperature, meaning the occurrence of one variable affects the probability of another.

Table 9 summarises the number of low-rainfall and high-temperature occurrences for the four drought categories. The values in set braces denote the number of occurrences followed by the probability of the occurrence. In square brackets are the probabilities from the logistic regression based on SOI as the covariate. The comparable probabilities suggest that SOI is a good predictor of the joint probabilities of the compound low rainfall and high temperatures.

Table 6. (**a**) Joint and conditional probabilities based on the Frank copula (Elevation 100–350 m); (**b**) $P(Y < y, X > x) = \frac{\sum_{\forall xy}[Y<y,X>x]}{n}$ and $P(Y < y|X > x) = \frac{\sum_{\forall xy}[Y<y,X>x]}{\sum_{\forall x}[X>x]}$ (Elevation 100–350 m).

Joint Probability	Conditional Probability	Drought
	(**a**)	
$P(Y < 74, X > 32) = 0.6378$	$P(Y < 74 \mid X > 32) = 0.7973$	Mild
$P(Y < 59, X > 35) = 0.3551$	$P(Y < 59 \mid X > 32) = 0.5959$	Moderate
$P(Y < 52, X > 37) = 0.1551$	$P(Y < 52 \mid X > 32) = 0.3959$	Severe
$P(Y < 44, X > 40) = 0.0378$	$P(Y < 44 \mid X > 32) = 0.1972$	Extreme
	(**b**)	
$P(Y < 74, X > 32) = 0.6316$	$P(Y < 74 \mid X > 32) = 0.7889$	Mild
$P(Y < 59, X > 35) = 0.3523$	$P(Y < 59 \mid X > 32) = 0.5996$	Moderate
$P(Y < 52, X > 37) = 0.1554$	$P(Y < 52 \mid X > 32) = 0.3688$	Severe
$P(Y < 44, X > 40) = 0.0487$	$P(Y < 44 \mid X > 32) = 0.1859$	Extreme

Table 7. (**a**) Joint and conditional probabilities based on the Frank copula (Elevation 350–700 m); (**b**) $P(Y < y, X > x) = \frac{\sum_{\forall xy}[Y<y,X>x]}{n}$ and $P(Y < y|X > x) = \frac{\sum_{\forall xy}[Y<y,X>x]}{\sum_{\forall x}[X>x]}$ (Elevation 100–350 m).

Joint Probability	Conditional Probability	Drought
	(a)	
$P(Y < 74, X > 32) = 0.6566$	$P(Y < 74 \| X > 32) = 0.8207$	Mild
$P(Y < 59, X > 35) = 0.3898$	$P(Y < 59 \| X > 32) = 0.6273$	Moderate
$P(Y < 52, X > 37) = 0.1898$	$P(Y < 52 \| X > 32) = 0.4261$	Severe
$P(Y < 44, X > 40) = 0.0566$	$P(Y < 44 \| X > 32) = 0.2180$	Extreme
	(b)	
$P(Y < 74, X > 32) = 0.6480$	$P(Y < 74 \| X > 32) = 0.8099$	Mild
$P(Y < 59, X > 35) = 0.3734$	$P(Y < 59 \| X > 32) = 0.6045$	Moderate
$P(Y < 52, X > 37) = 0.1916$	$P(Y < 52 \| X > 32) = 0.4181$	Severe
$P(Y < 44, X > 40) = 0.0663$	$P(Y < 44 \| X > 32) = 0.2094$	Extreme

Table 8. (**a**) Joint and conditional probabilities based on the Frank copula (Elevation 700–1200 m); (**b**) $P(Y < y, X > x) = \frac{\sum_{\forall xy}[Y<y,X>x]}{n}$ and $P(Y < y|X > x) = \frac{\sum_{\forall xy}[Y<y,X>x]}{\sum_{\forall x}[X>x]}$ (Elevation 700–1200 m).

Joint Probability	Conditional Probability	Drought
	(a)	
$P(Y < 74, X > 32) = 0.6509$	$P(Y < 74 \| X > 32) = 0.8136$	Mild
$P(Y < 59, X > 35) = 0.3835$	$P(Y < 59 \| X > 32) = 0.6197$	Moderate
$P(Y < 52, X > 37) = 0.1835$	$P(Y < 52 \| X > 32) = 0.4191$	Severe
$P(Y < 44, X > 40) = 0.0509$	$P(Y < 44 \| X > 32) = 0.2123$	Extreme
	(b)	
$P(Y < 74, X > 32) = 0.1144$	$P(Y < 74 \| X > 32) = 0.7873$	Mild
$P(Y < 59, X > 35) = 0.0683$	$P(Y < 59 \| X > 32) = 0.5991$	Moderate
$P(Y < 52, X > 37) = 0.0315$	$P(Y < 52 \| X > 32) = 0.3988$	Severe
$P(Y < 44, X > 40) = 0.0119$	$P(Y < 44 \| X > 32) = 0.2168$	Extreme

Table 9. (**a**) Number of occurrences for Cluster 1 (Elevation: 100–350 m ($n = 3366$)); (**b**) Number of occurrences for Cluster 2 (Elevation: 350–700 m ($n = 3366$)); (**c**) Number of occurrences for Cluster 3 (Elevation: 700–1200 m ($n = 612$)).

Indicator	Mild	Moderate	Severe	Extreme
		(a)		
	$P(Y < 74, X > 32)$	$P(Y < 59, X > 35)$	$P(Y < 52, X > 37)$	$P(Y < 44, X > 40)$
$I = 1$	(1464) 0.4349 [0.3039]	(99) 0.0294 [0.0023]	(0) 0 [0]	(0) 0 [0]
$I = 0$	(1902) 0.5651 [0.6961]	(3267) 0.9706 [0.9977]	(3366) 1 [1]	(3366) 1 [1]
		(b)		
	$P(Y < 74, X > 32)$	$P(Y < 59, X > 35)$	$P(Y < 52, X > 37)$	$P(Y < 44, X > 40)$
$I = 1$	(1088) 0.3232 [0.1982]	(72) 0.0214 [0.0027]	(0) 0 [0]	(0) 0 [0]
$I = 0$	(2278) 0.6768 [0.8018]	(3294) 0.9786 [0.9973]	(3366) 1 [1]	(3366) 1 [1]
		(c)		
	$P(Y < 74, X > 32)$	$P(Y < 59, X > 35)$	$P(Y < 52, X > 37)$	$P(Y < 44, X > 40)$
$I = 1$	(14) 0.0229 [0.0021]	(0) 0 [0]	(0) 0 [0]	(0) 0 [0]
$I = 0$	(598) 0.9771[0.9979]	(612) 1 [1]	(612) 1 [1]	(612) 1 [1]

The number of occurrences for mild drought $P(Y \leq 74, X \geq 32)$ (Cluster 1), $P(Y \leq 74, X \geq 32)$ (Cluster 2) and $P(Y \leq 74, X \geq 32)$ (Cluster 3) are shown in Figures 5–7, respectively. The monitoring system captured the major drought years in Southern Africa between 1970 and 2020, which are 1982/1983, 1991/1992, 2002/2003, 2015/2016, and the 2019/2020 seasons [26]. Out of the 1088 incidents of mild drought from cluster 2, 85 were for the 2015/2016 period, which was seen as the most severe drought for the sampling period 2000–2020. These results are consistent with those of [26].

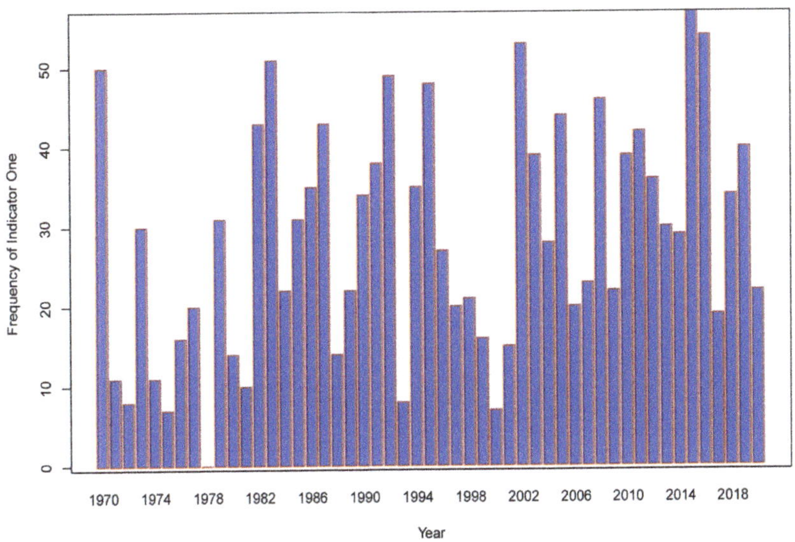

Figure 5. Frequency of indicator ones for mild drought $P(Y \leq 74, X \geq 32)$. Cluster 1 (Elevation: 100–350 m).

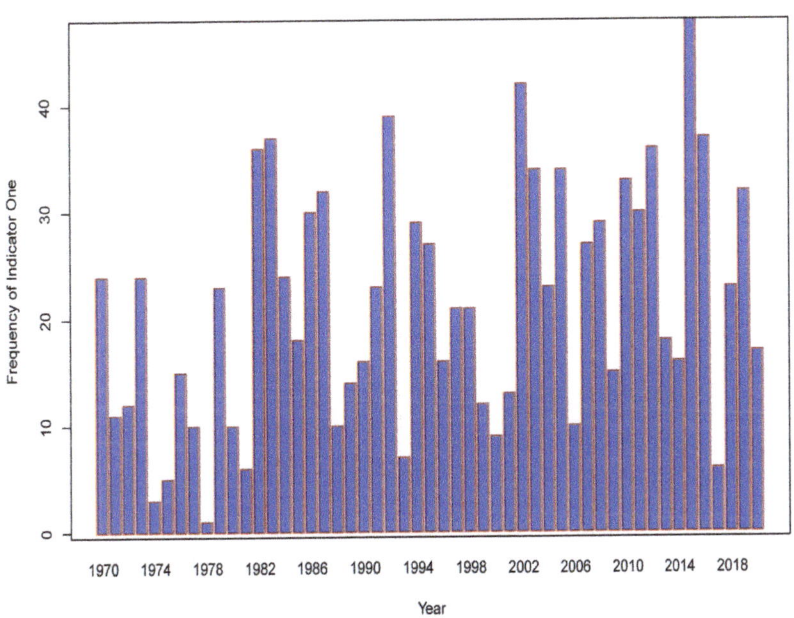

Figure 6. Frequency of indicator ones for mild drought $P(Y \leq 74, X \geq 32)$. Cluster 2 (Elevation: 350–700 m).

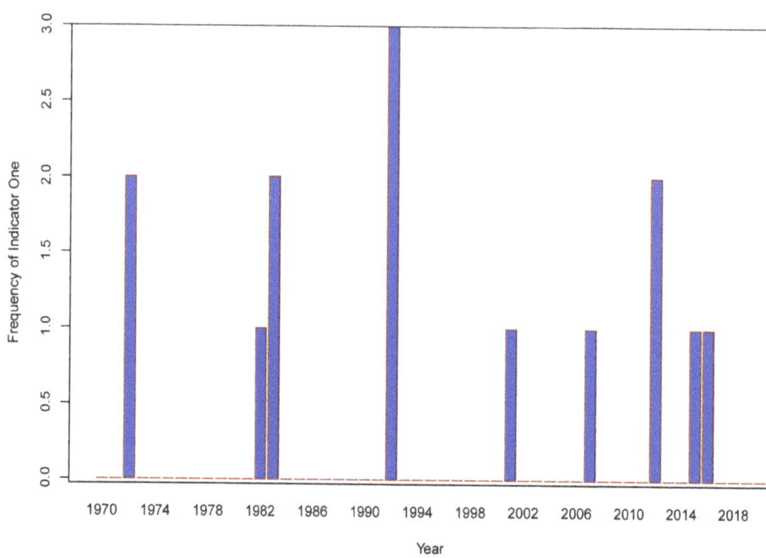

Figure 7. Frequency of indicator ones for mild drought $P(Y \leq 74, X \geq 32)$. Cluster 3 (Elevation: 700–1200 m).

4. Discussion

The current study was motivated by work discussed in [6]. It focused on analysing the concurrent occurrence of low rainfall and high temperatures in the Lowveld region of South Africa's Limpopo province. The primary aim was to evaluate how elevation influences the joint modelling of drought risk involving low rainfall and high temperatures. This involved using elliptical and Archimedean copulas for joint modelling and estimating the probabilities of getting low rainfall and high temperatures.

The findings of the study revealed that in elevations ranging from 100–350 m, 350–700 m, and 700–1200 m, there were approximately 64%, 66%, and 65% chances of experiencing mild droughts, with corresponding risks of moderate drought at around 36%, 39% and 38%, and severe drought at approximately 16%, 19%, and 18%, respectively. An analysis of mild and moderate drought occurrences for different elevation clusters indicated specific frequency probabilities, such as 0.43 and 0.03 for elevation 100–350 m and 0.32 and 0.02 for elevation 350–700 m, with zero probabilities for severe and extreme droughts. At 700–1200 m elevations, the probability was 0.02 for mild drought, with no occurrence of moderate, severe, or extreme droughts.

Furthermore, these frequency probabilities were compared with those obtained from a logistic regression model utilising the Southern Oscillation Index (SOI) as the sole covariate, demonstrating comparable results. The models exhibited strong predictive performance regarding compound low rainfall and high temperatures during the 2015/2016 season, as depicted in Figures 5–7. These results are consistent with those of [24].

One of the limitations of the study is that it did not include other indicators such as the standardised precipitation index (SPI), Palmer drought severity index (PDSI), normalised difference vegetation index (NDVI), drought severity and coverage index (DSCI), or the standardised precipitation evapotranspiration index (SPEI), among others. It is known that integrating multiple drought indicators provides a comprehensive assessment of drought risk and severity. This will be carried out in future research.

5. Conclusions

The study investigated the increasing occurrence of extreme compound events, focusing on concurrent low rainfall and high temperatures in South Africa's Lowveld region in the Limpopo province. It assessed how elevation influences drought risk, using copulas for joint modelling and estimating concurrent probabilities. Findings from the study show varying probabilities of mild, moderate, and severe droughts across elevation ranges, with comparisons to a logistic regression model using the southern oscillation index. The study's modelling framework provides insights into the complex relationship between high temperatures and low rainfall, offering valuable implications for disaster management and suggesting robust methodologies applicable globally.

Author Contributions: Conceptualization, C.S.; Methodology, C.S. and T.R.; Software, T.R.; Validation, C.S. and T.R.; Formal analysis, C.S. and T.R.; Investigation, C.S. and T.R.; Data curation, C.S.; Writing—original draft, C.S. and T.R.; Writing—review & editing, C.S. and T.R. These authors contributed equally to this work. All authors have read and agreed to the published version of the manuscript.

Funding: This research did not receive funding.

Institutional Review Board Statement: Not applicable.

Informed Consent Statement: Not applicable.

Data Availability Statement: The analytic data can be downloaded from https://github.com/csigauke (accessed on 31 October 2023).

Acknowledgments: The authors sincerely thank the anonymous reviewers for their helpful comments and suggestions on this paper.

Conflicts of Interest: There are no conflicts of interest.

Appendix A

Appendix A.1. Tables

Table A1 gives a summary of the metadata.

Table A1. Meta data. Source: Nemukula et al. [6]

Grid ID	r1c1	r1c2	r1c3	r1c4	r1c5	r1c6
Latitude Longitude	−22.5 30	−22.5 30.5	−22.5 31	−22.5 31.5	−22.5 32	−22.5 32.5
Grid ID	**r2c1**	**r2c2**	**r2c3**	**r2c4**	**r2c5**	**r2c6**
Latitude Longitude	−23 30	−23 30.5	−23 31	−23 31.5	−23 32	−23 32.5
Grid ID	**r3c1**	**r3c2**	**r3c3**	**r3c4**	**r3c5**	**r3c6**
Latitude Longitude	−23.5 30	−23.5 30.5	−23.5 31	−23.5 31.5	−23.5 32	−23.5 32.5
Grid ID	**r4c1**	**r4c2**	**r4c3**	**r4c4**	**r4c5**	**r4c6**
Latitude Longitude	−24 30	−24 30.5	−24 31	−24 31.5	−24 32	−24 32.5

Appendix A.2. Figures

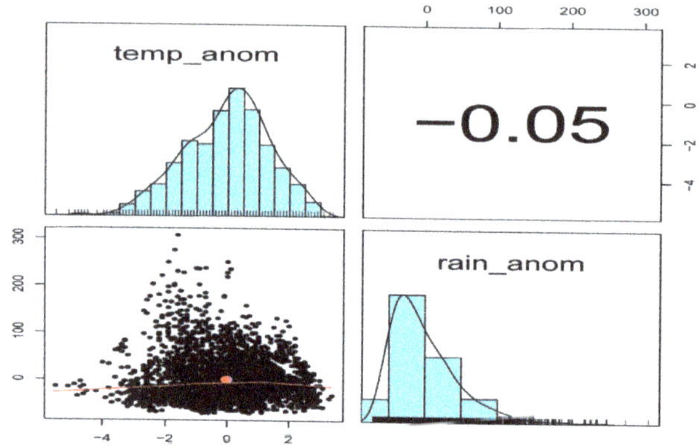

Figure A1. Histograms (diagonal), a scatter plot of the rainfall and temperature (**bottom left**). The red dot represents the reference point (0,0). Kendall's rank correlation coefficient (**top right**) of temperature and rainfall data for Cluster 1 (elevation 100–350 m).

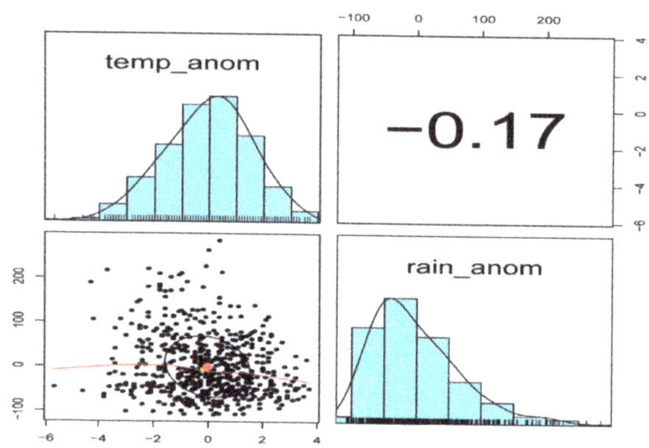

Figure A2. Histograms (diagonal), a scatter plot of the rainfall and temperature (**bottom left**). The red dot represents the reference point (0,0). Kendall's rank correlation coefficient (**top right**) of temperature and rainfall data for Cluster 3 (elevation 700–1200 m).

References

1. Sedlmeier, K.; Mieruch, S.; Schädler, G.; Kottmeier, C. Compound extremes in a changing climate—A Markov chain approach. *Nonlinear Process. Geophys.* **2016**, *23*, 375–390. [CrossRef]
2. Miao, C.; Sun, Q.; Duan, Q.; Wang, Y. Joint analysis of changes in temperature and rainfall on the Loess Plateau during the period 1961–2011. *Clim. Dyn.* **2016**, *47*, 3221–3234. [CrossRef]
3. Genest, C.; Favre, A.-C. Everything You Always Wanted to Know about Copula Modeling but Were Afraid to Ask. *J. Hydrol. Eng.* **2007**, *12*, 347–368. [CrossRef]
4. Gudendorf, G.; Segers, J. *Extreme-Value Copulas*; Springer: Berlin/Heidelberg, Germany, 2010; pp. 1–21. [CrossRef]
5. Serrano, J.F. Semiparametric bivariate extreme-value copulas. *arXiv* **2022**, arXiv:2109.11307. [CrossRef]
6. Nemukula, M.M.; Sigauke, C.; Chikoore, H.; Bere, A. Modelling Drought Risk Using Bivariate Spatial Extremes: Application to the Limpopo Lowveld Region of South Africa. *Climate* **2023**, *11*, 46. [CrossRef]
7. Serinaldi, F. Can we tell more than we can know? The limits of bivariate drought analyses in the United States. *Stoch. Environ. Res. Risk Assess.* **2016**, *30*, 1691–1704. [CrossRef]
8. Sharma, S.; Mujumdar, P. Increasing frequency and spatial extent of concurrent meteorological droughts and heatwaves in India. *Sci. Rep.* **2017**, *7*, 15582. [CrossRef] [PubMed]
9. Zscheischler, J.; Seneviratne, S.I. Dependence of drivers affects risks associated with compound events. *Sci. Adv.* **2017**, *3*, e1700263. [CrossRef]
10. Hao, Z.; Singh, V.P.; Hao, F. Compound Extremes in Hydroclimatology: A Review. *Water* **2018** , *10*, 718. [CrossRef]

11. Tavakol, A.; Rahmani, V.; Harrington, J., Jr. Probability of compound climate extremes in a changing climate: A copula-based study of hot, dry, and windy events in the central United States. *Environ. Res. Lett.* **2020**, *15*, 104058. [CrossRef]
12. Liu, Y.; Cheng, Y.; Zhang, X.; Li, X.; Cao, S. Combined Exceedance Probability Assessment of Water Quality Indicators Based on Multivariate Joint Probability Distribution in Urban Rivers. *Water* **2018**, *10*, 971. [CrossRef]
13. McKee, T.B.; Doesken, N.J.; Kleist, J. The Relationship of Drought Frequency and Duration to Time Scales. In Proceedings of the 8th Conference on Applied Climatology, Anaheim, CA, USA, 17–22 January 1993; pp. 179–184.
14. Esit, M.; Yuce, M.I. Copula-based bivariate drought severity and duration frequency analysis considering spatial–temporal variability in the Ceyhan Basin, Turkey. *Theor. Appl. Climatol.* **2023**, *151*, 1113–1131. [CrossRef]
15. Carrillo, J.; Hernández-Barrera, S.; Expósito, F.J.; Díaz, J.P.; González, A.; Pérez, J.C. The uneven impact of climate change on drought with elevation in the Canary Islands. *NPJ Clim. Atmos. Sci.* **2023**, *6*, 31. [CrossRef]
16. Mbiriri, M.; Mukwada. G.; Manatsa, D. Influence of Altitude on the Spatiotemporal Variations of Meteorological Droughts in Mountain Regions of the Free State Province, South Africa (1960–2013). *Adv. Meteorol.* **2018**, *2018*, 5206151. [CrossRef]
17. Ndlovu, M.S.; Demlie, M. Assessment of Meteorological Drought and Wet Conditions Using Two Drought Indices Across KwaZulu-Natal Province, South Africa. *Atmosphere* **2020**, *11*, 623. [CrossRef]
18. Naifar, N. Modelling dependence structure with Archimedean copulas and applications to the iTraxx CDS index. *J. Comput. Appl. Math.* **2011**, *235*, 2459–2466. [CrossRef]
19. Embrechts, P.; Lindskog, F.; McNeil, A. Modelling dependence with copulas. *Rapp. Tech. Dep. Math. Inst. Fed. Technol. Zur. Zur.* **2001**, *14*, 1–50.
20. Zhang, Q.; Li, J.; Singh, V.P. Application of Archimedean copulas in the analysis of the rainfall extremes: Effects of rainfall changes. *Theor. Appl. Climatol.* **2012**, *107*, 255–264. [CrossRef]
21. Nelsen, R.B. *An Introduction to Copulas*, 2nd ed.; Springer Series in Statistics Springer; Springer: Berlin/Heidelberg, Germany, 2006.
22. Frank, M.J. On the simultaneous associativity of F(x,y) and x + y—F(x,y). *Aequ. Math.* **1979**, *19*, 194–226. [CrossRef]
23. Gumbel, E.J. Distributions des valeurs extremes en plusiers dimensions. *Publ. Inst. Statist. Univ. Paris* **1960**, *9*, 171–173. Available online: https://hal.science/hal-04092830 (accessed on 2 February 2024).
24. Hao, Z.; Hao, F.; Xia, Y.; Singh, V.P.; Zhang, X. A monitoring and prediction system for compound dry and hot events. *Environ. Res. Lett.* **2019**, *14*, 114034. [CrossRef]
25. Chikoore, H.; Jury, M.R. South African drought, deconstructed. *Weather Clim. Extrem.* **2021**, *33*, 100334. [CrossRef]
26. Bhaga, T.D.; Dube, T.; Shekede, M.D.; Shoko, C. Impacts of Climate Variability and Drought on Surface Water Resources in Sub-Saharan Africa Using Remote Sensing: A Review. *Remote Sens.* **2020**, *12*, 4184. [CrossRef]

atmosphere

MDPI

Article

Spatiotemporal Patterns of Typhoon-Induced Extreme Precipitation in Hainan Island, China, 2000–2020, Using Satellite-Derived Precipitation Data

Mengyu Xu [1,2], Yunxiang Tan [1,2], Chenxiao Shi [3], Yihang Xing [1,2], Ming Shang [4], Jing Wu [5], Yue Yang [1,2], Jianhua Du [3] and Lei Bai [1,2,*]

[1] School of Ecology, Hainan University, Haikou 570228, China; xumy@hainanu.edu.cn (M.X.); yunxiangtan@hainanu.edu.cn (Y.T.); panda@hainanu.edu.cn (Y.X.); yueyang@hainanu.edu.cn (Y.Y.)
[2] Center for Eco-Environment Restoration Engineering of Hainan Province, Hainan University, Haikou 570228, China
[3] Hainan Meteorological Information Center, Hainan Key Laboratory of South China Sea Meteorological Disaster Prevention and Mitigation, Haikou 570203, China; shichenxiao127421@cma.cn (C.S.); dujianhua843623@cma.cn (J.D.)
[4] School of Earth Science and Engineering, Hebei University of Engineering, Handan 056038, China; shangming@hebeu.edu.cn
[5] Lanzhou Central Meteorological Observatory, Lanzhou 730020, China; wujing10@mails.ucas.ac.cn
* Correspondence: caecar@hainanu.edu.cn

Abstract: Extreme precipitation events induced by tropical cyclones have increased frequency and intensity, significantly impacting human socioeconomic activities and ecological environments. This study systematically examines the spatiotemporal characteristics of these events across Hainan Island and their influencing factors using GsMAP satellite precipitation data and tropical cyclone track data. The results indicate that while the frequency of typhoon events in Hainan decreased by 0.3 events decade^{-1} from 1949 to 2020, extreme precipitation events have increased significantly since 2000, especially in the eastern and central regions. Different typhoon tracks have distinct impacts on the island, with **Track 1** (**Northeastern track**) and **Track 2** (**Central track**) primarily affecting the western and central regions and **Track 3** (**Southern track**) impacting the western region. The impact of typhoon precipitation on extreme events increased over time, being the greatest in the eastern region, followed by the central and western regions. Incorporating typhoon precipitation data shortened the recurrence interval of extreme precipitation in the central and eastern regions. Diurnal peaks occur in the early morning and late evening, primarily affecting coastal areas. Typhoon duration (CC_max = 0.850) and wind speed (CC_max = 0.369) positively correlated with extreme precipitation, while the pressure was negatively correlated. High sea surface temperature areas were closely associated with extreme precipitation events. The atmospheric circulation indices showed a significant negative correlation with extreme precipitation, particularly in the western and central regions. ENSO events, especially sea surface temperature changes in the Niño 1 + 2 region (−0.340 to −0.406), have significantly influenced typhoon precipitation characteristics. These findings can inform region-specific disaster prevention and mitigation strategies for Hainan Island.

Keywords: precipitation intensity; orographic effect; tropical cyclones; remote sensing; climate change

1. Introduction

Typhoon disasters pose a major threat to socioeconomic development and human security, often causing extreme precipitation upon landfall that can lead to secondary disasters such as floods, landslides, and mudslides [1–4]. The southeastern coastal regions of China are particularly vulnerable to these impacts due to frequent tropical cyclone activities, which are the primary contributors to extreme precipitation events [3–5]. Since 2020, Hainan Island has been impacted by several powerful typhoons, including Kompasu (2118), Nasha (2220),

Rai (2122), and Lionrock (2117). These storms have brought heavy rainfall and destructive winds, resulting in extreme precipitation and severely affecting sustainable development. The urbanization process on the island increases the surface roughness, which in turn amplifies the strength of typhoon-induced extreme precipitation (TIEP) [3–6]. This problem is particularly acute in the coastal areas of Hainan where the population and infrastructure are concentrated, further exacerbating the adverse effects of TIEPs [3,4,7]. This situation shows significant challenges to the development of the Hainan Free Trade Port. Therefore, it is essential to systematically investigate the spatial and temporal patterns and formation mechanisms of extreme precipitation events on Hainan Island.

With global warming, there has been a notable increase in the frequency of tropical cyclones over the past three decades [3,4,8]. Simultaneously, the impact of these cyclones on extreme precipitation events in China's southeastern coastal regions has become more pronounced [3,4,9]. Notably, more than 40 percent of such events are attributable to tropical cyclones [1,3,4,10]. With global warming, both the severity of tropical cyclones and the magnitude of associated extreme precipitation events have consistently risen [1–4,11]. This trend indicates a complex evolution in the spatiotemporal dynamics of typhoon activities, potentially increasing the risk of extreme precipitation in the future.

In recent years, there has been a significant decrease in the number of tropical cyclones affecting Hainan Island. Paradoxically, the frequency of days with extreme heavy rainfall events influenced by these cyclones, as well as the total precipitation from these events, has increased [3,4,12]. The average annual frequency, precipitation amount, and the proportion of typhoon precipitation to the total annual precipitation on Hainan Island all demonstrate a downward trend. Yet, the maximum precipitation during typhoon events and the maximum daily precipitation present a slight upward trend [3,4,13]. In contrast, another study reports a mild downward trend in maximum daily precipitation across different precipitation intensity levels [3,4,14]. These conflicting findings may result from the limited number of in situ observations, which may not adequately represent conditions across Hainan Island's complex topography. Additionally, most studies have relied on daily precipitation data, which offers relatively low temporal resolution and fails to capture the detailed spatiotemporal structure of extreme typhoon precipitation. Consequently, there is an urgent need for systematic analysis of the spatiotemporal patterns of extreme precipitation events induced by typhoons on Hainan Island, using high-resolution gridded precipitation data.

Previous studies on extreme precipitation have largely relied on indices from meteorological stations, such as daily precipitation amounts [3,4,15]. These indices often fail to capture the detailed spatiotemporal patterns of extreme typhoon precipitation due to the sparse and uneven distribution of stations on Hainan Island, especially in regions with complex terrain. However, recent advances in satellite remote sensing technology compensate for these shortcomings by providing high-resolution precipitation data with more comprehensive and continuous spatial and temporal coverage [3,4,14,16]. This improvement is crucial in areas like oceans where traditional surface observation coverage is sparse. Global analyses using data from the Global Precipitation Measurement (GPM) satellite have uncovered distinct diurnal peaks in tropical cyclone precipitation, which are typically elusive in station-based observations [17]. Therefore, compared with the station, higher temporal and spatial resolution data can be used to conduct a multi-scale study on the extreme precipitation pattern of the typhoon on Hainan Island.

While previous studies have examined extreme precipitation changes globally, the unique topographic and climatic features of Hainan Island, a typical tropical island, significantly influence the spatial distribution of typhoon-induced extreme precipitation. The factors influencing the intensity and spatial distribution of TIEP include the dynamics of typhoon movement, such as stalling, local terrain, and moisture availability [3,4]. As typhoons make landfall on Hainan Island, their forward motion typically decelerates, extending their duration over the region. This deceleration is especially pronounced in the central and western mountainous terrain, where orographic effects further intensify precipitation. Furthermore, Hainan Island's maritime environment provides a continuous supply of water vapor, unlike inland regions where typhoons rapidly lose their moisture source after landfall. This sustained moisture availability, combined with the aforementioned factors, contributes to the intensification of typhoon-induced extreme precipitation events on the island. Radar observations and numerical modeling experiments focused on specific cases have corroborated the enhancement of typhoon precipitation in the central and western mountainous areas of Hainan Island [18,19]. In China, the precipitation associated with landfalling typhoons is strongly influenced by the storm's intensity and translational speed [20,21]. However, the characterization of these relationships for Hainan Island is hindered by the lack of long-term observational data. Concurrently, rising sea surface temperatures (SSTs), associated with global climate change, are expected to increase the likelihood of severe precipitation events related to typhoons. These elevated SSTs significantly influence the occurrence and magnitude of extreme precipitation events [22,23]. Analyses of historical climate data reveal that typhoon activity frequency along the southern coast of Hainan Island is significantly modulated by large-scale climate patterns, particularly the El Niño-Southern Oscillation (ENSO) [24]. Therefore, from a climatological perspective, it is essential to consider the influence of large-scale SST patterns on extreme typhoon precipitation. This approach allows for a more comprehensive assessment of the spatial extent and intensity of extreme typhoon precipitation on Hainan Island, accounting for both local and remote climate drivers.

Recent studies have advanced our understanding of the spatiotemporal characteristics of extreme typhoon-induced precipitation on Hainan Island. However, significant knowledge gaps persist, particularly regarding spatiotemporal resolution, multi-scale analysis, and comprehensive quantification of influencing factors. To address these limitations, this study aims for: (1) Utilize high spatiotemporal resolution satellite-retrieved precipitation data to reveal the fine spatiotemporal pattern of extreme typhoon precipitation on Hainan Island; (2) Explore the distribution patterns and variation characteristics of extreme typhoon precipitation across multiple spatiotemporal scales; (3) Comprehensively analyze the interactions between typhoon characteristics and environmental fields and other factors and quantitatively assess their impacts on extreme typhoon precipitation; and (4) Integrate typhoon track information to examine regional differences in extreme typhoon precipitation across Hainan Island and explore their underlying causes. This study utilizes high-resolution long-term gridded precipitation data in conjunction with optimal typhoon track data to investigate the distribution characteristics of extreme precipitation events associated with typhoons on Hainan Island across multiple scales.

2. Materials and Methods

2.1. Overview of the Study Area

Hainan Island, China's second-largest island (≈34,000 km^2), is located in the northwestern South China Sea (18°10′–20°10′ N, 108°37′–111°03′ E). The island's topography exhibits a distinctive annular structure, with the central highlands, notably Wuzhishan (1840 m a.s.l.) and Yinggeling (1812 m a.s.l.), descending to the surrounding lowlands (Figure 1a). Its tropical monsoon climate has average annual temperatures of 22.5–25.6 °C and precipitation of 1500–2500 mm, with distinct dry and wet seasons [25].

Rainfall distribution across Hainan Island is heterogeneous, governed by the tropical monsoon circulation. The rainy season, spanning from April to October, contributes over 70% of the annual precipitation. Hainan's location makes it highly typhoon-prone, with tropical cyclones from the northwestern Pacific typically affecting the island from May through November.

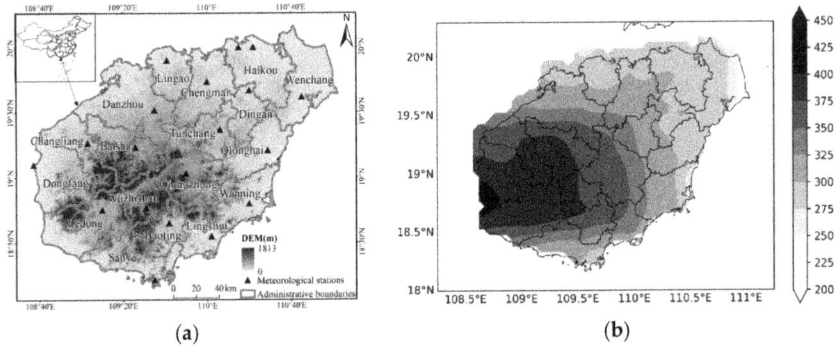

Figure 1. (**a**) Topographic map of Hainan Island and spatial distribution of meteorological stations; (**b**) multi-year average TIP.

Figure 1b shows the multi-year average typhoon-induced precipitation (TIP), affecting virtually all of Hainan Island. TIP varies spatially, averaging 250–450 mm annually. Western and central mountainous regions receive the highest rainfall (>350 mm/year), while precipitation decreases towards coastal areas. The northeastern region (such as Wenchang) experiences the lowest TIP (<300 mm/year), while western parts (such as Dongfang and Ledong) receive the highest (>400 mm/year).

Combining Figures 1a and 1b reveal that high TIP areas encompass the entire mountainous region and western plains. Exceptions include certain mountainous areas around Baoting and Sanya, which do not exhibit elevated precipitation values. This pattern is likely due to the orographic effects in central and western Hainan Island, where the forced ascent of moist air associated with typhoons enhances rainfall, making these regions more vulnerable to direct impacts from typhoons.

2.2. Datasets

This study uses historical tropical cyclone track data (1949–2020) from the Shanghai Typhoon Institute of the China Meteorological Administration. The dataset includes tropical cyclone metrics such as center coordinates, 2-min average maximum wind speeds (m s^{-1}), and minimum central pressure (hPa). Data from 2000–2020 were selected to examine the recent distribution of TIP over Hainan Island.

Precipitation data is from the Global Satellite Mapping of Precipitation (GsMAP) Reanalysis product by the Japan Aerospace Exploration Agency (JAXA). The GsMAP Reanalysis dataset features a fine spatial resolution of 0.1° and a temporal resolution of 1 h. It spans the period from April 2000 to December 2020 and provides extensive geographic coverage from 60° N to 60° S. Previous research validates its reliability for Hainan Island, showing correlations up to 0.870 on a monthly scale, with slight overestimation [26].

Sea surface temperature (SST) data is sourced from the National Centers for Environmental Information (NCEI) NOAA Optimum Interpolation Sea Surface Temperature (OISST) platform. This dataset combines satellite, ship, and buoy observations, using optimal interpolation at 0.25° × 0.25° resolution from 1981 to 2024. This study utilizes essential atmospheric circulation indices, such as the Niño SST (available at https://www.cpc.ncep.noaa.gov/data/indices/sstoi.indices, last accessed on 1 May 2024) and the Northern Hemisphere Subtropical High Area Index, among others. The latter indices are obtained from the National Climate Center of the China Meteorological Administration (available at https://cmdp.ncc-cma.net/Monitoring/cn_index_130.php, last accessed on 1 May 2024). The monthly averages of these indices are used to investigate their correlations with tropical cyclone activity. Comprehensive results are in Appendix A, while the main text focuses on significantly correlated indices.

2.3. Methodology

Statistical analyses of return period events commonly assume that extreme events follow the generalized extreme value (GEV) distribution [27]. This assumption is based on the premise that typhoon-induced extreme precipitation is independent and identically distributed (i.i.d.), providing a foundation for applying the GEV model with these datasets. The GEV probability density function (PDF) is [28]:

$$H_i(x|\mu,\sigma,\xi) = \frac{1}{\sigma}[1+\xi(\frac{x-\mu}{\sigma})]^{-(1+\frac{1}{\xi})}\exp(-[1+\xi(\frac{x-\mu}{\sigma})]^{-\frac{1}{\xi}}),$$

where μ is the location parameter, σ is the scale parameter, and ξ is the shape parameter. In Section 3.3 we use data from different time scales to calculate recurrence intervals for 25, 50, 100, 200, and 300-year return period events.

For trend analysis, the Theil–Sen regression method was used due to its robustness to outliers [29]. Data were smoothed using a three-point average method. Pearson correlation coefficients assessed relationships between typhoon extreme precipitation characteristics and total precipitation amounts.

2.4. Definition of Typhoon Precipitation and Typhoon-Induced Precipitation Extremes

Typhoon-induced precipitation (TIP) was defined as the precipitation occurring within an area covered by a 2.5° influence radius (≈260–277 km on Hainan Island) centered on the typhoon's track [26]. Typhoons directly impacting Hainan (2000–2020) were identified using best track data and GsMAP precipitation data. GsMAP data (0.1° spatial, 1-h temporal resolution) were clipped to Hainan's terrestrial boundaries using an official map [Map Approval Number: GS(2019)1822]. Extreme precipitation events were identified using 95th and 99th percentile values (C95 and C99) of non-typhoon precipitation as thresholds, calculated on a monthly scale for multiple temporal scales (1 h, 3 h, 6 h, 12 h and 24 h). Section 3.4 uses the 99th percentile for spatial distribution analysis.

Based on typhoon precipitation data, the K-means clustering algorithm was used to partition Hainan Island into three distinct regions (Figure 2): 1. Eastern region: plains with uniform typhoon precipitation due to minimal topographic variation. 2. Central region: Mountainous with heterogeneous precipitation due to orographic effects. 3. Western region: Northwestern and southwestern plains with lower precipitation intensities.

2.5. Typhoon-Induced Extreme Precipitation's Regionalization and the Typhoon Track Cluster

This study uses the K-means clustering algorithm to objectively delineate the Hainan Island's terrestrial regions. Based on multi-year average typhoon precipitation relative to total precipitation (Figure 2a), three distinct regions were identified on Hainan Island: the eastern region, central region, and western region (Figure 2b).

The eastern region, situated on the island's eastern side, primarily comprises plains with relatively flat terrain. This area experiences weaker typhoon precipitation intensity due to minimal topographic variation, which does not favor orographic lifting. Consequently, the spatial distribution of typhoon precipitation in this region is notably uniform, exhibiting minimal local variability. Conversely, the central region, located in the center of Hainan Island, features a rugged and varied landscape dominated by mountains. This region harbors the island's principal mountain ranges with high elevations that enhance orographic lifting on the windward slopes, frequently resulting in intense precipitation events. The complex topography introduces significant spatial heterogeneity in typhoon precipitation, with pronounced disparities in rainfall between mountainous and valley areas. The western region, encompassing the northwestern and southwestern plains, is positioned on the leeward side of the monsoonal currents and exhibits less topographic relief. While typhoons impact this region, the absence of substantial orographic lift leads to comparatively lower precipitation intensities during such events. This regional classification provides a framework for analyzing typhoon-induced precipitation patterns across Hainan Island's diverse topography.

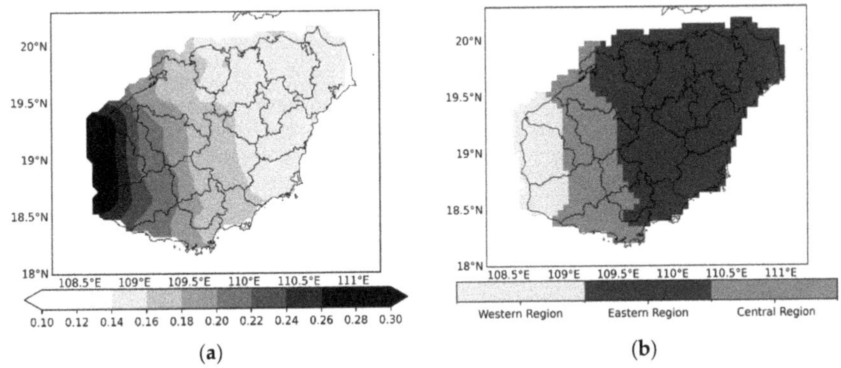

Figure 2. (**a**) Multi-year average typhoon precipitation ratio; (**b**) Spatial distribution zones of typhoon-induced extreme precipitation (TIEP).

To identify extreme precipitation events induced by specific typhoons, K-means clustering analysis was used to categorize the typhoon trajectories affecting Hainan Island from 2000 to 2020. Typhoon metrics including the typhoon center positions, typhoon movement speeds, and typhoon intensities were examined to classify them into three distinct groups (Figure 3). Subsequent analyses of these categorized typhoons were based on these classifications.

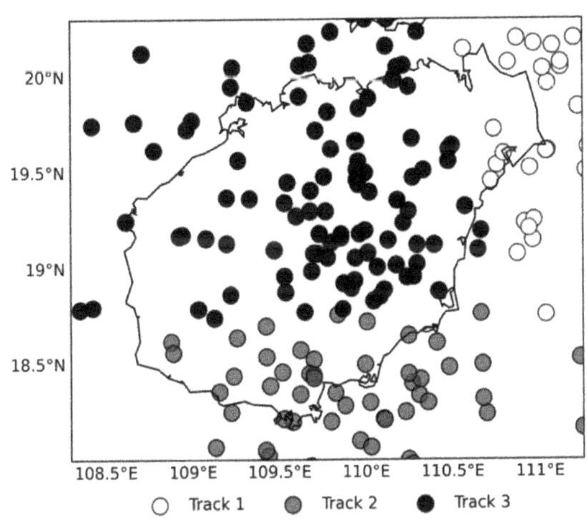

Figure 3. Spatial patterns of clustered typhoon tracks over Hainan Island.

The cluster analysis yielded three typhoon track groups (Figure 3): Typhoons in the first group (**Track 1**) primarily originated from the southeast of Hainan Island, traversing the central region of the island from east to west, characterized by relatively high movement speeds and moderate intensities.

The second group (**Track 2**) predominantly arose from the southern sea area of Hainan Island, progressing from south to north across the island's southern tip, with rapid movement speeds but lesser intensities compared to Track 1.

The third group (**Track 3**) exhibited more varied trajectories, including typhoons forming near Hainan Island and moving northward, and those passing east to west over the northern part. Slowest speeds but generally stronger intensities, are associated with higher frequency of extreme precipitation events. Detailed characteristics of each track group, including average speeds, intensities, and associated precipitation patterns, are presented in Appendix A, Table A2.

Following the classification of these typhoon tracks, this study examined the differences in extreme precipitation characteristics events triggered by each typhoon category. Typhoon-induced precipitation data were correlated with their respective track categories, to quantify the frequency and average intensity of extreme precipitation events across various temporal scales (e.g., 1 h, 6 h, and 12 h durations).

3. Results

3.1. Spatiotemporal Analysis

3.1.1. Change in Typhoon Frequency

Figure 4 presents the number of typhoons affecting Hainan Island from 1949 to 2020. Over this period, typhoon frequency shows a statistically significant declining trend of 0.3 events decade^{-1} (p = 0.00) with notable inter-decadal variability. Specifically, the mid-1950s and mid-1960s experienced higher frequencies of typhoon events, averaging around 6 events year^{-1}, with some years reaching up to ten events. The years from 1949 to 1959 were marked by a significant increase in typhoon activity, culminating in 1960 with 11 events, the highest annual count observed in the past seventy years. From 1960 to 1989, although the overall number of typhoons began to decline, the rate of decrease slowed, with most years recording fewer than four events per year. The period from the 1990s to the early 2010s witnessed larger fluctuations, with typhoon counts in some years approaching eight, indicating a pattern of initial increase followed by a decrease. Over the last two decades (2000–2020), typhoon events affecting Hainan have fluctuated between three to six events annually. This analysis aligns with Wu's findings, indicating a significant reduction in typhoon frequency impacting Hainan Island from 1962 to 2005 [12].

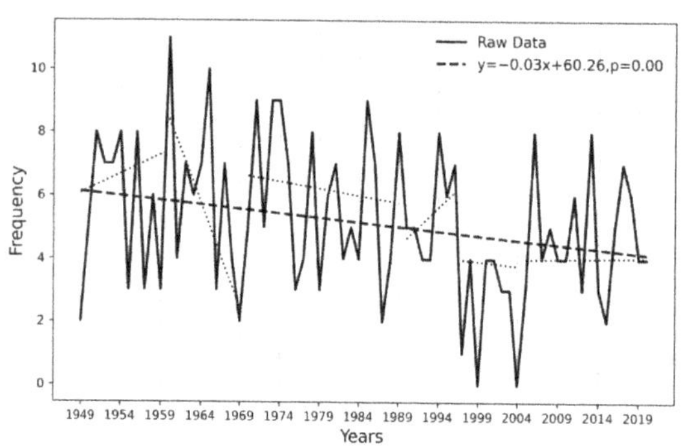

Figure 4. Number of typhoons affecting Hainan Island from 1950 to 2020. Notes: The dashed trend line is derived from Theil–Sen regression after three-point average smoothing. Dotted lines represent segmented regression after breakpoint detection, which is used to analyze the trend changes in different periods.

3.1.2. Change in the Frequency of Extreme Precipitation

Figure 5 shows the temporal variability of extreme precipitation events associated with typhoons affecting Hainan Island from 2000 to 2020. These events were identified using GsMAP precipitation data, employing the 95th and 99th percentile as hourly thresholds. The results reveal an upward trend in the frequency of these extreme events at both thresholds: a 1.5 events decade^{-1} increase at the 95th percentile threshold and a 1.8 events decade^{-1} increase at the 99th percentile threshold. The higher rate of increase at the 99th percentile threshold suggests an intensification in the severity of extreme precipitation events. This trend indicates that the more extreme rainfall events are escalating more substantially compared to less extreme ones.

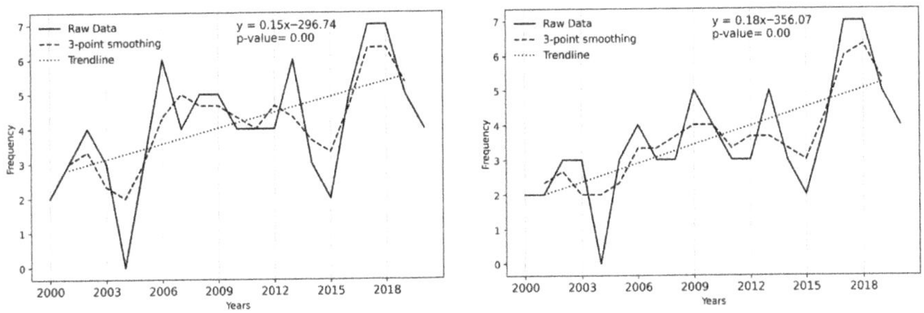

(**a**) (**b**)

Figure 5. Frequency of extreme precipitation events distinguished with the 95th (**a**) and 99th (**b**) percentile threshold, respectively.

3.1.3. Change in Extreme Precipitation and Contribution

Figure 6 shows the time series of the total extreme precipitation associated with typhoons affecting Hainan Island from 2000 to 2020, based on GsMAP data at the 95th and 99th percentile thresholds on an hourly scale. Excluding 2004, which had no typhoon impacts, the data shows significant interannual variability in extreme precipitation. The years 2001, 2011, 2016, and 2018 experienced notably higher levels of extreme precipitation. 2001 recorded exceptionally high values, with 686 mm at the 95th percentile and 378 mm at the 99th percentile in the western region. The total extreme precipitation (TEP) across the regions of Hainan Island shows a gradient from the western region (most severe impacts), to the central and eastern regions, respectively.

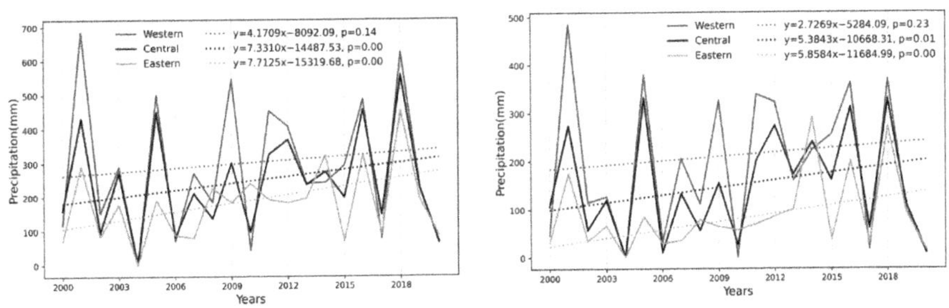

(**a**) (**b**)

Figure 6. Typhoon-induced extreme precipitation in three sub-regions of Hainan Island: Analysis using 1-hour precipitation thresholds at the 95th (**a**) and 99th (**b**) percentiles.

Trend analysis of typhoon-associated extreme precipitation in Hainan Island reveals an increasing pattern across the sub-regions. Extreme precipitation events at both the 95th and 99th percentile thresholds show a consistent upward trend in the central and eastern regions.

At the 95th percentile threshold, the eastern region shows the most pronounced upward trend, with an increase of 77.1 mm decade^{-1} in extreme precipitation. This trend suggests an escalating frequency and intensity of extreme rainfall during typhoon events in this area. Extreme precipitation in the central region shows marked increases at both percentile thresholds: 73.3 mm decade^{-1} (95th percentile) and 58.6 mm decade^{-1} (99th percentile). This consistent rise suggests a uniform intensification of extreme rainfall events during typhoons across different intensity levels. In contrast, the western region demonstrates a more modest change, with a slight upward trend of 27.3 mm decade^{-1} observed only at the 99th percentile threshold, indicating that only the most extreme events are intensifying in this area.

Figure 7 presents a time series of the ratio between extreme precipitation and total typhoon-induced precipitation (TIP) over Hainan Island from 2000 to 2020. The analysis uses the 95th and 99th percentiles as hourly thresholds, based on GsMAP satellite-retrieved precipitation data. The extreme precipitation proportion exhibits significant inter-annual variability, influenced by tropical depressions and monsoon systems. Years 2003, 2005, 2012, 2014, and 2018 recorded higher ratios, exceeding 80% at the 95th percentile threshold. 2014 marked the peak contribution with extreme precipitation comprising 96.9% (95th percentile threshold) and 88.1% (99th percentile threshold) of the typhoon-induced precipitation.

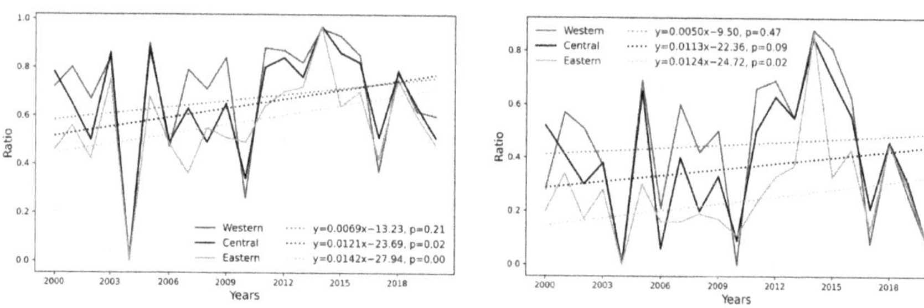

(**a**) (**b**)

Figure 7. Temporal evolution of the ratio between typhoon-induced extreme precipitation (TIEP) and total typhoon rainfall, based on 1-h 95th (**a**) and 99th (**b**) percentile thresholds.

The trend analysis indicates a general increase in the proportion of extreme precipitation relative to the total typhoon precipitation across Hainan Island. The eastern region shows a pronounced increase of 14.2% decade^{-1} at the 95th percentile, suggesting a rising frequency of extreme precipitation events. The central region exhibits an increase of 12.1% decade^{-1} and 11.3% decade^{-1} at the 95th and 99th percentiles, respectively. The western region displays a more moderate growth, with trends of 6.9% decade^{-1} and 5.0% decade^{-1} at these thresholds.

3.1.4. Intra-Day Changes in Extreme Precipitation

Figure 8 depicts the diurnal cycle of extreme precipitation across three sub-regions. The data shows a bimodal distribution, with maxima during the early morning hours (0200–0700 h at UTC) and late afternoon to evening (1800–2000 h at UTC), and minima around 0000 and 1200 h UTC. Hainan Island's tropical monsoon climate is characterized by intense solar radiation, which enhances surface heating and promotes convective activity. This convective peak in the late afternoon often triggers significant precipitation events. Conversely, by the early morning, surface cooling increases atmospheric stability, promoting water vapor condensation and precipitation formation. The eastern region experiences the highest frequency of extreme precipitation, with over 50 occurrences at the 99th percentile during several hours. The western region records these events less frequently. These findings align with research by Manuel F. Rios Gaona [17], who reported that the highest rainfall during typhoon events typically occurs at 0200, 0500, 1400, and 1800 h UTC [3].

3.1.5. Spatial Distribution of the Maximum Precipitation

Figure 9 illustrates the intensity patterns of the maximum hourly and daily precipitation from typhoons following different trajectories. The analysis reveals significant spatial variability, highlighting the critical role of the typhoon paths in shaping precipitation distribution. For **Track 1** high hourly precipitation rates in Baisha, Danzhou, and Qionghai, while other areas experience more moderate precipitation. **Track 2** is associated with lower hourly precipitation, peaking in Qionghai and Wenchang. **Track 3** produces the most intense typhoon-induced precipitation (TIP), especially in the western and central regions, with Sanya recording an extreme hourly maximum of 91 mm/h.

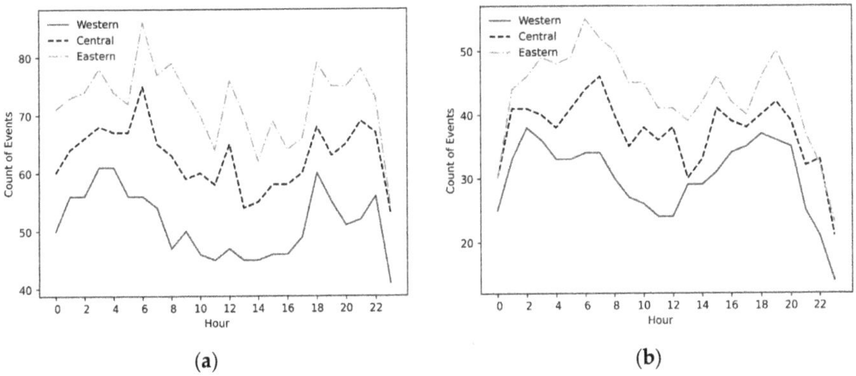

Figure 8. Intra-day temporal pattern of extreme precipitation based on 1-h data with 95th percentile (**a**) and 99th percentile (**b**) thresholds.

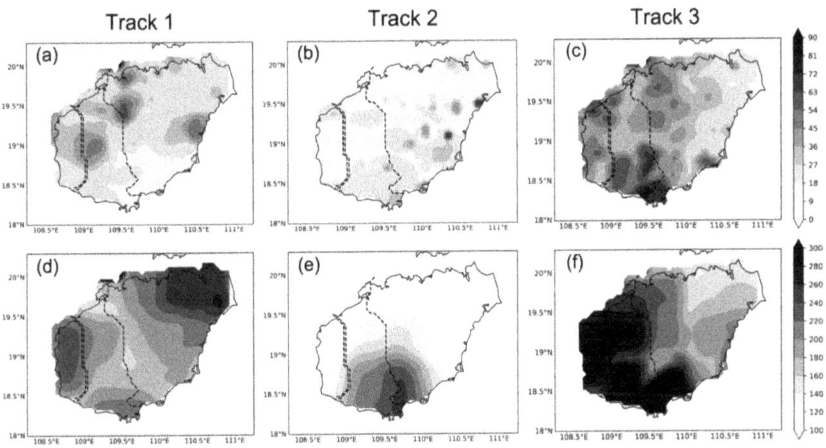

Figure 9. The maximum hourly precipitation and maximum daily precipitation in different typhoon tracks. Notes: The subplots of (**a–c**) are the maximum hourly precipitation in the typhoon Tracks 1–3, respectively, while the subplots of (**d–f**) show the maximum daily precipitation in the typhoon Tracks 1–3, respectively.

Daily precipitation patterns also vary by track. **Track 1** typhoons bring increased daily precipitation to eastern regions (Haikou and Wenchang) and western areas, with lower amounts in central mountainous regions. **Track 2** typhoons show prominent daily precipitation maxima in central and eastern regions, especially near Sanya and Lingshui, decreasing westward and northward. This pattern suggests weakening typhoon intensity or moisture content as systems move inland. **Track 3** typhoons consistently generate higher daily precipitation rates across western and central regions, reaching a maximum of 354 mm day^{-1}, indicating severe impacts.

3.2. *Typhoon Contributions to Extreme Precipitation*

3.2.1. Spatial Distribution of the Contribution Rates

With hourly precipitation data from 2000 to 2020, this study applies the 95th and 99th percentile values as thresholds to define typhoon-induced extreme precipitation (TIEP) events. We quantify the contribution of typhoons to extreme precipitation at various temporal scales (1 h, 3 h, 6 h, 12 h, and 1 day), denoted as C95 and C99 (Figures 10 and 11). Results reveal that as the thresholds increase, typhoon contribution to extreme precipitation events decreases, with the C95 values exceeding those of C99. This suggests that typhoons exert a more substantial influence on the lower tail of the extreme precipitation distribution.

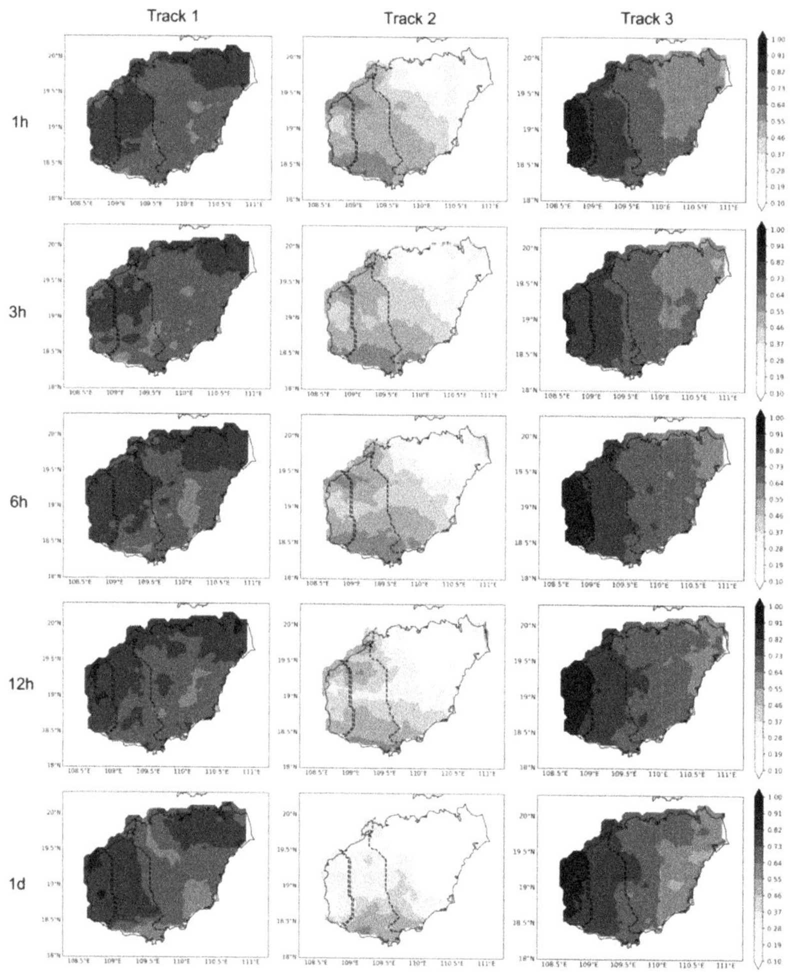

Figure 10. Spatial distribution of typhoon-induced extreme precipitation (TIEP) at a threshold of the 95th percentile.

Distinct spatial and temporal variations in contribution rates are observed across different typhoon tracks. For **Track 1**, the highest C95 values are recorded in the western, central, and eastern regions, especially around Haikou, with a noticeable decline from coastal to inland areas. C99 values show variations across different temporal scales, with higher rates progressively shifting from southeast to northwest as the temporal scale increases. For **Track 2**, elevated C95 values are primarily noted in the western and central regions, exhibiting a gradient decrease from southwest to northeast. C99 values similarly affect these areas, but the peak values gradually shift toward the northern sectors of the central and western regions as the temporal scale extends. **Track 3** demonstrates a significant west-to-east decreasing trend in both C95 and C99 distribution, with the western regions identified as zones of high extreme precipitation contribution rates.

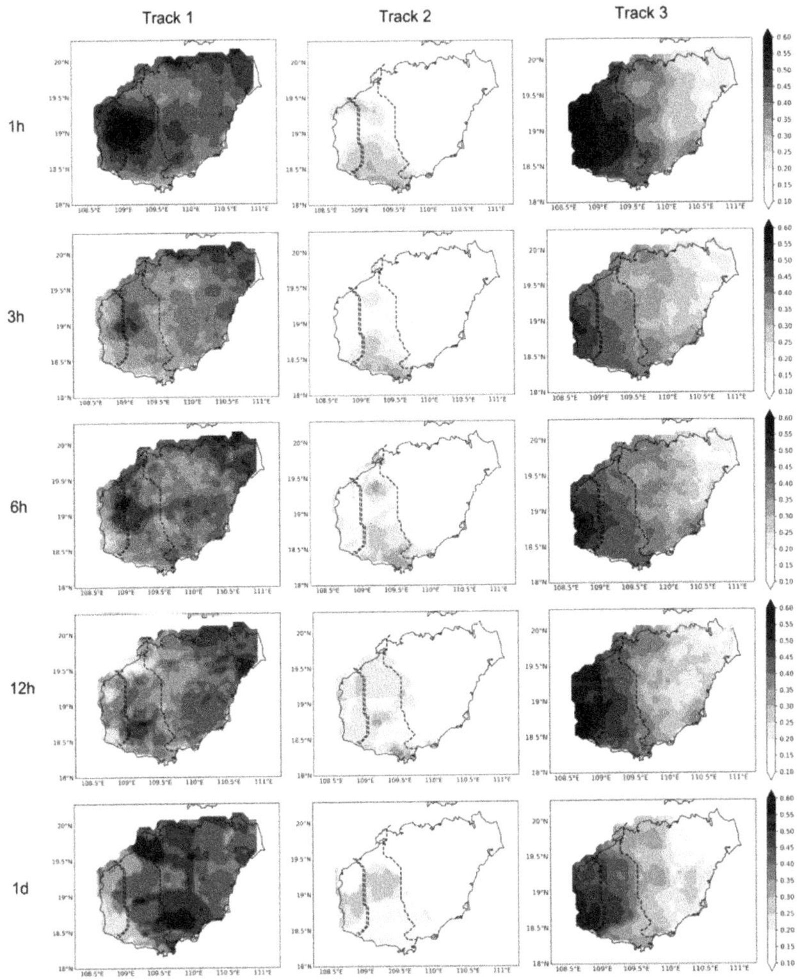

Figure 11. Spatial distribution of TIEP at a threshold of the 99th percentile.

3.2.2. Intra-Annual Change in the Contribution Rate

This study delineates variations in the contribution rates of typhoon precipitation to Hainan Island across different tracks on a monthly scale (Figure 12). Results reveal an increasing trend in the overall typhoon contribution rates over expansion time scales, with C95 consistently exceeding C99.

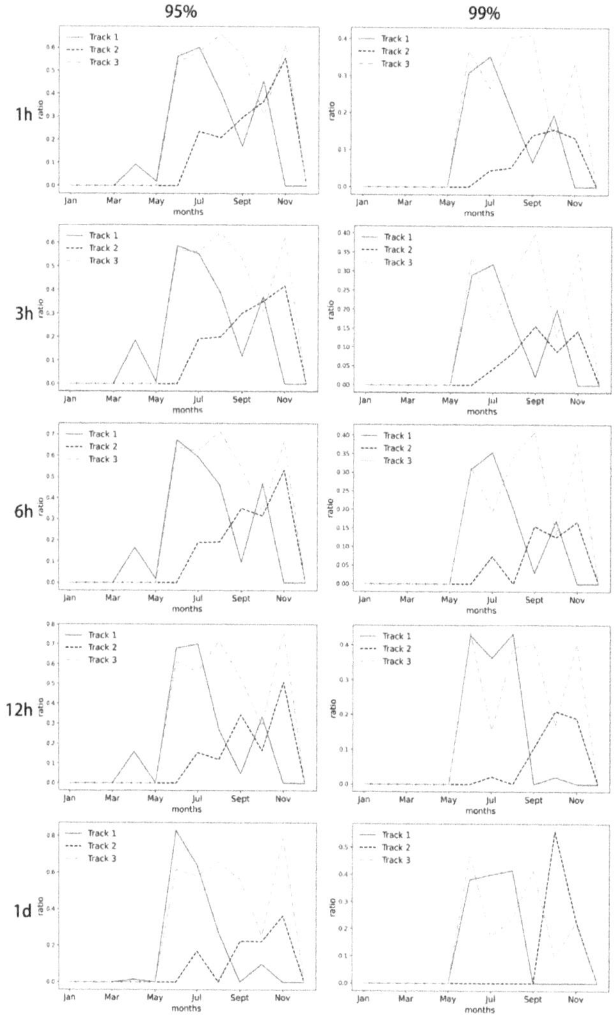

Figure 12. Temporal pattern of typhoon contribution to extreme precipitation for different tracks.

There are notable differences in the temporal contribution rates among the typhoons following different tracks. **Track 1** typhoons exhibit peak contribution rates between June and August, significantly impacting the terrestrial regions of Hainan Island during this period. High contribution rates are also observed in October, but gradually decline with increasing time scales, indicating that typhoons in October generally exert a relatively brief and weaker impact before swiftly relocating. **Track 2** typhoons display varied seasonal patterns. At an hourly scale, a notable peak occurs from September to November. As time scales increase, contribution rates significantly amplify from June to August, albeit with a decline in the 1-day scale contribution rates during these months. This pattern indicates a broader temporal impact of these typhoons on Hainan Island, with extreme precipitation events primarily concentrated between September and November. **Track 3** typhoons show a more consistent pattern, affecting Hainan Island from June to November. However, the impacts are comparatively moderate in July and October, potentially linked to the specific climatic dynamics inherent to this track.

3.3. Statistical Analysis of Extreme Values

This study employs the GEV distribution model to estimate and simulate the return levels of precipitation under varying return periods (P) on Hainan Island (Figure 13). Percentiles are categorized into three groups: "TC" for typhoon periods, "nTC" for non-typhoon periods, and "Mix" for both periods combined. The analysis reveals distinct spatial variations in these percentiles across different regions and time scales.

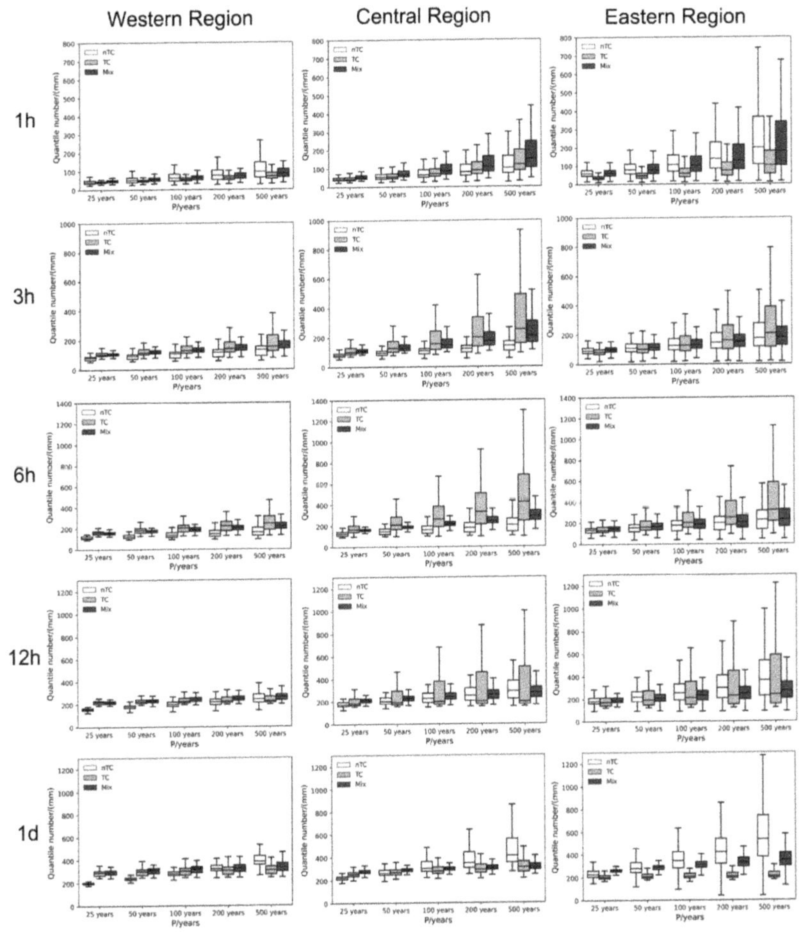

Figure 13. TIEP in different return periods and different temporal scales.

For any given return period, the eastern region of Hainan Island experiences a higher percentile for nTC compared to Mix, suggesting a dominant influence of non-typhoon precipitation. Conversely, in the central and western regions, Mix percentiles are higher than those for nTC, indicating that the inclusion of typhoon data significantly alters the statistical characteristics of extreme precipitation events, potentially shortening their return periods.

On shorter return periods, both the western and eastern regions exhibit a higher Mix percentile compared to nTC. However, this relationship reverses in favor of nTC on longer return periods (12 h or 1 day). Additionally, the percentiles for TC initially increase and subsequently decrease, a pattern that likely reflects the short-duration impacts of typhoons, which typically affect the island for 2 to 3 days and may not persist throughout an entire day. The transient nature of typhoon-induced precipitation is critical for accurate statistical analysis. When examining longer time periods, these brief but intense rainfall events can lead to underestimation of percentiles if not properly accounted for. This potential bias highlights the importance of careful methodology in resampling and statistical assessments of extreme precipitation events.

3.4. Factors Influencing the TIEP

3.4.1. Extreme Precipitation in Relation to Air Pressure, Wind Speed, and Precipitation

This study examines the determinants of extreme precipitation events by analyzing the relationships between typhoon characteristics (duration, pressure, and wind speed) and the precipitation amounts across different regions of Hainan Island from 2000 to 2020 (Table 1).

Table 1. Correlation coefficients between the TIEP and typhoon features in different sub-regions with different typhoon track paths.

Region	Track	Air Pressure	Wind Speed	Duration
All Regions	Track 1	0.246	−0.327	0.589 **
	Track 2	−0.052	0.134	0.292
	Track 3	−0.190	0.257	0.850 ***
Western Region	Track 1	0.202	−0.235	0.530 **
	Track 2	0.024	0.031	−0.036
	Track 3	−0.320 *	0.369 *	0.674 ***
Central Region	Track 1	0.263	−0.279	0.539 **
	Track 2	−0.117	0.143	0.101
	Track 3	−0.287	0.336 *	0.691 ***
Eastern Region	Track 1	0.272	−0.358	0.575 **
	Track 2	0.092	0.022	0.305
	Track 3	−0.064	0.109	0.612 ***

Notes: *, **, and ***, denote the CC with $p < 0.05$, $p < 0.01$ and $p < 0.001$, respectively.

The analysis reveals a consistent positive correlation between the duration of typhoon activities and total island extreme precipitation (TIEP) across Hainan Island. This suggests that longer-lasting typhoons tend to produce a greater amount of TIEP. In contrast, atmospheric pressure and precipitation are generally negatively correlated, except for the **Track 1** region; lower pressure and increased precipitation during typhoons. Wind speed exhibits a positive correlation with precipitation intensity, likely due to enhanced moisture transport from maritime sources to land and facilitated vertical convective transport of heat and moisture.

The **Track 1** region on Hainan Island displays unique extreme precipitation characteristics. In this region, extreme precipitation is positively correlated with atmospheric pressure and negatively correlated with wind speed. This atypical relationship may be attributed to the low-pressure systems prevalent in **Track 1**, which are often accompanied by substantial moisture accumulation and robust convective activity conducive to extreme precipitation events. In these low-pressure environments, the ascending motion facilitates moisture condensation, thereby enhancing precipitation intensity. Moreover, an elevated pressure gradient can amplify the transport of moisture, further intensifying the precipitation dynamics. However, strong winds in this region might hinder the effective transport of moisture and disrupt convective processes, leading to a distinct impact pattern from typhoons that follow other trajectories.

3.4.2. Relationship between the Extreme Precipitation and the SST

To elucidate the relationship between extreme precipitation events and the SST, we calculated the correlation coefficients (CCs) for 2000–2020 between extreme precipitation amounts in various Hainan Island regions and SST variations in the surrounding marine areas (Figure 14). Regions traversed by typhoon centers exhibit notably higher CCs. This robust coupling between SST variations and the extreme precipitation events occurrence may be attributed to the typhoons' transport of substantial moisture over these warmer sea areas. Typhoon path and intensity are critical determinants of local precipitation patterns. Furthermore, warmer SSTs in proximity to typhoon centers provide a rich source of heat and moisture, significantly enhancing the precipitation capacity of the typhoons.

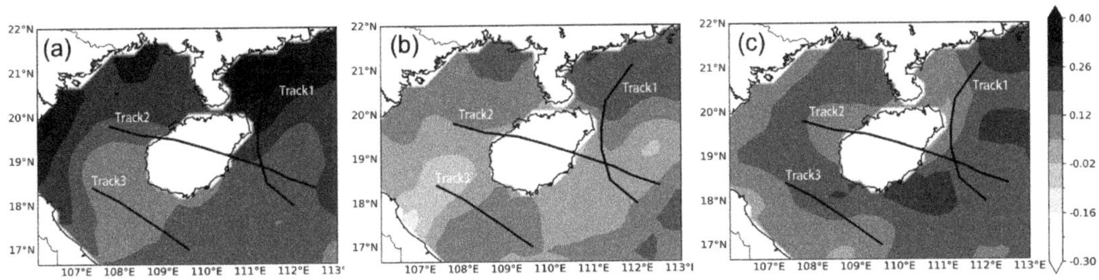

Figure 14. Spatial distribution of correlations between extreme precipitation and SST in the western, central, and eastern regions, respectively. Notes: (1) Subplots of **(a–c)** show the correlation coefficients between extreme precipitation and SST in the western, central and eastern regions, respectively. (2) The black lines in the left diagram represent the representative paths of the three typhoon paths.

3.4.3. Extreme Precipitation in Relation to the Atmospheric Circulation and ENSO

This study examines the relationship between the TIEP and atmospheric circulation indices (Table 2). The analysis reveals a significant negative correlation between the extreme precipitation and atmospheric circulation indices, particularly strong in the western (CCs ranging from −0.303 to −0.385) and central (CCs ranging from −0.277 to −0.368) regions of Hainan Island. The correlation is comparatively weaker in the eastern region (CCs ranging from −0.172 to −0.260).

Table 2. Correlation relationships between extreme precipitation and atmospheric circulation indices and SST indices.

Index	All Regions	Western Region	Central Region	Eastern Region
Asia Polar Vortex Area Index	−0.240 **	−0.314 **	−0.284 **	−0.185 *
Pacific Polar Vortex Area Index	−0.311 **	−0.373 **	−0.358 **	−0.252 **
North American Polar Vortex Area Index	−0.306 **	−0.362 **	−0.344 **	−0.255 **
Northern Hemisphere Polar Vortex Area Index	−0.320 **	−0.385 **	−0.368 **	−0.260 **
Asia Polar Vortex Intensity Index	−0.287 **	−0.351 **	−0.327 **	−0.232 **
Pacific Polar Vortex Intensity Index	−0.228 **	−0.303 **	−0.277 **	−0.172 *
Northern Hemisphere Polar Vortex Intensity Index	−0.266 **	−0.331 **	−0.308 **	−0.212 **
Pacific Subtropical High Ridge Position Index	0.487 **	0.484 **	0.506 **	0.446 **
Northern Hemisphere Subtropical High Northern Boundary Position Index	0.420 **	0.438 **	0.446 **	0.375 **
North African Subtropical High Northern Boundary Position Index	0.362 **	0.387 **	0.386 **	0.320 **
Mid-Eastern Pacific 200 mb Zonal Wind Index	−0.352 **	−0.350 **	−0.366 **	−0.323 **
Niño 1 + 2 SST	−0.406 **	−0.340 **	−0.398 **	−0.398 **
Niño 3 SST	−0.274 **	−0.191 *	−0.252 **	−0.286 **
Niño 4 SST	0.064	0.117	0.095	0.033
Niño 3.4 SST	−0.107	−0.026	−0.076	−0.134

Notes: *, and **, denote the CC with $p < 0.01$ and $p < 0.001$, respectively.

Intensification and expansion of the northern hemisphere's polar vortex are observed to inhibit typhoon development and restrict their northward movement, consequently diminishing the precipitation associated with these systems. Furthermore, SST changes in the Niño 1 + 2 region (CCs ranging from −0.340 to −0.406) exert a significant influence on the activity of typhoon systems in the Northwest Pacific. These factors critically modulate typhoon-associated precipitation in Hainan Island. Notably, when the Western Pacific Subtropical High is elevated, a corresponding increase in extreme precipitation in Hainan is typically observed.

To delineate temporal relationships between extreme precipitation events and climatic indices, this study analyzes a time series of extreme precipitation data from Hainan Island alongside representative indices (Figure 15). The analysis demonstrates a discernible negative correlation between the TIEP and both the Northern Hemisphere Polar Vortex Area Index and the Niño 1 + 2 SST Index. Lower values of these indices are frequently associated with heightened occurrences of intense extreme precipitation.

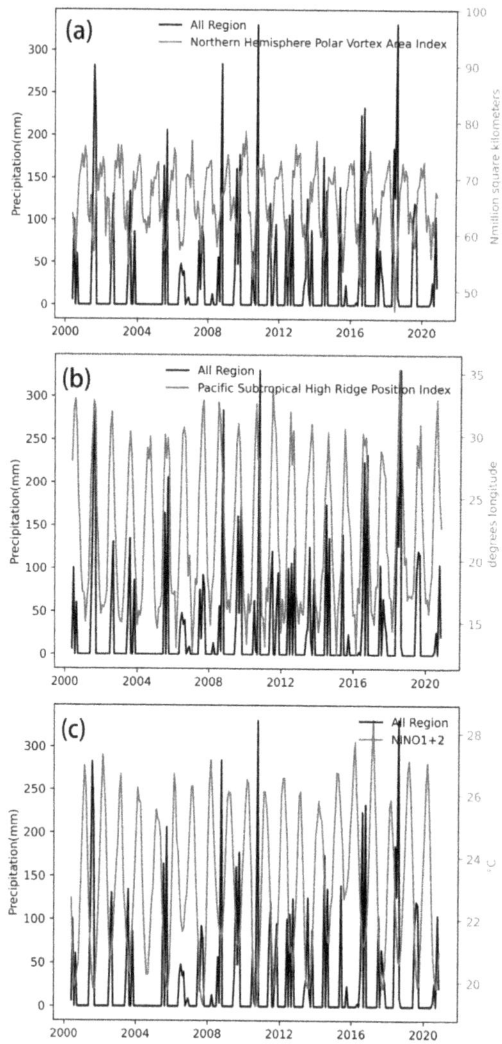

Figure 15. Comparisons between the TIEP and Northern Hemisphere Polar Vortex Area Index (**a**), Pacific Subtropical High Ridge Position Index (**b**), and Niño 1 + 2 SST Index (**c**).

A reduced Northern Hemisphere Polar Vortex Area Index correlates with increased frequency and intensity of polar cold air descending toward lower latitudes, instigating atmospheric circulation anomalies that precipitate heavy rainfall events. Concurrently, a lower Niño 1 + 2 SST Index, indicative of a cooler SST in the equatorial Eastern Pacific, influences global atmospheric circulation patterns. This alteration amplifies the strength and alters the position of the subtropical high, thereby elevating the potential for TIEP.

Conversely, there is a positive correlation between extreme precipitation and the Pacific Subtropical High Ridge Position Index. When the subtropical high is positioned at higher latitudes, it often results in the formation of a robust convergence zone on its southern flank. This convergence, coupled with ascending moist air, fosters conditions conducive to heavy precipitation. The intensity of the TIEP is notably enhanced under these circumstances.

4. Discussion

Typhoon intensity exhibited significant interannual variability, with 2001, 2016, and 2018 experiencing particularly elevated levels of extreme precipitation. This variability underscores the influence of natural climatic oscillations on TC-associated extreme precipitation events. Large-scale climate phenomena, including ENSO, significantly modulate typhoon frequency and intensity [10,23]. During El Niño years, studies show a substantial decrease in the passage rate and extreme wind speeds of tropical cyclones along China's southeastern coast [30–32], corroborating our findings. The correlation coefficients between the Niño 1 + 2 SST Index and typhoon activity on Hainan Island range from −0.340 to −0.406, indicating a negative relationship that supports the broader impacts of Pacific Ocean SST variations on typhoon dynamics.

Global TC precipitation rates are projected to increase due to enhanced tropical water vapor from anthropogenic warming, with a median increase of 14% [1]. This projection suggests more frequent extreme typhoon-related precipitation events in the future. Studies have shown an increase in both the number of days with extreme heavy precipitation and the total precipitation volume associated with TCs [12]. These findings align with trends observed on Hainan Island, where a general increase in extreme precipitation has been noted. In the central region, extreme precipitation (95th percentile threshold) has risen by 77.1 mm decade^{-1}, likely corresponding with increases in TC intensity and duration at landfall in recent decades [33]. Additionally, the continuous rise in global temperatures since 1960 [34] has led to higher temperatures and evaporation rates, thereby increasing the atmospheric water vapor content. This change requires more energy for water vapor condensation, resulting in prolonged energy accumulation in cloud formations before precipitation events. Consequently, the intensification of extreme precipitation acts as a significant indicator of increased heavy rainfall events [35]. Moreover, a study by Zhao, N. [36] indicates divergent trends in the maximum daily rainfall (TMDR) from tropical cyclones on Hainan Island. The frequencies for TMDR reaching thresholds of 50 mm, 100 mm, and 250 mm show a trend of decreasing by 0.7 decade^{-1}, a slight decrease of 0.2 decade^{-1}, and a slight increase of 0.1 decade^{-1}, respectively. At higher thresholds, these trends suggest that extreme precipitation events are becoming more frequent, with typhoons having an increasingly severe impact on extreme precipitation patterns on Hainan Island.

Spatial analyses reveal significant variability in how different typhoon tracks contribute to extreme precipitation events on Hainan Island (Figures 10 and 11). Notably, Tracks 1 and 3 significantly influence precipitation primarily in the western regions and at the northern end of the eastern regions. These observations support earlier studies [14,37,38], which show that the typical paths of these typhoons pass just north of Wuzhishan Mountain. As these typhoons approach this area, the northwesterly wind flow on the southwest flank of the typhoon center experiences orographic lifting and blocking by Wuzhishan Mountain, leading to substantial increases in precipitation. This interaction between typhoon trajectories and orographic features explains why the northwestern region of Hainan Island is particularly prone to intense rainfall. Furthermore, typhoon centers tend to slow down on the windward slopes of Wuzhishan Mountain, extending their duration in the area. Simultaneously, a weaker Western Pacific Subtropical High may diminish the steering currents at 500 hPa, further slowing typhoon movement [36]. These dynamics contribute to an increased rate of extreme precipitation in this area.

Recent studies have established a robust correlation between the inter-annual frequency of extreme summer precipitation events and SST variations [39]. Our research supports these findings, demonstrating a strong linkage between SST fluctuations and the occurrence of extreme precipitation events. Elevated SSTs provide critical thermal energy and moisture, which significantly enhance both typhoon development and precipitation capabilities, thereby directly contributing to extreme precipitation events. Additionally, increased SSTs in regions frequented by typhoon centers can intensify moisture convergence and precipitation processes through the positive feedback mechanisms of local atmospheric circulations.

5. Conclusions

This study analyzes the spatiotemporal distribution and contributing factors of typhoon-induced extreme precipitation events on Hainan Island using GsMAP satellite-retrieved precipitation data and best track data from the China Meteorological Administration. The main findings are as follows:

(1) From 1949 to 2020, typhoon events affecting Hainan decreased by 0.3 events decade^{-1} ($p < 0.01$). However, extreme precipitation events increased by 1.5 and 1.8 events decade^{-1} at the 95th and 99th percentiles, respectively, from 2000 to 2020. Extreme precipitation surged by 77.1 mm decade^{-1} in the eastern region, 73.3 mm decade^{-1} in the central region, and 58.6 mm decade^{-1} in the western region, respectively.

(2) Typhoon paths significantly influence the spatial distribution of extreme precipitation events, with Track 1 and 2 affecting the western and central regions. While Track 3 primarily impacts the western region. The contribution rates of typhoons to extreme precipitation vary spatiotemporally and seasonally.

(3) Incorporating typhoon precipitation data reduces the return period of extreme precipitation events in the central and eastern regions but extends it in the eastern region.

(4) Typhoon duration, wind speed, and SST positively correlate with extreme precipitation volumes, while atmospheric pressure and circulation indices show negative correlations. ENSO events, particularly SST variations in the Niño 1 + 2 region (correlation coefficients ranging from −0.340 to −0.406), notably impact typhoon activity.

Author Contributions: M.X. conceived and designed the experiments; analyzed and interpreted the data; contributed reagents, materials, analysis tools, or data; and wrote the paper. Y.T. conceived and designed the experiments; performed the experiments; analyzed and interpreted the data; and wrote the paper. C.S. performed the experiments and contributed reagents, materials, analysis tools, or data. Y.X. performed the experiments and contributed analysis tools or data. M.S. assisted in the analysis and interpretation of the data and contributed materials. J.W. contributed to the acquisition and analysis of data and assisted in the preparation of the manuscript. Y.Y. contributed to experimental setup and data collection. J.D. assisted in the design of the methodology and provided critical feedback on the manuscript. L.B. oversaw the project direction and planning; contributed to the conception and design of the experiments; provided final approval of the version to be published; and ensured questions related to all aspects of the work were appropriately resolved. All authors have read and agreed to the published version of the manuscript.

Funding: This project is funded by the National Natural Science Foundation of China (No. 32260294), the Hainan University Research Fund (KYQD(ZR)-22083), and the Hainan Provincial Natural Science Foundation of China (No. 423QN208).

Institutional Review Board Statement: Not applicable.

Informed Consent Statement: Informed consent was obtained from all subjects involved in the study.

Data Availability Statement: The typhoon track data is sourced from the Shanghai Typhoon Institute under the China Meteorological Administration. Precipitation data is derived from the Global Satellite Mapping of Precipitation (GSMaP) Reanalysis product. Sea Surface Temperature (SST) data comes from multiple sources: the Niño SST indices provided by the Climate Prediction Center (CPC) of the National Centers for Environmental Prediction (NCEP) are available at https://www.cpc.ncep.noaa.gov/data/indices/sstoi.indices (accessed on 16 May 2024) additional SST data provided by the National Centers for Environmental Information (NCEI) in the United States can be accessed at https://www.ncdc.noaa.gov/oisst (accessed on 16 May 2024); and the National Climate Center of the China Meteorological Administration also offers SST indices, available at https://cmdp.ncc-cma.net/Monitoring/cn_index_130.php (accessed on 16 May 2024).

Acknowledgments: We would like to express our sincere gratitude to Weijie Liao and Shixi Li for their invaluable assistance in the preliminary work and data preparation for this study. Their contributions have been instrumental in laying the foundation for our research and analysis.

Conflicts of Interest: The authors declare no conflicts of interest.

Appendix A

Table A1. Correlation coefficients between the TIEP and typhoon features in different sub-regions with different typhoon track paths.

Index	All Region	Western Region	Central Region	Eastern Region
Northern Hemisphere Subtropical High Area Index	0.362 ***	0.415 ***	0.403 ***	0.306 ***
North African Subtropical High Area Index	0.342 ***	0.408 ***	0.385 ***	0.282 ***
North African–North Atlantic–North American Subtropical High Area Index	0.376 ***	0.442 ***	0.422 ***	0.313 ***
Indian Subtropical High Area Index	−0.104	−0.098	−0.110	−0.096
Western Pacific Subtropical High Area Index	0.164 *	0.152 *	0.171 **	0.152 *
Eastern Pacific Subtropical High Area Index	0.370 ***	0.415 ***	0.408 ***	0.316 ***
North American Subtropical High Area Index	0.384 ***	0.446 ***	0.430 ***	0.321 ***
Atlantic Subtropical High Area Index	0.392 ***	0.455 ***	0.438 ***	0.329 ***
South China Sea Subtropical High Area Index	−0.113	−0.129 *	−0.119	−0.099
North American–Atlantic Subtropical High Area Index	0.394 ***	0.458 ***	0.441 ***	0.329 ***
Pacific Subtropical High Area Index	0.313 ***	0.335 ***	0.340 ***	0.274 ***
Northern Hemisphere Subtropical High Intensity Index	0.381 ***	0.446 ***	0.429 ***	0.317 ***
North African Subtropical High Intensity Index	0.357 ***	0.435 ***	0.407 ***	0.291 ***
North African–North Atlantic–North American Subtropical High Intensity Index	0.389 ***	0.468 ***	0.442 ***	0.318 ***
Indian Subtropical High Intensity Index	−0.101	−0.093	−0.105	−0.095
Western Pacific Subtropical High Intensity Index	0.210 ***	0.189 **	0.213 ***	0.199 **
Eastern Pacific Subtropical High Intensity Index	0.347 ***	0.395 ***	0.388 ***	0.293 ***
North American Subtropical High Intensity Index	0.402 ***	0.475 ***	0.455 ***	0.332 ***
North Atlantic Subtropical High Intensity Index	0.401 ***	0.485 ***	0.456 ***	0.327 ***
South China Sea Subtropical High Intensity Index	−0.103	−0.117	−0.108	−0.090
North American–North Atlantic Subtropical High Intensity Index	0.405 ***	0.485 ***	0.460 ***	0.332 ***
Pacific Subtropical High Intensity Index	0.315 ***	0.332 ***	0.340 ***	0.278 ***
Northern Hemisphere Subtropical High Ridge Position Index	0.475 ***	0.486 ***	0.500 ***	0.428 ***
North African Subtropical High Ridge Position Index	0.439 ***	0.464 ***	0.470 ***	0.389 ***
Indian Subtropical High Ridge Position Index	0.432 ***	0.460 ***	0.464 ***	0.382 ***
Indian Subtropical High Ridge Position Index	−0.176 **	−0.127 *	−0.179 **	−0.175 **
Western Pacific Subtropical High Ridge Position Index	0.489 ***	0.489 ***	0.508 ***	0.447 ***
Eastern Pacific Subtropical High Ridge Position Index	0.072	0.069	0.074	0.068
North American Subtropical High Ridge Position Index	0.066	0.066	0.069	0.061
Atlantic Sub Tropical High Ridge Position Index	0.149*	0.137 *	0.151 *	0.141 *
South China Sea Subtropical High Ridge Position Index	−0.030	0.013	−0.035	−0.037
North American–North Atlantic Subtropical High Ridge Position Index	0.065	0.065	0.068	0.059
Pacific Subtropical High Ridge Position Index	0.487 ***	0.484 ***	0.506 ***	0.446 ***
Northern Hemisphere Subtropical High Northern Boundary Position Index	0.111	0.104	0.113	0.104
North African Subtropical High Northern Boundary Position Index	0.195 **	0.191 **	0.197 **	0.181 **

Table A1. *Cont.*

Index	All Region	Western Region	Central Region	Eastern Region
North African–North Atlantic–North American Subtropical High Northern Boundary Position Index	0.135 *	0.135 *	0.138 *	0.125
Indian Subtropical High Northern Boundary Position Index	0.078	0.079	0.073	0.075
Western Pacific Subtropical High Northern Boundary Position Index	0.059	0.044	0.062	0.058
Eastern Pacific Subtropical High Northern Boundary Position Index	0.310 ***	0.316 ***	0.319 ***	0.284 ***
North American Subtropical High Northern Boundary Position Index	0.185 **	0.178 **	0.188 **	0.172 **
Atlantic Subtropical High Northern Boundary Position Index	0.252 ***	0.255 ***	0.260 ***	0.230 ***
South China Sea Subtropical High Northern Boundary Position Index	−0.035	−0.041	−0.026	−0.035
North American–Atlantic Subtropical High Northern Boundary Position Index	0.175 **	0.170 **	0.179 **	0.163 *
Pacific Subtropical High Northern Boundary Position Index	0.129 *	0.121	0.132 *	0.121
Western Pacific Sub Tropical High Western Ridge Point Index	0.177 **	0.179 **	0.183 **	0.162 *
Asia Polar Vortex Area Index	−0.240 ***	−0.314 ***	−0.284 ***	−0.185 **
Pacific Polar Vortex Area Index	−0.311 ***	−0.373 ***	−0.358 ***	−0.252 ***
North American Polar Vortex Area Index	−0.306 ***	−0.362 ***	−0.344 ***	−0.255 ***
Atlantic–European Polar Vortex Area Index	−0.201 **	−0.227 ***	−0.230 ***	−0.167 **
Northern Hemisphere Polar Vortex Area Index	−0.320 ***	−0.385 ***	−0.368 ***	−0.260 ***
Asia Polar Vortex Intensity Index	−0.287 ***	−0.351 ***	−0.327 ***	−0.232 ***
Pacific Polar Vortex Intensity Index	−0.228 ***	−0.303 ***	−0.277 ***	−0.172 **
North American Polar Vortex Intensity Index	−0.242 ***	−0.292 ***	−0.274 ***	−0.198 **
Atlantic–European Polar Vortex Intensity Index	−0.216 ***	−0.268 ***	−0.254 ***	−0.171 **
Northern Hemisphere Polar Vortex Intensity Index	−0.266 ***	−0.331 ***	−0.308 ***	−0.212 ***
Northern Hemisphere Polar Vortex Central Longitude Index	−0.118	−0.128 *	−0.137 *	−0.099
Northern Hemisphere Polar Vortex Central Latitude Index	0.175 **	0.167 **	0.185 **	0.161 *
Northern Hemisphere Polar Vortex Central Intensity Index	0.342 ***	0.375 ***	0.371 ***	0.297 ***
Eurasian Zonal Circulation Index	−0.140 *	−0.223 ***	−0.188 **	−0.088
Eurasian Meridional Circulation Index	−0.126 *	−0.125 *	−0.131 *	−0.114
Asian Zonal Circulation Index	−0.120	−0.215 ***	−0.170 **	−0.066
Asian Meridional Circulation Index	−0.106	−0.116	−0.113	−0.093
East Asian Trough Position Index	−0.056	−0.035	−0.056	−0.057
East Asian Trough Intensity Index	0.399 ***	0.423 ***	0.427 ***	0.354 ***
Tibet Plateau Region 1 Index	0.411 ***	0.406 ***	0.426 ***	0.377 ***
Tibet Plateau Region 2 Index	0.410 ***	0.415 ***	0.428 ***	0.372 ***
India–Burma Trough Intensity Index	−0.203 **	−0.253 ***	−0.229 ***	−0.165 **
Arctic Oscillation, AO	−0.027	−0.016	−0.017	−0.032
Antarctic Oscillation, AAO	0.007	−0.033	−0.019	0.028
North Atlantic Oscillation, NAO	0.002	−0.013	0.000	0.007
Pacific/ North American Pattern, PNA	0.094	0.056	0.075	0.106
East Atlantic Pattern, EA	0.217 ***	0.234 ***	0.242 ***	0.187 **
West Pacific Pattern, WP	−0.108	−0.087	−0.107	−0.106
North Pacific Pattern, NP	0.130 *	0.120	0.132 *	0.123
East Atlantic–West Russia Pattern, EA/WR	−0.167 **	−0.156 *	−0.179 **	−0.154 *
Tropical–Northern Hemisphere Pattern, TNH	−0.251 ***	−0.230 ***	−0.253 ***	−0.238 ***
Polar–Eurasia Pattern, POL	0.066	0.112	0.091	0.039
Scandinavia Pattern, SCA	−0.100	−0.126 *	−0.108	−0.083
Pacific Transition Pattern, PT	0.311 ***	0.347 ***	0.328 ***	0.273 ***
30 hPa zonal wind Index	−0.040	−0.075	−0.083	−0.009
50 hPa zonal wind Index	−0.112	−0.158 *	−0.162 *	−0.071
Mid-Eastern Pacific 200 mb Zonal Wind Index	−0.352 ***	−0.350 ***	−0.366 ***	−0.323 ***
West Pacific 850 mb Trade Wind Index	0.111	0.096	0.110	0.107
Central Pacific 850 mb Trade Wind Index	−0.110	−0.115	−0.118	−0.098
East Pacific 850 mb Trade Wind Index	−0.051	−0.031	−0.049	−0.054
Atlantic–European Circulation W Pattern Index	0.217 ***	0.235 ***	0.235 ***	0.189 **
Atlantic–European Circulation C Pattern Index	−0.204 **	−0.225 ***	−0.224 ***	−0.175 **

Notes: *, **, and ***, denote the CC with $p < 0.05$, $p < 0.01$ and $p < 0.001$, respectively.

Table A2. Typhoon classification and extreme precipitation caused.

Start Time	End Time	Extreme Precipitation (mm)	Track
2000/6/1 6:00	2000/6/1 12:00	0.00	2
2000/7/15 18:00	2000/7/17 0:00	197.98	1
2000/7/19 0:00	2000/7/19 6:00	0.00	3
2000/9/8 12:00	2000/9/9 18:00	184.99	2
2001/7/1 6:00	2001/7/2 6:00	203.01	1
2001/7/25 12:00	2001/7/25 18:00	0.00	1
2001/8/9 18:00	2001/8/10 12:00	0.00	2
2001/8/28 18:00	2001/9/11 12:00	937.60	3
2002/8/18 6:00	2002/8/19 12:00	114.33	1
2002/9/14 6:00	2002/9/15 12:00	30.65	1
2002/9/24 12:00	2002/9/27 0:00	294.46	3
2002/9/27 12:00	2002/9/28 6:00	0.00	3
2003/7/21 0:00	2003/7/22 0:00	147.98	3
2003/8/24 12:00	2003/8/25 12:00	184.52	3
2003/11/17 12:00	2003/11/19 12:00	210.65	3
2005/7/29 0:00	2005/7/30 18:00	392.91	3
2005/9/17 18:00	2005/9/18 0:00	49.44	2
2005/9/25 6:00	2005/9/26 18:00	435.40	3
2006/6/28 0:00	2006/6/29 0:00	108.62	1
2006/7/3 6:00	2006/7/4 0:00	133.55	3
2006/8/3 6:00	2006/8/3 12:00	21.47	1
2006/8/22 6:00	2006/8/24 18:00	0.00	1
2006/9/13 6:00	2006/9/13 18:00	0.00	1
2006/9/24 12:00	2006/9/25 0:00	63.36	2
2006/11/13 18:00	2006/11/14 18:00	0.00	2
2006/12/13 6:00	2006/12/14 0:00	0.00	2
2007/7/3 12:00	2007/7/5 12:00	196.83	3
2007/8/6 6:00	2007/8/7 6:00	0.00	2
2007/9/24 0:00	2007/9/25 6:00	134.11	3
2007/10/2 0:00	2007/10/3 6:00	78.92	2
2008/4/17 18:00	2008/4/19 0:00	0.00	1
2008/8/6 12:00	2008/8/7 6:00	106.45	3
2008/9/29 12:00	2008/9/29 18:00	0.00	2
2008/10/3 0:00	2008/10/4 12:00	214.92	1
2008/10/13 6:00	2008/10/14 18:00	260.86	3
2009/7/11 18:00	2009/7/12 6:00	59.24	3
2009/8/6 12:00	2009/8/9 6:00	426.45	3
2009/9/10 18:00	2009/9/11 12:00	117.02	3
2009/10/11 18:00	2009/10/14 0:00	275.94	3
2009/10/19 6:00	2009/10/20 12:00	43.72	2
2010/7/15 18:00	2010/7/17 6:00	51.34	2
2010/7/21 12:00	2010/7/22 6:00	58.46	1
2010/8/23 18:00	2010/8/24 0:00	7.68	2
2010/10/4 12:00	2010/10/10 6:00	674.45	3
2011/6/22 18:00	2011/6/24 6:00	189.08	3
2011/7/29 0:00	2011/7/30 0:00	257.60	3
2011/10/3 18:00	2011/10/5 12:00	0.00	2
2011/11/8 0:00	2011/11/9 0:00	142.30	2
2012/6/16 6:00	2012/6/18 0:00	230.20	1
2012/8/17 0:00	2012/8/17 12:00	152.55	3
2012/10/27 6:00	2012/10/28 0:00	229.19	2
2012/10/29 0:00	2012/10/29 12:00	0.00	3
2013/6/22 0:00	2013/6/23 0:00	73.55	3
2013/7/1 12:00	2013/7/2 0:00	66.20	1

Table A2. *Cont.*

Start Time	End Time	Extreme Precipitation (mm)	Track
2013/8/2 0:00	2013/8/3 0:00	138.10	3
2013/8/6 18:00	2013/8/7 6:00	0.00	2
2013/8/14 0:00	2013/8/14 6:00	20.91	1
2013/9/29 12:00	2013/9/30 6:00	0.00	2
2013/11/4 0:00	2013/11/4 12:00	0.00	2
2013/11/10 6:00	2013/11/11 0:00	120.60	3
2014/7/18 0:00	2014/7/18 18:00	243.65	1
2014/9/7 12:00	2014/9/8 6:00	26.80	1
2014/9/16 0:00	2014/9/16 12:00	213.28	3
2015/6/21 12:00	2015/6/23 18:00	413.86	3
2015/10/3 18:00	2015/10/4 6:00	72.06	1
2016/5/27 0:00	2016/5/27 6:00	0.00	1
2016/7/26 6:00	2016/7/27 6:00	133.02	3
2016/8/17 18:00	2016/8/19 0:00	370.73	3
2016/10/12 18:00	2016/10/13 0:00	38.16	2
2016/10/17 12:00	2016/10/19 6:00	276.80	3
2017/7/15 0:00	2017/7/16 6:00	156.10	2
2017/7/22 0:00	2017/7/25 6:00	59.15	2
2017/9/14 12:00	2017/9/15 0:00	76.33	2
2017/9/24 3:00	2017/9/25 0:00	255.08	3
2017/10/9 12:00	2017/10/9 18:00	45.70	2
2017/10/15 12:00	2017/10/16 6:00	113.44	1
2017/11/12 6:00	2017/11/13 6:00	265.96	2
2018/6/5 0:00	2018/6/7 12:00	412.57	1
2018/7/17 15:00	2018/7/18 6:00	97.43	3
2018/7/21 21:00	2018/7/23 21:00	315.96	3
2018/8/9 0:00	2018/8/11 0:00	581.26	1
2018/8/11 15:00	2018/8/12 15:00	105.57	1
2018/8/14 21:00	2018/8/16 12:00	178.04	3
2018/9/12 18:00	2018/9/13 12:00	122.99	1
2019/7/2 6:00	2019/7/3 12:00	294.38	3
2019/7/31 12:00	2019/8/2 18:00	317.07	3
2019/8/29 0:00	2019/8/29 12:00	95.44	2
2019/9/1 18:00	2019/9/2 12:00	173.95	2
2019/9/3 12:00	2019/9/6 6:00	224.51	2
2020/7/31 12:00	2020/8/1 12:00	58.60	2
2020/10/13 0:00	2020/10/14 0:00	80.71	3
2020/10/24 12:00	2020/10/25 12:00	297.82	2
2020/11/14 18:00	2020/11/15 0:00	20.40	2

References

1. Knutson, T.; Camargo, S.J.; Chan, J.C.L.; Emanuel, K.; Ho, C.H.; Kossin, J.; Mohapatra, M.; Satoh, M.; Sugi, M.; Walsh, K.; et al. Tropical Cyclones and Climate Change Assessment: Part II: Projected Response to Anthropogenic Warming. *Bull. Am. Meteorol. Soc.* **2020**, *101*, E303–E322. [CrossRef]
2. Shan, K.; Lin, Y.; Chu, P.; Yu, X.; Song, F. Seasonal advance of intense tropical cyclones in a warming climate. *Nature* **2023**, *623*, 83–89. [CrossRef] [PubMed]
3. Lianshou, C.; Zhiyong, M.; Chunhua, C. An overview on the research of typhoon rainfall distribution. *J. Mar. Meteorol.* **2017**, *37*, 1–7.
4. Lianshou, C.; Yinglong, X. Review of Typhoon Very Heavy Rainfall in China. *Meteorol. Environ. Sci.* **2017**, *40*, 3–10.
5. Du, X.; Chen, H.; Li, Q.; Ge, X. Urban Impact on Landfalling Tropical Cyclone Precipitation: A Numerical Study of Typhoon Rumbia (2018). *Adv. Atmos. Sci.* **2023**, *40*, 988–1004. [CrossRef]
6. Zhang, W.; Villarini, G.; Vecchi, G.A.; Smith, J.A. Urbanization exacerbated the rainfall and flooding caused by hurricane Harvey in Houston. *Nature* **2018**, *563*, 384–388. [CrossRef] [PubMed]
7. Meng, C.; Xu, W.; Qiao, Y.; Liao, X.; Qin, L. Quantitative Risk Assessment of Population Affected by Tropical Cyclones Through Joint Consideration of Extreme Precipitation and Strong Wind—A Case Study of Hainan Province. *Earth's Future* **2021**, *9*, e2021EF002365. [CrossRef]
8. Dake, C. Global intensification of weak tropical cyclones in the past 30 years. *Sci. Sin.* **2024**, *54*, 659–660.

9. Yongmei, W.; Fumin, R.; Weijing, L.; Wang, X.L. Climatic characteristics of typhoon precipitation over China. *J. Trop. Meteorol.* **2008**, *14*, 233–238.
10. Zhang, Q.; Lai, Y.; Gu, X.; Shi, P.; Singh, V.P. Tropical Cyclonic Rainfall in China: Changing Properties, Seasonality, and Causes. *J. Geophys. Res. Atmos.* **2018**, *123*, 4476–4489. [CrossRef]
11. Gori, A.; Lin, N.; Xi, D.; Emanuel, K. Tropical cyclone climatology change greatly exacerbates US extreme rainfall-surge hazard. *Nat. Clim. Change* **2022**, *12*, 171–178. [CrossRef]
12. Wu, Y.; Wu, S.; Zhai, P. The impact of tropical cyclones on Hainan Island's extreme and total precipitation. *Int. J. Climatol.* **2007**, *27*, 1059–1064. [CrossRef]
13. Lu, G.; Wang, W.; Zheng, M.; Cai, Q.B. Spatial and temporal distribution characteristics of typhoon precipitation in Hainan. *Trans. Atmos. Sci.* **2015**, *38*, 710–715.
14. Jiang, X.; Ren, F.; Li, Y.; Qiu, W.; Ma, Z.; Cai, Q. Characteristics and Preliminary Causes of Tropical Cyclone Extreme Rainfall Events over Hainan Island. *Adv. Atmos. Sci.* **2018**, *5*, 580–591. [CrossRef]
15. Chu, P.; Chen, D.; Lin, P. Trends in precipitation extremes during the typhoon season in Taiwan over the last 60 years. *Atmos. Sci. Lett.* **2014**, *15*, 37–43. [CrossRef]
16. Yu, C.; Hu, D.; Di, Y.; Wang, Y. Performance evaluation of IMERG precipitation products during typhoon Lekima (2019). *J. Hydrol.* **2021**, *597*, 126307. [CrossRef]
17. Rios Gaona, M.F.; Villarini, G. Characterization of the diurnal cycle of maximum rainfall in tropical cyclones. *J. Hydrol.* **2018**, *564*, 997–1007. [CrossRef]
18. Yan, Z.; Renyong, Y.; Shouchang, C.; Cai, Q. Observation and Simulation of Effect of Terrain on Rainstorm Amplification in Typhoon Haiyan. *Meteorol. Sci. Technol.* **2018**, *46*, 1147–1153.
19. Yang, R.Y.; Min, J.Z.; Feng, W. Numerical simulations of the hainan island topographicefects on the rainfall of west-going typhoons. *J. Trop. Meteorol.* **2013**, *29*, 474–480.
20. Yu, Z.; Wang, Y.; Xu, H.; Davidson, N.; Chen, Y.; Chen, Y.; Yu, H. On the Relationship between Intensity and Rainfall Distribution in Tropical Cyclones Making Landfall over China. *J. Appl. Meteorol. Clim.* **2017**, *56*, 2883–2901. [CrossRef]
21. Lai, Y.; Li, J.; Gu, X.; Chen, Y.D.; Kong, D.; Gan, T.Y.; Liu, M.; Li, Q.; Wu, G. Greater flood risks in response to slowdown of tropical cyclones over the coast of China. *Proc. Natl. Acad. Sci. USA* **2020**, *117*, 14751–14755. [CrossRef] [PubMed]
22. Li, X.; Han, G.; Yang, J.; Wang, C. Remote Sensing Analysis of Typhoon-Induced Storm Surges and Sea Surface Cooling in Chinese Coastal Waters. *Remote Sens.* **2023**, *15*, 1844. [CrossRef]
23. Kawase, H.; Yamaguchi, M.; Imada, Y.; Hayashi, S.; Murata, A.; Nakaegawa, T.; Miyasaka, T.; Takayabu, I. Enhancement of Extremely Heavy Precipitation Induced by Typhoon Hagibis (2019) due to Historical Warming. *Sola* **2021**, *17A*, 7–13. [CrossRef]
24. Zhou, L.; Gao, S.; Jia, J.; Yang, Y.; Tong, C.; Wang, A. Paleo-Typhoon Events as Indicated by Coral Reef Boulder Deposits on the Southern Coast of Hainan Island, China. *Front. Mar. Sci.* **2021**, *8*, 746773. [CrossRef]
25. Chou, J.; Dong, W.; Tu, G.; Xu, Y. Spatiotemporal distribution of landing tropical cyclones and disaster impact analysis in coastal China during 1990–2016. *Phys. Chem. Earth Parts A/B/C* **2020**, *115*, 102830. [CrossRef]
26. Xu, M.; Tan, Y.; Li, S.; Shang, M.; Shi, C.; Bai, L. Characterisation of Multi-Scale Spatial and Temporal Distribution of Typhoon Precipitation on Hainan Island Based on Remotely Sensed Precipitation Data. *Chin. J. Trop. Agric.* **2025**, *45*. *accepted in press*.
27. Fisher, R.A.; Tippett, L.H.C. Limiting forms of the frequency distribution of the largest or smallest member of a sample. *Math. Proc. Camb.* **1928**, *24*, 180–190. [CrossRef]
28. Jenkinson, A.F. The frequency distribution of the annual maximum (or minimum) values of meteorological elements. *Q. J. R. Meteor. Soc.* **1955**, *81*, 158–171. [CrossRef]
29. Ohlson, J.A.; Kim, S. Linear valuation without OLS: The Theil-Sen estimation approach. *Res. Account Stud.* **2015**, *20*, 395–435. [CrossRef]
30. Chen, Y.; Duan, Z. Impact of ENSO on typhoon wind hazard in the coast of southeast China. *Nat. Hazards* **2018**, *92*, 1717–1731. [CrossRef]
31. Chu, P.; Zhang, H.; Chang, H.; Chen, T.L.; Tofte, K. Trends in return levels of 24-hr precipitation extremes during the typhoon season in Taiwan. *Int. J. Climatol.* **2018**, *38*, 5107–5124. [CrossRef]
32. Liu, L.; Wang, Y. Trends in Landfalling Tropical Cyclone–Induced Precipitation over China. *J. Clim.* **2020**, *33*, 2223–2235. [CrossRef]
33. Zhang, P.; Ren, G.; Qin, Y.; Zhai, Y.; Zhai, T.; Tysa, S.K.; Xue, X.; Yang, G.; Sun, X. Urbanization Effects on Estimates of Global Trends in Mean and Extreme Air Temperature. *J. Clim.* **2021**, *34*, 1923–1945. [CrossRef]
34. Su, J.; Ren, G.; Zhang, Y.; Yang, G.; Xue, X.; Lee, R. Changes in extreme rainfall over mainland China induced by landfalling tropical cyclones. *Environ. Res. Commun.* **2022**, *4*, 101004. [CrossRef]
35. Zhao, N.; Hong, J.; Lau, K.H. Impact of supply chain digitalization on supply chain resilience and performance: A multi-mediation model. *Int. J. Prod. Econ.* **2023**, *259*, 108817. [CrossRef] [PubMed]
36. Duan, L.; Chen, L. Diagnostic Analysis and Numerical Study of Torrential Rain Associated with the Tropical Storm Fitow (0114). *Chin. J. Atmos. Sci.* **2005**, *29*, 343–353. [CrossRef]
37. Liu, H.; Huang, X.; Fei, J.; Zhang, C.; Cheng, X. Characteristics and Preliminary Causes of Extremely Persistent Heavy Rainfall Generated by Landfalling Tropical Cyclones Over China. *Earth Space Sci.* **2022**, *9*, e2022EA002238. [CrossRef]

38. Hatsuzuka, D.; Sato, T. Impact of SST on Present and Future Extreme Precipitation in Hokkaido Investigated Considering Weather Patterns. *J. Geophys. Res. Atmos.* **2022**, *127*, e2021JD036120. [CrossRef]
39. Fujibe, F. Relationship between Interannual Variations of Extreme Hourly Precipitation and Air/Sea-Surface Temperature in Japan. *Sola* **2015**, *11*, 5–9. [CrossRef]

atmosphere

MDPI

Article

Projection of Extreme Summer Precipitation over Hubei Province in the 21st Century

Abrar Mubark [1,2], Qian Chen [1,*], Mohamed Abdallah [3,4], Awad Hussien [2] and Monzer Hamadalnel [5]

1 Collaborative Innovation Center on Forecast and Evaluation of Meteorological Disasters, Key Laboratory for Aerosol-Cloud-Precipitation of the China Meteorological Administration, Nanjing University of Information Science and Technology, Nanjing 210044, China; 202252010020@nuist.edu.cn
2 School of Atmospheric Sciences, Nanjing University of Information Science and Technology, Nanjing 210044, China; 202252010028@nuist.edu.cn
3 Sudan Meteorological Authority (SMA), Khartoum P.O. Box 574, Sudan; m_alriah@nuist.edu.cn
4 School of Geographical Sciences, Nanjing University of Information Science and Technology, Nanjing 210044, China
5 Department of Astronomy and Meteorology, Faculty of Science and Technology, Omdurman Islamic University, Oumdurman 14415, Sudan; mnzr@oiu.edu.sd
* Correspondence: chenq_05@nuist.edu.cn

Abstract: The link between the escalation of global warming and the increase in extreme precipitation events necessitates a deeper understanding of future trends. This study focused on the dynamics of extreme rainfall in Hubei Province throughout the 21st century, a region already sensitive to climatic shifts and extreme weather occurrences. Using the high-resolution global climate model RegCM4 driven by another high-resolution model, HadGEM2-ES, and based on the representative concentration pathway (RCP8.5) emissions scenario, this research predicted the changes in rainfall patterns in Hubei Province during the summer of the 21st century. The accuracy of the adjusted model was confirmed through the use of five extreme rainfall indices (EPIs), namely maximum 5-day amount of precipitation (RX5day), number of heavy rain days (R10), the simple daily intensity index (SDII), consecutive dry days (CDD), and consecutive wet days (CWD), that measured the intensity and frequency of such events. In particular, excluding the index for continuous dry days (CDD), there was an anticipated increase in extreme rainfall during the summer in the mid-21st century. The number of heavy rain days (R10mm) increased significantly ($p < 0.05$) in the southeastern parts, especially for Wuhan, Xiantao, Qianjiang, Jinzhou, and Ezhou. The EPI values were higher in southeastern Hubei. Consequently, areas such as Wuhan, Xiantao, and Qianjiang in Hubei Province are projected to face more frequent and severe extreme rainfall episodes as the century progresses.

Keywords: extreme climate indices; climate projections; precipitation; bias correction; RegCM4

1. Introduction

The environment and human life are now at risk due to climate change, which has caused widespread alarm in recent times. Climate change has become a threat to people's lives and property and has been linked to increased disasters [1]. Extreme rainfall is considered a double-edged sword, as it can lead to the stagnation of dams and rivers and, on the other hand, causes numerous floods at global and regional levels. Given the impact of extreme climate events on society, it is crucial to conduct in-depth investigations to understand the long-term fluctuations and the factors that contribute to fluctuations in exceptional precipitation levels in historical and future periods [2,3].

Reports from the Intergovernmental Panel on Climate Change (IPCC) have indicated that severe impacts on ecosystems, people, settlements, and infrastructure have significantly increased with the frequency and intensity of extreme weather and climate events [4]. The rises in extreme rainfall rates and the occurrence of extreme precipitation are likely to increase in the future as the average global temperature continues to rise [5–9]. The precipitation trend tends to exhibit varying patterns across China, with increases observed in winter and summer in the majority of areas, while significant decreases are observed in autumn in many regions of central and eastern China [10]. Zhai et al. [8] found that precipitation extremes trends differed from one season to another in eastern China. The summer precipitation trend is very similar to that of annual totals. Spring precipitation has increased in southern northeast China and north China but decreased significantly in the mid-reaches of the Yangtze River. Autumn precipitation has generally decreased throughout eastern China. In winter, precipitation has significantly decreased over the northern part of eastern China but increased in the south. Chen et al. [9] suggested that annual rainfall in China will increase by the end of the 21st century compared with current levels.

Recent research has indicated that recurrent droughts are caused by climate variability, leading to changes in the intensity and frequency of precipitation at regional and local levels [11]. Models' outputs can be used to investigate the changes in extreme precipitation in the future, such as simulations from the RegCM4 model. Previous studies indicated that RegCM4 can accurately replicate China's spatial distribution of extreme climate events [12,13], and performs well in climate predictions, illustrating the interannual variability in China's river basins and demonstrating improved summer performance [14].

The extreme precipitation indices may exhibit distinct variations across various regions [15]. Although many studies have investigated the spatial–temporal variations in extreme precipitation over different regions [16,17], few studies have focused on extreme precipitation and its impacts under future scenarios in Hubei Province, China [18,19]. Hubei Province is highly susceptible to changes in climate and has already experienced severe weather incidents, with extreme precipitation events mainly occurring in spring and summer [20]. Using extreme precipitation indices (EPIs), this study looked at how the characteristics of extreme precipitation in Hubei Province will change over time in the 21st century under the RCP8.5 scenario, due to its alignment with high greenhouse gas emissions scenarios and its representation of the upper limit of the representative concentration pathways (RCPs) [21].

This study's objectives were (1) to evaluate the efficiency of the model in simulating precipitation over Hubei Province, (2) to analyze the spatial–temporal changes in the annual extreme precipitation indices in Hubei Province through the reference period (1979–2005), and (3) to analyze the spatial–temporal changes in the extreme precipitation indices in summer within Hubei Province through the mid-21st century and late 21st century.

2. Study Area

Hubei Province is located at the center of the Yangtze River in central China. The province is known for its plentiful natural resources, varied mountainous landscapes, and diverse terrain. The climate of the region is characterized by a subtropical monsoon zone, which experiences an average annual rainfall of 800–1600 mm. Most of this precipitation occurs during the rainy season, which lasts from mid-June to mid-July [22]. The landform of Hubei Province represents an incomplete watershed with three high sides, a middle depression opening to the south, and a gap to the north [23]. The Dabie and Wudang Mountains in north Hubei are also considered to be among the natural landscapes of the province [24]. Figure 1 shows the location of Hubei Province.

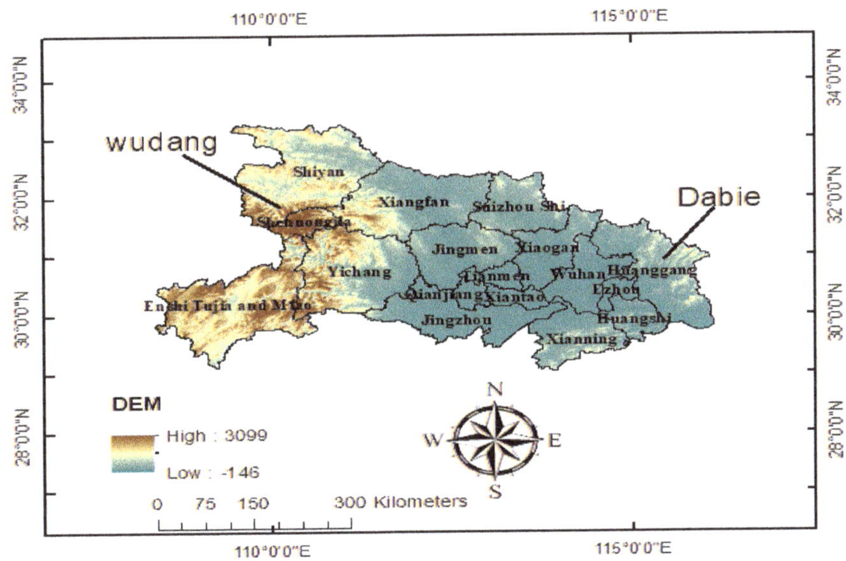

Figure 1. Location of Hubei Province.

3. Data and Methods

3.1. Reanalysis of ERA5 Datasets

ERA5, the fifth-generation reanalysis data created by the European Center for Medium-Range Weather Forecasts (ECMWF) [25], were used in this study as reference data. The quality of the input data in ERA5 has been significantly enhanced compared with its earlier versions [26]. It has also effectively captured the annual and seasonal patterns of precipitation over several parts of China, with continuous updates to the present day. The spanning period has been covered since 1979, and some versions have been extended to 1950 [27]. Because of its good performance in representing actual rainfall observations in different climatic zones around the globe [28,29], we used ERA5 as reference data to evaluate the historical extremes in precipitation for a simulation of a climate model and adjusted future projections in this study. A daily precipitation dataset with a spatial resolution of $0.25° \times 0.25°$ latitude/longitude covering the period from 1979 to 2005 was utilized to evaluate the model's performance in simulating rainfall over Hubei Province in China. The grid dataset is freely available at https://cds.climate.copernicus.eu/cdsapp#!/dataset/reanalysis-era5-single-levels (Accessed on 15 October 2023).

3.2. The Reginal Climate Model (RegCM)

To assess future changes in extreme precipitation in Hubei, we relied on RegCM4, a model developed by the Abdus Salam International Centre for Theoretical Physics (ICTP) during the specified timeframe of 1979–2005. Established at the National Center for Atmospheric Research (NCAR), the regional climate model system is formally referred to as RegCM. Furthermore, it has been modified to accurately represent the specific weather patterns of Hubei Province, leading to the development of precise predictions regarding climate change at a local level. The regional model RegCM4, designed for specific areas, utilized the initial and lateral boundary conditions from HadGEM2-ES, the Earth system model used by the Met Office Hadley Centre for CMIP5 20th century simulations.

As mentioned in the study of Giorgi et al. [30], RegCM4 is the latest version compared with previous versions. Recently, a large research community has made significant enhancements to its structure, and additional modern physical parameters have been developed. It can be readily implemented in any geographical area worldwide, exhibiting varying levels of precision across a diverse array of research disciplines. The model was run using a grid spacing of 0.25° × 0.25° (latitude by longitude) throughout CORDEX-EA (East Asia), including continental China and centered at (35 °N, 115 °E). The study utilized the annual mean precipitation calculated for the simulated RegCMs using both the corrected and raw data from the historical period from 1979 to 2005 and a two-period bias correction for future precipitation. According to the IPCC Fifth Assessment Report (AR5), the analysis was performed for the mid-21st century (2046–2065) and the late 21st century (2078–2097) (i.e., the far future). To verify the performance of the model and to obtain high accuracy, a bias correction process was used for the simulated RegCMs using both the corrected and raw data from the historical period for reference data and two-period bias correction for future precipitation. the software used is ArcMap (version 10.4.1), which is a popular GIS software that enables users to generate, modify, and assess geospatial data.

The bias correction technique was carried out, utilizing the quantile mapping technique, which is particularly efficient in correcting outliers, applying this method to both observed and simulated data. This approach involves mapping the simulated data onto the observed data based on the cumulative probability distribution, as described in detail by Park et al. [31].

3.3. *Climate Indices*

EPIs were constructed using the Joint WMO Expert Group on Climate Change Detection and Indices (CCl/CLIVAR/JCOMM) (http://etccdi.pacificclimate.org/) (Accessed on 10 November 2023). This is a collection of indicators consisting mainly of data on rainfall extremes, as they allow for the assessment of various aspects related to the variations in the strength, occurrence rate, and length of severe occurrences and have been extensively used in different research investigations [32].

We determined the specific values of the absolute indices based on a fixed threshold for the recorded rainfall. We used five different EPIs in this study, including the maximum 5-day amount of precipitation (RX5day), the number of heavy rain days (R10), consecutive wet days (CWD), consecutive dry days (CDD), and a simple daily intensity index (SDII), to capture various aspects of the extreme precipitation events in Hubei Province. Table 1 provides a summary of each index. All meteorological indices evaluate either the severity or frequency of precipitation; however, the CDD index measures the duration of dry spells.

Table 1. Definitions of five EPIs used in the study.

Indicator Units	Index Name	Definitions	Units
Rx5day	Maximum 5-day precipitation	Annual maximum consecutive 5-day precipitation	mm
R10mm	Number of heavy precipitation days	Annual number of days when the daily precipitation is >l0 mm	days
CWD	Consecutive wet days	Average daily precipitation on wet days, mm/day	days
CDD	Consecutive dry days	Average daily precipitation on dry days, mm/day	days
SDII	Simple daily intensity index	Daily precipitation (≥1 mm) in the year	mm/day

3.4. *Methods of Analysis*

In this study, the correlation coefficient was used to evaluate the reliability of the rainfall data. It was necessary to evaluate both the model's outputs and the precipitation data from the ERA5 reanalysis. For this purpose, CC was used to evaluate the degree of linear correlation between the model's outputs and reanalyzed data. When the CC is 1 or close to 1, it indicates a strong correlation between the model outputs and ERA5 data.

$$CC = \frac{\sum_{n=1}^{n}(Gi - \overline{G})(Si - \overline{S})}{\sum_{n=1}^{n}(Gi - \overline{G})^2 \sum_{n=1}^{1}(Si - \overline{S})^2}, \quad (1)$$

Moreover, the root mean square error (RMSE) method was applied to assess the overall accuracy and error level of the rainfall products from the model's outputs compared with the ERA5 reanalyzed data.

$$RMSE = \sqrt{\frac{1}{N}\sum_{n=1}^{N}(S_n - S_n)^2} \quad (2)$$

The Mann–Kendall (MK) test was used to analyze the significance of the trend in the time series of EPIs for the three periods: the reference period (1979–2005), the mid-21st century (2046–2065), and the late 21st century (2078–2097). The World Meteorological Organization has suggested this approach for analyzing trends in specific time series [33]. The purpose of the analysis was to demonstrate the statistical significance of upward and downward trends. The strength of this method depends on the amount, size, and variability of the data being analyzed, as well as the magnitude of the data.

4. Results

4.1. Validation of the Model Simulation and ERA5

Initially, we assessed the model's accuracy in predicting precipitation patterns in Hubei Province by comparing the mean simulated annual precipitation with the corresponding observed data from 1979 to 2005. We used several statistical tests (Table 2) to determine how much the ERA5 reanalyzed rainfall product and the RegCM model output differed in accuracy. We also adopted the Taylor diagram, widely acknowledged as one of the most suitable techniques for the graphical representation of model-matched performance against observations. This diagram displays the centered RMSE, correlation coefficient (CC), and ratio of standard deviations. Figure 2 shows a Taylor diagram illustrating a comparison between the daily mean rainfall for the reanalysis of ERA5 and the output of RegCM.

Table 2. Statistical summary of daily historical precipitation over Hubei in an evaluation of RegCM against ERA5.

	Mean	Bias	RMSD	RMSE	STD	R
ERA5	3.85	0	0	0	1	1
RegCM	3.69	0.17	4.34	4.34	0.99	027

Table 2 and Figure 2, which demonstrate the use of CC to assess the reliability of rainfall data for both the reanalyzed data and the model's outputs, provided the relative statistical metrics. The ERA5 product showed a better correlation (CC = 1.0) and an increased standard deviation compared with the RegCM. We used the RMSE method to evaluate the overall accuracy and magnitude of error of the rainfall products derived from the model's outputs against the ERA5 reanalyzed data. The model outputs exhibited a higher level of error compared with the ERA5 product.

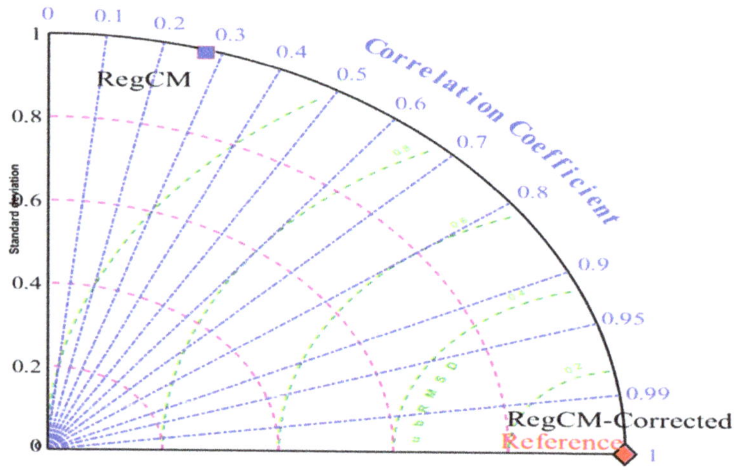

Figure 2. Taylor diagram comparing daily ERA5 products with the output of RegCM.

4.2. Climatology of Rainfall over Hubei

The study region was analyzed to investigate the temporal distribution of rainfall. The precipitation of Hubei Province shows annual cycles of variability in daily precipitation (see Figure 3). The daily precipitation in Hubei Province for both ERA5 and RegCM was analyzed before and after bias correction during the flood season, which lasts from May to August. In contrast, the winter months of January, November, and December are the driest times of the year, with average daily precipitation below 3 mm. The outcomes of the study suggested that RegCM was capable of accurately capturing the daily amount of rainfall for all regions, and tended to overestimate it during wet months and underestimate the rainfall during dry months.

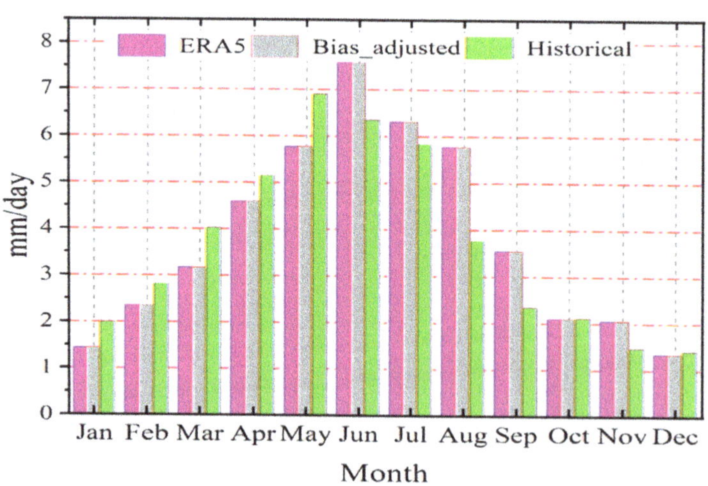

Figure 3. Annual cycle of daily mean precipitation over 1979–2005. The pink box represents the annual cycle for ERA5, the green box represents the annual cycle before bias correction (BC), and the gray box represents the annual cycle after BC.

4.3. Changes in Annual Mean Precipitation

The following section presents a comparison among the ERA5 reanalyzed data, the RegCM4 simulation data, and the bias-corrected simulated precipitation. It is important to note that there was spatial variation in both the ERA5 and bias-corrected simulated precipitation, indicating that RegCM4 responded effectively to the process of bias correction. With regards to the reanalyzed data, the average precipitation was primarily concentrated in the northwestern region. The intensity of precipitation increased in these areas and decreased towards the center and south. However, the annual mean of bias-corrected simulated precipitation also exhibited an increase in the northwestern parts and a decrease towards the center and some northern regions. ERA5 products accurately captured the spatial amount of rainfall, as shown in Figure 4a and as demonstrated in a report by Jiao et al. [34]. Figure 4 shows that most areas affected by extreme precipitation events were located in western and northern part, with the average annual rainfall reaching 1500 mm/year. The frequency of heavy rain decreased towards Shiyan and Suizhou, with values below 400 mm/year of rainfall. This variation in rainfall is attributed to the presence of hills and mountainous terrain that prevail in the western and eastern parts. In contrast, the central part of the province is dominated by low-level mountains and hills, with little difference in elevation, leading to minimal variation in the mean annual rainfall according to the topographic units.

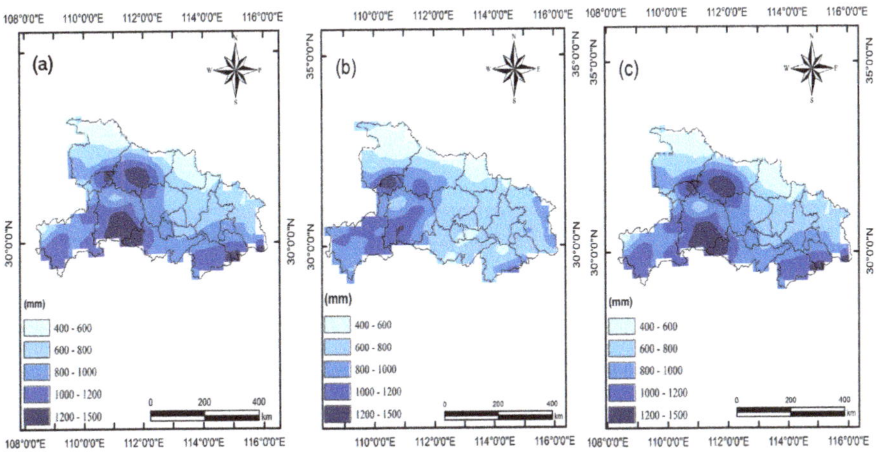

Figure 4. Spatial variations in annual mean precipitation over Hubei Province during 1979–2005: (**a**) precipitation from ERA5 products, (**b**) simulated precipitation before bias correction, and (**c**) simulated precipitation after bias correction.

The center and some of the northern parts of Hubei Province recorded less rainfall, with an annual average up to 1000 mm. These areas correspond to the driest areas of the province. Consequently, we expect a relative reduction in the frequency of extreme precipitation events in these areas.

Figure 4b shows the average historical annual rainfall from the model's simulations. We found that there was not much difference from that in Figure 4a, indicating that the model has good potential for spatially representing rainfall over the region. Figure 4c illustrates the variations in the annual mean precipitation after bias correction. The spatial variation in average rainfall after the process of bias correction is depicted in this figure. The amount of annual rainfall was also concentrated in the western and northern parts, and decreased in the central and southern regions.

The outputs of RegCM4 coincided with ERA5 (see Figure 5), which means the correction method was effective in the past; for this reason, its application was adopted in future projections. The results showed obvious variations in the spatial patterns of the average annual precipitation for both the ERA5 products and the RegCM model's outputs before and after bias correction, which indicated that the RegCM4 responded to the process of bias correction, so it was necessary to perform the process of bias correction to obtain more accurate and reliable results.

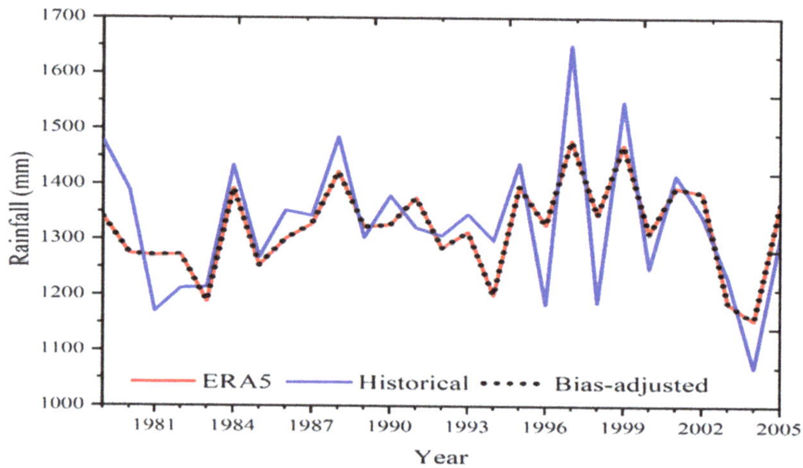

Figure 5. Time series of annual mean precipitation for the ERA5 product and RegCM4 before and after bias correction.

4.4. Spatial Changes in EPIs

Figure 6 shows the spatial changes in the five observed average annual EPIs over Hubei Province from 1979 to 2005. As depicted in the figure, the distributions of Rx5day and SDII ranged across 20–104 mm and 2.6–11.2 mm, respectively. The annual distribution of R10mm, CWD, and CDD ranged across 20–128 mm, 5–22.8 days, and 1.2–8.8 days, respectively. According to the observed extreme wet indicators (all indicators except for CDD), the highest values were found in the western and northern parts, while the lowest values were in the eastern and central parts. The distributions of Rx5day and SDII were similar to those of the average annual rainfall, which increased in Yichang, Shiyan, and Xiangyang, and were lower in the eastern to central part of Hubei.

The Rx5day values increased in the western and northern region, mainly in Yichang and Xiangyang (also known as Xiangfan) (see Figure 6a). The maximum value of R10mm was found in Yichang and Shennongjia, suggesting these areas experienced frequent and intense rainfall events during this period (see Figure 6b).

CWD decreased in the eastern and central parts, and increased in the western and northern parts and some southern areas (see Figure 6c). In contrast, CDD increased in the southern–central parts and decreased in the western part (see Figure 6c). CWD increased distinctly in western Hubei, while CDD decreased in western Hubei compared with other regions. This may be due to differences in the terrain. Generally, SDII corresponded to changes in total annual precipitation in China. Increases in SDII were observed in most parts of the region (see Figure 6e). The spatial distribution of these indices was very similar, indicating that the ERA5 reanalyzed data are effective in capturing spatial patterns. The ERA5 reanalyzed data indicated that almost all EPIs, except for CDD, showed higher values in the western and southern regions, while the most severe drought events occurred in the southern and central parts of Hubei between 1979 and 2005. Consequently, extreme precipitation events in the western and northern regions were the most hazardous, while drought events occurred in the southern and central parts of the province between 1979 and 2005.

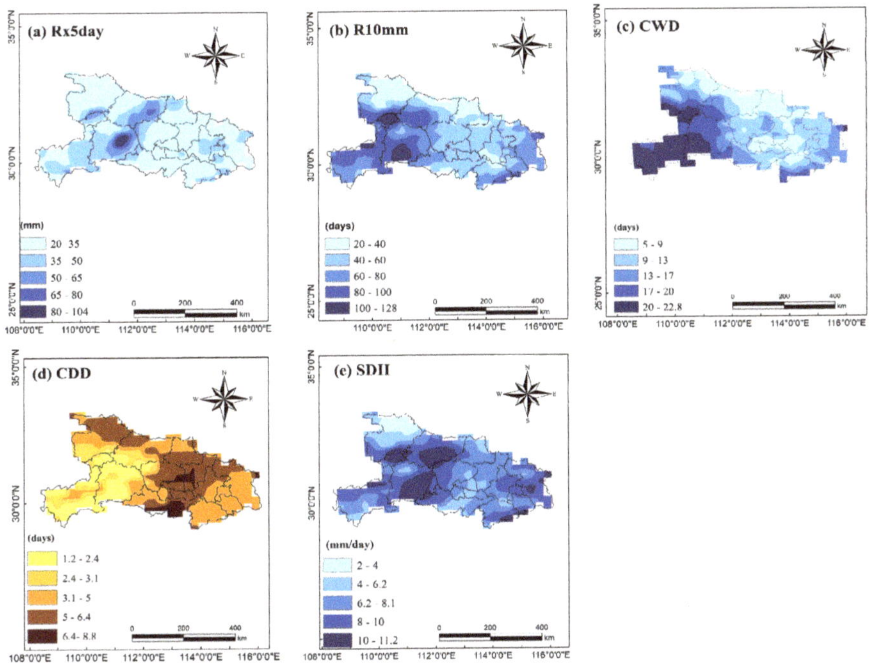

Figure 6. Spatial distribution of five annual mean EPIs in Hubei Province for 1979–2005: (**a**) Rx5day, (**b**) R10 mm, (**c**) CWD, (**d**) SDD, and (**e**) SDII.

4.5. Temporal Changes in EPIs

Figure 7 and Table 3 depicts the temporal precipitation patterns of the EPIs from 1979 to 2005 in Hubei Province. Rx5day showed a positive trend during this period, indicating a non-significant increase in the 1980s in both Yichang and Xiangfan, with a value of 57 mm (see Table 2). Among the strength indicators, the frequency index R10mm increased non-significantly in the 1990s in both Yichang and Shennongjia, showing a positive trend of 30 days, which suggests that heavy precipitation was more intense and frequent during this period. As for the consecutive dry index (CDD), it was found to increase non-significantly, showing a positive trend of 41 days in Xiaogan, Jinmen, and Jinzhou, with a rise observed in the early 1980s. As shown in Table 3 there were no notable patterns in any of the five EPIs. However, there was a gradual decrease since the mid-1980s, followed by an increase in the 1990s, indicating an overall rise in the trend of drought in these areas.

Table 3. Mann–Kendall test for five extreme precipitation indices.

EPI	S	Z	p	Trend
Rx5day	75	1.5427	0.12291	
R10mm	30	0.60636	0.54427	Based on the available data, no statistically significant trend was observed
CWD	−35	0.70879	0.47845	
CDD	41	0.88052	0.37858	
SDII	19	0.37524	0.70748	

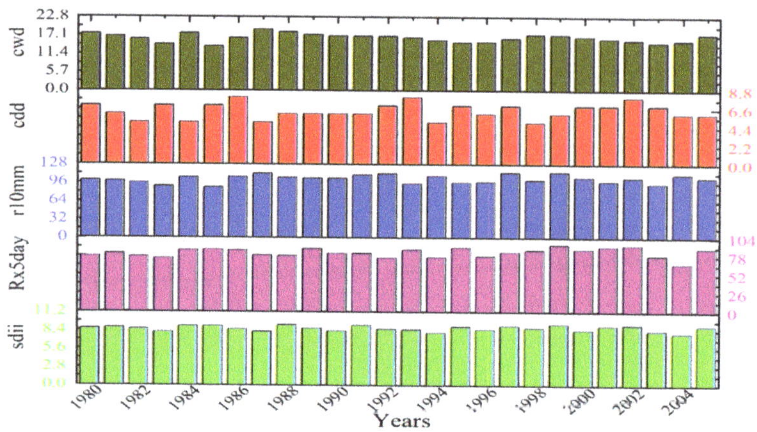

Figure 7. Temporal patterns for five EPIs over Hubei Province from 1979 to 2005.

The CWD showed a negative tendency of 35 days, indicating a decrease in heavy rainfall on wet days. Therefore, this index has decreased in the eastern and northern parts. The daily intensity (SDII) increased non-significantly by 19 mm/day in Yichang, Shennongjia, Xiangfan, and some of the southeastern parts.

4.6. Seasonal Analysis of Extreme Precipitation Indices in Summer for the Mid-21st Century (2046–2065)

This study discussed the EPIs' characteristics in Hubei Province during the summer. The concentration of precipitation, including extreme events, during this time, is a noteworthy aspect.

4.6.1. Spatial Patterns

As shown in Figure 8, spatial changes occurred in the five EPIs in the mid-21st century (2046–2065) compared with the reference period. Rx5day increased in the central and northern part of Hubei, mainly in Wuhan, Qianjiang, Jingzhou, and Suizhou, with values larger than 65 mm, and it decreased in Yichang and Xianyang, with values below −30 mm (see Figure 8a). This indicated that this region will witness a rise in the intensity of rainfall through summer. A significant increase in R10mm was seen in the south of Hubei with values larger than 60, especially in Wuhan, Qianjiang, Jingzhou, and Suizhou, indicating that these areas are more at risk of extreme precipitation during summer. It decreased in Enshi and Yichang in west Hubei, with values below −15 days, indicating less rainfall in these areas (see Figure 8b). CWD also increased in Xiantao, Qianjiang, and Jingzhou, with a value exceeding 40 days (see Figure 8c). Summer CDD decreased in middle and southern Hubei, with values up to −15 days, and increased in most parts of Hubei, with values larger than 45 days, implying that the middle region will witness less drought during summer (see Figure 8d). The change in the SDII also increased in Shyian, Qianjiang, Jingzhou, and Suizhou in central and eastern Hubei, with values up to 45 mm/days, and decreased in Yichang and Xianyang. This indicated that these regions will experience a rise in the intensity of rainfall throughout summer (see Figure 8e). Generally, all the EPIs increased, except CDD during the summer in the 21st century. Therefore, the southern and central regions such as Wuhan, Qianjiang, Jingzhou, and Suizhou will be highly exposed to the risk of floods and extreme events.

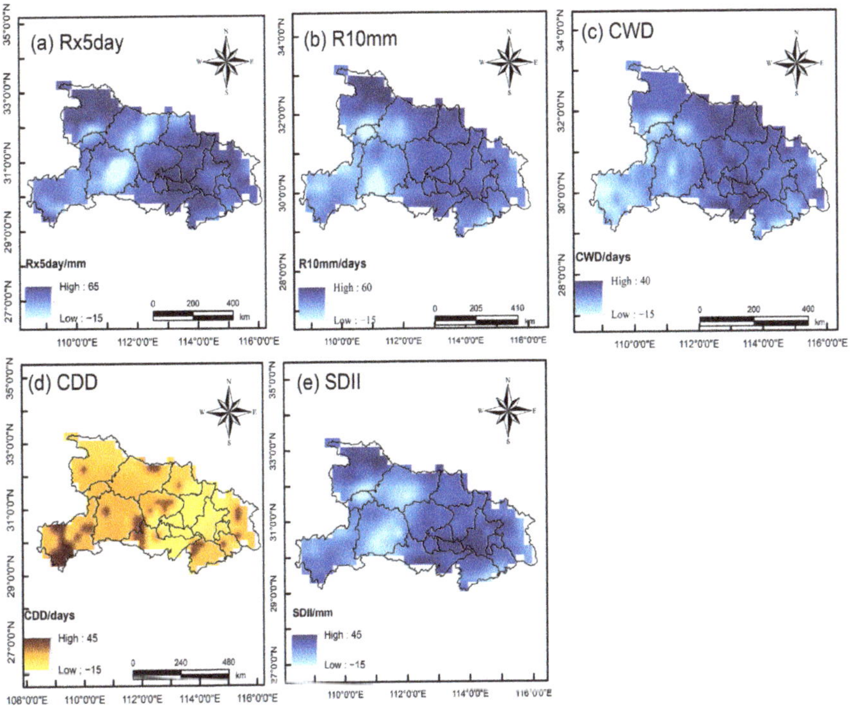

Figure 8. Differences in the spatial patterns of the extreme precipitation and climatic indices during the summer between the mid-21st century (2046–2065) and the period of 1986–2005.

4.6.2. Temporal Patterns

In summer, as shown in Figure 9 and Table 4, Rx5day increased non-significantly at a rate of 28 mm in Wuhan, Ezhou, and Shyian, and decreased in western Hubei. R10mm increased significantly in Wuhan, Xiantao, Qianjiang, Tianmen, and Jingzhou at a rate of 64 days ($p < 0.05$) but decreased in Enshi, Yichang, and Xianyang (see Table 4). This indicated that these regions face a higher risk of extreme precipitation in the summer. Conversely, CDD showed a negative trend of 54 days, indicating less severe drought over the central parts and more heavy precipitation days in this area. CWD increased mainly in the northern and southern parts at a rate of 43 days, and decreased in Enshi and Yichang. That means those areas would be less to extreme events. In general, most of the EPIs examined (except CDD) rose during the summer of the mid-21st century. SDII increased mainly in the southern northern of Hubei at a rate of 12 days and decreased in the western parts. Precipitation increased in the southern and central parts, and decreased in the western parts. Therefore, the southern and central regions will face a higher risk of extreme events by the middle of the 21st century.

Table 4. The M-K test for summer over the mid-21st century.

EPI	High	Low	S	Z	*p*	Trend
Rx5day	65	−30	28	0.876	0.38103	Non-significant increasing
R10mm	15	−5	64	2.044	0.040955	Significant increasing
CWD	40	−25	43	1.7195	0.16175	Non-significant increasing
CDD	45	−15	−54	1.7195	0.085515	Non-significant decreasing
SDII	45	−15	12	0.35689	0.77029	Non-significant increasing

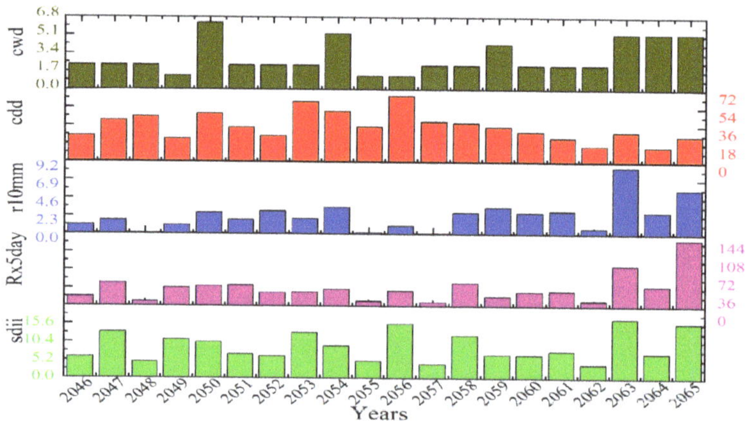

Figure 9. Temporal patterns of extreme precipitation and climatic indices during the summer season in the mid-21st century (2046–2065).

4.7. Spatial–Temporal Changes of EPIs in Summer for the Late 21st Century (2078–2097)

4.7.1. Spatial Changes

The spatial changes in five EPIs in the late 21st century (2078–2097) compared with the reference period are shown in Figure 10. the consecutive dry days CDD showed a reversed distribution, for wet extreme indices (all indices except CDD) increased in northern and southern Hubei.

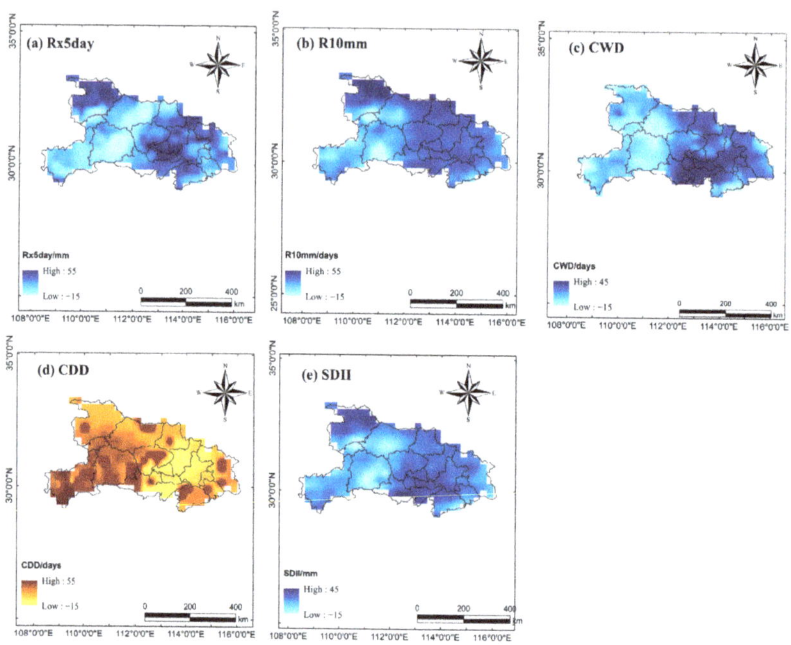

Figure 10. Differences in the spatial changes of extreme precipitation and climatic indices during the summer season between the late 21st century and the period of 1986–2005.

Increases in Rx5day were found in the southern and northern parts for values of up to 55 mm, and a decrease in the western parts of Hubei around −15 mm (see Figure 10a). This may indicate increased Rx5day in those areas during the summer season. R10mm increased in the southern parts with values of up to 55 but decreased towards Enshi, Shennongjia, Shiyan, and Xiangyang (see Figure 10b). This suggests that these areas may be less at risk of heavy rainfall during the summer. Figure 10c displays an increase in CWD in the southern parts of Hubei, with values up to 45 days, indicating that these areas could be wetter. The CDD shows increases in most parts of Hubei of up to 55 days, except the middle parts, and it decreased in Wuhan, Xiagan, and Xianto, which indicated that these areas will be exposed to heavy rains and that more attention should be paid to these areas (see Figure 10d). According to Figure 10e, a decrease in SDII can be observed in the western parts, with values up to 45 mm/day, while Xiantao, Qianjiang, and Jingzhou will experience an increase in SDII.

4.7.2. Temporal Changes

The temporal changes for the five EPIs over Hubei for the summer of the late 21st century are shown in Figure 11 and Table 5. Rx5day had a negative tendency of −2 mm in Yichang and Xiangyang. This may indicate a decrease in the intensity of rainfall on wet days in those areas during the summer in the late 21st century. Accordingly, R10mm also showed a negative tendency of −49 days, indicating a non-significant decrease in the north and west of Hubei. CWD had a positive non-significant increasing trend of 24 days in Xiantao, Tianmen, and Jinzhou, and decreased in the western and northern parts. The southern part will experience an increase in CWD during the summer, suggesting that these regions will be wetter. An increase in CDD is expected in most region of Hubei at a rate of 54 days, except the middle areas. This indicates that EPIs will increase in these areas. SDII also showed a negative tendency of −38 days in the west and north parts, and decreased in Yichang and Enshi. This means that these regions will suffer less from extreme precipitation.

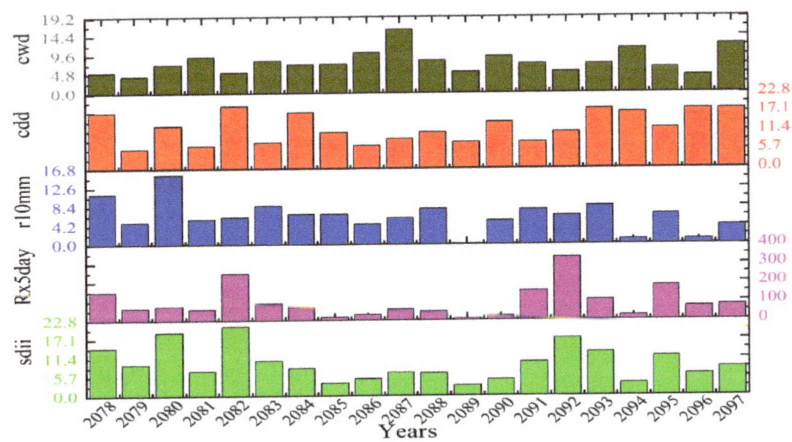

Figure 11. Temporal patterns of extreme precipitation and climatic indices during summer in the late 21st century (2078–2097).

Most EPIs over Hubei Province should decrease in the summer during the late 21st century, except for CDD and CWD, which are projected to increase. The projected seasonal analysis of precipitation differed between the summer of the late 21st century and the summer of the mid-21st century. Specifically, the EPIs, except for CDD, should increase in the summer of the mid-21st century and decrease in the summer of the late 21st century.

Table 5. The M-K test for summer over the late 21st century.

EPI	High	Low	S	Z	*p*	Trend
Rx5day	70	−55	−2	0.032444	0.97412	There is no statistically significant trend
Rx10mm	15	−5	−49	1.5581	0.1192	
CWD	20	−10	24	0.74622	0.45554	
CDD	25	−15	54	1.7195	0.085515	
SDII	20	−10	−38	1.2004	0.22997	

5. Discussions and Conclusions

Utilizing ERA5 reanalyzed data from ECMWF and the outcomes of high-resolution climate change projections for Hubei via RegCM4, this study investigated future extreme precipitation in Hubei Province. Based on RegCM4's output, five EPIs (RX5day, R10mm, CWD, SDII, and CDD) were selected to analyze the spatial–temporal fluctuations of extreme precipitation during the summer season in Hubei. The period from 1979 to 2005 was used as a reference point for future changes, with the model's efficacy assessed by comparing its outcomes with ERA5 data.

Generally, shifts in precipitation extremes can significantly affect overall yearly precipitation. A significant increase in R10mm was detected during the summer of the mid-21st century, particularly in the southern and northern parts of Hubei, indicating a higher risk of extreme events in regions such as Wuhan, Xiantao, Qianjiang, Jinzhuo, and Ezhou. These results are consistent with research conducted by Wang and Li [18]. It has been reported that the amount of heavy rainfall has increased compared with previous times, potentially worsening the strain on drainage systems in Wuhan City [35]. In the summer of the late 21st century, RX5day, R10mm, and SDII showed a slight decline towards the western parts of Hubei, indicating a decrease in extreme precipitation in these areas. In contrast, CDD increased in most parts of Hubei, suggesting a rise in drought rates during the summer of the late 21st century. There are factors other than rainfall that contribute to climate change and the presence of extreme precipitation indicators in the summer of the 21st century, such as prevailing winds and mountains. Menezes et al. [36] analyzed numerous years of surface wind data from the Quick Scatterometer (QuikSCAT) satellite to elucidate the spatial and temporal variability of westerly wind jets in the mountain gap of the northern Red Sea. He found that these jets were relatively cold and dry air from the Arabian Desert.

Based on these results, the general trend of variation in frequency indices in Hubei Province during the summer of the mid-21st century is increasing, indicating that precipitation and intensity will rise. Extreme precipitation events will occur more frequently in the southern and northern regions of Hubei, potentially leading to increased flood disasters in the southeastern part of the province. The primary goal of this research was to investigate the regional patterns of future changes as well as the temporal progression of extreme indices between 2046–2065 and 2078–2097, using the RCP8.5 scenario. The findings showed that the model can accurately replicate the spatial and temporal patterns of extreme climate events in Hubei Province. The major conclusions include the following.

1. The R10mm will increase significantly ($p < 0.05$) during the summer of the mid-21st century in the south northern parts, especially for Wuhan, Xiantao, Qianjiang, Jinzhuo, and Ezhou. The majority of EPIs, except CDD, are expected to rise. This indicates the need for more focus on disaster prevention during summer.
2. Rx5day, R10mm, and SDII will all gradually decrease in the summer of the late 21st century toward the western parts. However, they may climb throughout other parts of Hubei, indicating an increase in the rates of drought during the summer of the 21st century.
3. More focus needs to be placed on the rainstorms that are expected in Wuhan, Xiantao, Qianjiang, and Tianmen, particularly during the summer of the mid-21st century, where there may be a significant increase in the R10mm. This could lead to more floods and financial losses during the periods of intense rainfall.

In conclusion, this study enhances our understanding of severe precipitation incidents in Hubei Province during the 21st century in the context of the RCP8.5 scenario. However, analyzing the uncertainty involved in the simulation and the projection of extreme rainfall events is essential for further research, given the complexity of rainfall patterns. Although the good efficiency of the bias correction method is beneficial for the simulation of the observed precipitation, especially in the temporal scale, there is potential room for improvement spatially. Therefore, for future work, the study recommends more investigation on bias correction techniques, as well as using datasets of high-resolution observations.

Author Contributions: Conceptualization, A.M.; methodology, A.H.; software, M.A.; validation, Q.C. and A.M.; formal analysis, A.M.; investigation, A.M. and Q.C.; resources, A.M.; data curation, A.M.; writing—original draft preparation, A.M.; writing—review and editing, Q.C.; visualization, M.H.; supervision, Q.C.; project administration, A.M.; funding acquisition, Q.C. All authors have read and agreed to the published version of the manuscript.

Funding: This study was supported by the National Science Foundation of China (grant 42275078, 41775136).

Institutional Review Board Statement: Not applicable.

Informed Consent Statement: Not applicable.

Data Availability Statement: The data presented in this study are available on request from the corresponding author. The DEM data used in this study have been deposited in NASA Shuttle Radar Topography Mission (SRTM) (2013). Shuttle Radar Topography Mission (SRTM) Global. Distributed by Open Topography. https://doi.org/10.5069/G9445JDF Accessed: 10 August 2024.

Acknowledgments: The authors thank the National Key Scientific and Technological Infrastructure project "Earth System Science Numerical Simulator Facility" (EarthLab) for the support of this work. Finally, we are grateful for the comments from two anonymous reviewers who helped improved the quality of this manuscript.

Conflicts of Interest: The authors declare no conflict of interest.

References

1. Chen, Y.J.; Lin, H.J.; Liou, J.J.; Cheng, C.T.; Chen, Y.M. Assessment of Flood Risk Map under Climate Change RCP8.5 Scenarios in Taiwan. *Water* **2022**, *14*, 207. [CrossRef]
2. Alexander, L.V. Global Observed Long-Term Changes in Temperature and Precipitation Extremes: A Review of Progress and Limitations in IPCC Assessments and Beyond. *Weather Clim. Extrem.* **2016**, *11*, 4–16. [CrossRef]
3. Alexander, L.V.; Fowler, H.J.; Bador, M.; Behrangi, A.; Donat, M.G.; Dunn, R.; Funk, C.; Goldie, J.; Lewis, E.; Rogé, M.; et al. On the Use of Indices to Study Extreme Precipitation on Sub-Daily and Daily Timescales. *Environ. Res. Lett.* **2019**, *14*, 125008. [CrossRef]
4. Intergovernmental Panel on Climate Change. *Climate Change 2014: Synthesis Report: Longer Report*; Intergovernmental Panel on Climate Change: Geneva, Switzerland, 2014; ISBN 9789291691432.
5. Boo, K.O.; Kwon, W.T.; Baek, H.J. Change of Extreme Events of Temperature and Precipitation over Korea Using Regional Projection of Future Climate Change. *Geophys. Res. Lett.* **2006**, *33*, L01701. [CrossRef]
6. Khoi, D.N.; Quan, N.T.; Nhi, P.T.T.; Nguyen, V.T. Impact of Climate Change on Precipitation Extremes over Ho Chi Minh City, Vietnam. *Water* **2021**, *13*, 120. [CrossRef]
7. Yang, X.-Y.; Zhang, S.-B.; Lyu, Y.-Q.; Zhao, Y.; Lyu, S.-H. Characteristics and Future Projections of Summer Extreme Precipitation in Sichuan Province, China. *J. Mt. Sci.* **2020**, *17*, 1696–1711. [CrossRef]
8. Zhai, P.; Zhang, X.; Wan, H.; Pan, X. Trends in Total Precipitation and Frequency of Daily Precipitation Extremes over China. *J. Clim.* **2005**, *18*, 1096–1108. [CrossRef]
9. Chen, H.P. Projected Change in Extreme Rainfall Events in China by the End of the 21st Century Using CMIP5 Models. *Chin. Sci. Bull.* **2013**, *58*, 1462–1472. [CrossRef]
10. Zou, L.; Zhou, T. Near Future (2016-40) Summer Precipitation Changes over China as Projected by a Regional Climate Model (RCM) under the RCP8.5 Emissions Scenario: Comparison between RCM Downscaling and the Driving GCM. *Adv. Atmos. Sci.* **2013**, *30*, 806–818. [CrossRef]
11. Alriah, M.A.A.; Bi, S.; Nkunzimana, A.; Elameen, A.M.; Sarfo, I.; Ayugi, B. Assessment of Observed Changes in Drought Characteristics and Recent Vegetation Dynamics over Arid and Semiarid Areas in Sudan. *Theor. Appl. Clim.* **2024**, *155*, 3541–3561. [CrossRef]
12. Li, L.; Xiao, Z.; Luo, S.; Yang, A. Projected Changes in Precipitation Extremes over Shaanxi Province, China, in the 21st Century. *Adv. Meteorol.* **2020**, *2020*, 1808404. [CrossRef]

13. Ji, Z.; Kang, S. Evaluation of Extreme Climate Events Using a Regional Climate Model for China. *Int. J. Climatol.* **2015**, *35*, 888–902. [CrossRef]
14. Gao, X.; Shi, Y.; Han, Z.; Wang, M.; Wu, J.; Zhang, D.; Xu, Y.; Giorgi, F. Performance of RegCM4 over Major River Basins in China. *Adv. Atmos. Sci.* **2017**, *34*, 441–455. [CrossRef]
15. Fu, G.; Yu, J.; Yu, X.; Ouyang, R.; Zhang, Y.; Wang, P.; Liu, W.; Min, L. Temporal Variation of Extreme Rainfall Events in China, 1961–2009. *J. Hydrol.* **2013**, *487*, 48–59. [CrossRef]
16. Zhou, X.; Huang, G.; Wang, X.; Cheng, G. Future Changes in Precipitation Extremes Over Canada: Driving Factors and Inherent Mechanism. *J. Geophys. Res. Atmos.* **2018**, *123*, 5783–5803. [CrossRef]
17. Grimm, A.M.; Tedeschi, R.G. ENSO and Extreme Rainfall Events in South America. *J. Clim.* **2009**, *22*, 1589–1609. [CrossRef]
18. Wang, W.; Tang, H.; Li, J.; Hou, Y. Spatial-Temporal Variations of Extreme Precipitation Characteristics and Its Correlation with El Niño-Southern Oscillation during 1960–2019 in Hubei Province, China. *Atmosphere* **2022**, *13*, 1922. [CrossRef]
19. Chen, W.; Huang, C.; Wang, L.; Li, D. Climate Extremes and Their Impacts on Interannual Vegetation Variabilities: A Case Study in Hubei Province of Central China. *Remote Sens.* **2018**, *10*, 477. [CrossRef]
20. Wang, R.; Li, C. Spatiotemporal Analysis of Precipitation Trends during 1961–2010 in Hubei Province, Central China. *Theor. Appl. Clim.* **2016**, *124*, 385–399. [CrossRef]
21. Riahi, K.; Rao, S.; Krey, V.; Cho, C.; Chirkov, V.; Fischer, G.; Kindermann, G.; Nakicenovic, N.; Rafaj, P. RCP 8.5-A Scenario of Comparatively High Greenhouse Gas Emissions. *Clim. Change* **2011**, *109*, 33–57. [CrossRef]
22. Che, L.; Yin, S.; Guo, Y. *Flood Risk Assessment Based on the Historical Disaster Statistics and the Index System Method: A Case Study of Hubei Province, China*; PREPRINT (Version 1); Research Square Platform LLC: Durham, NC, USA, 2023. [CrossRef]
23. Tu, Y.; Zhao, Y.; Dong, R.; Wang, H.; Ma, Q.; He, B.; Liu, C. Study on Risk Assessment of Flash Floods in Hubei Province. *Water* **2023**, *15*, 617. [CrossRef]
24. Zhu, K.; Cheng, Y.; Zang, W.; Zhou, Q.; El Archi, Y.; Mousazadeh, H.; Kabil, M.; Csobán, K.; Dávid, L.D. Multiscenario Simulation of Land-Use Change in Hubei Province, China Based on the Markov-FLUS Model. *Land* **2023**, *12*, 744. [CrossRef]
25. Wu, G.; Lv, P.; Mao, Y.; Wang, K. ERA5 Precipitation over China: Better Relative Hourly and Daily Distribution than Absolute Values. *J. Clim.* **2023**, *37*, 1581–1596. [CrossRef]
26. Hoffmann, L.; Günther, G.; Li, D.; Stein, O.; Wu, X.; Griessbach, S.; Heng, Y.; Konopka, P.; Müller, R.; Vogel, B.; et al. From ERA-Interim to ERA5: The Considerable Impact of ECMWF's next-Generation Reanalysis on Lagrangian Transport Simulations. *Atmos. Chem. Phys.* **2019**, *19*, 3097–3214. [CrossRef]
27. Hersbach, H.; Bell, B.; Berrisford, P.; Hirahara, S.; Horányi, A.; Muñoz-Sabater, J.; Nicolas, J.; Peubey, C.; Radu, R.; Schepers, D.; et al. The ERA5 Global Reanalysis. *Q. J. R. Meteorol. Soc.* **2020**, *146*, 1999–2049. [CrossRef]
28. Alriah, M.A.A.; Bi, S.; Nkunzimana, A.; Elameen, A.M.; Sarfo, I.; Ayugi, B. Multiple Gridded-Based Precipitation Products' Performance in Sudan's Different Topographical Features and the Influence of the Atlantic Multidecadal Oscillation on Rainfall Variability in Recent Decades. *Int. J. Climatol.* **2022**, *42*, 9539–9566. [CrossRef]
29. Wu, G.; Qin, S.; Mao, Y.; Ma, Z.; Shi, C. Validation of Precipitation Events in ERA5 to Gauge Observations during Warm Seasons over Eastern China. *J. Hydrometeorol.* **2022**, *23*, 807–822. [CrossRef]
30. Giorgi, F.; Pal, J.S.; Bi, X.; Sloan, L.; Elguindi, N.; Solmon, F. Introduction to the TAC Special Issue: The RegCNET Network. *Theor. Appl. Clim.* **2006**, *86*, 1–4. [CrossRef]
31. Park, J.; Kang, M.S.; Song, I. Assessment of Flood Vulnerability Based on CMIP5 Climate Projections in South Korea. *J. Am. Water Resour. Assoc.* **2015**, *51*, 859–876. [CrossRef]
32. Gong, X.; Wang, X.; Li, Y.; Ma, L.; Li, M.; Si, H. Observed Changes in Extreme Temperature and Precipitation Indices on the Qinghai-Tibet Plateau, 1960–2016. *Front. Environ. Sci.* **2022**, *10*, 888937. [CrossRef]
33. Alriah, M.A.A.; Bi, S.; Shahid, S.; Nkunzimana, A.; Ayugi, B.; Ali, A.; Bilal, M.; Teshome, A.; Sarfo, I.; Elameen, A.M. Summer Monsoon Rainfall Variations and Its Association with Atmospheric Circulations over Sudan. *J. Atmos. Sol. Terr. Phys.* **2021**, *225*, 105751. [CrossRef]
34. Jiao, D.; Xu, N.; Yang, F.; Xu, K. Evaluation of Spatial-Temporal Variation Performance of ERA5 Precipitation Data in China. *Sci. Rep.* **2021**, *11*, 17956. [CrossRef] [PubMed]
35. Xiong, L.; Yan, L.; Du, T.; Yan, P.; Li, L.; Xu, W. Impacts of Climate Change on Urban Extreme Rainfall and Drainage Infrastructure Performance: A Case Study in Wuhan City, China. *Irrig. Drain.* **2019**, *68*, 152–164. [CrossRef]
36. Menezes, V.V.; Farrar, J.T.; Bower, A.S. Westward Mountain-Gap Wind Jets of the Northern Red Sea as Seen by QuikSCAT. *Remote Sens. Environ.* **2018**, *209*, 677–699. [CrossRef]

atmosphere

MDPI

Article

Assessment of the Drought Risk in Constanta County, Romania

Carmen Elena Maftei [1], Alina Bărbulescu [1,*] and Amela Osman [2]

[1] Faculty of Civil Engineering, Transilvania University of Brașov, 5 Turnului Str., 500152 Brașov, Romania; carmen.maftei@unitbv.ro
[2] RAJA S.A., 900590 Constanța, Romania; amela.osman22@gmail.com
* Correspondence: alina.barbulescu@unitbv.ro

Abstract: Drought poses a significant risk in many parts of the world, especially in regions reliant on agriculture. Evaluating this risk is an essential step in preventing and reducing its impact. In this context, we assess the drought intensity at six sites in Constanța County (Romania) using the de Martonne aridity index. The risk of aridity and vulnerability to drought were evaluated by the Drought Hazard Index (*DHI*) and Drought Risk Index (*DRI*), computed based on the Standardized Precipitation Index (*SPI*). The de Martonne index indicates a variation between the slightly arid and semi-arid climates for Adamclisi station, with periodic changes from semi-arid to arid. At Cernavodă station, we notice a passage from an arid period towards a moderately humid one (in 2005), followed by a movement in the opposite direction to the limit of the arid zone (in 2011), and a return inside the "limits" of the semi-arid to moderately arid climate. A similar variation for 2000–2018 is noticed at Medgidia, Hârșova, and Mangalia. *DRI* classifies two stations in the low risk to drought category and one in the moderate risk to drought class. The other two locations experience a high or very high risk of drought. The drought intensities varied in the intervals 0.503–1.109 at Constanța, 0.473–1.363 at Mangalia, 0.511–1.493 at Adamclisi, 0.438–1.602 at Hârșova, 0.307–1.687 at Medgidia, and 0.463–1.307 at Cernavodă, and the prolonged drought periods were over 99 months at all stations.

Keywords: hazard; vulnerability; *SPI*; *DVI*; *DHI*; *DRI*

1. Introduction

The growing occurrence of extreme events, such as drought, has become a major global concern. Drought, a complex hydro-meteorological phenomenon characterized by prolonged and abnormal moisture deficiency, significantly impacts various sectors, including agriculture, water resources, and ecosystems. The substantial challenges it poses to economies, human welfare, and the environment are evident in diverse geographical locations. Scientists have classified it into hydrological, meteorological, agricultural, and socio-economic drought, further emphasizing its global reach [1–7]. The accumulated water deficit and drought demand immediate attention from governments due to their significant impact on food security and population welfare. The European Commission (EC) and other organizations have taken action by preparing assessments of water losses. These documents underscore the importance of comprehensive management plans for each member of the European Union (EU) in order to conserve water resources. The Water Frame Directive (WFD) is the EU's primary regulation in this regard [8]. In 2007, the EC proposed the "Blueprint to Safeguard Europe's Water Resources" [9], which includes the analysis of the main challenges related to climate change and water scarcity and outlines actions to prevent and mitigate these phenomena's effects. In response to the EU members' need for significant progress in these areas, the EC proposed including the drought risk management plans in the River Basin Management Plans (RBMP) designed by each country [10].

Drought assessment can be conducted using suitable indices based on hydro-meteorological series and their analysis [11]. The Handbook of Drought Indicators and Indices [12], issued by the World Meteorological Organization (WMO) and the Global Water Partnership (GWP), is a comprehensive resource that provides the instruments for the assessment of drought severity. Other sources are also of interest. For example, 74 indices are reviewed in [3] out of 150 known [13].

The SPI is among the most used indices to assess drought, indicating the total rainfall departure from the mean for different periods and study sites based on a comparison with the historical long-term precipitation [14–18]. Given its advantages, which will be emphasized in the next sections, the SPI can be used to develop hazard risk indicators.

Hazard involves climate anomalies impacting drought, including temperature variability, rainfall, and evaporation [19]. Vulnerability represents the extent to which a system can be affected following the impact of a hazard. It includes all the physical, social, economic, and environmental conditions that increase the susceptibility of the respective system. Like hazard, vulnerability is an indicator of the future state of a system, defining the degree of ability or inability of the system to cope with the expected stress [20]. Risk is defined as the probability of the appearance of harmful outcomes arising from natural or anthropic-induced hazards interacting with vulnerable populations [21]. Given that the ultimate consequences of the drought are socio-economic [22,23], its monitoring is essential.

According to [24], about 30% of Romania experiences desertification and is characterized by a humid/dry to arid climate. Using the Expert Team on Climate Change Detection and Indices (ETCCDI), Birsan et al. [25] found that the number of summer days and tropical nights increased. The same behavior was noticed for warm spells, while in the frost season, it decreased. A decrease in agricultural production was noticed in 2010–2019 compared to the average values from the previous decade due to drought episodes. Dobrogea, the region investigated in this study, was the most affected by the temperature increase, shortage of precipitation, and lack of water in the soil [26]. Other studies [27–29] indicate that the temperature increased by 1.7 °C during 1965–2021 in Dobrogea, which has faced severe drought over the last 20 years. The evaluation of the trend of the series span from 1961 to 2018 recorded at Constanța meteorological station (situated in Constanța County, Dobrogea) indicates that the annual, quarterly, monthly, and seasonal minima and maxima data, except for those recorded in autumn, exhibit an increasing trend [29]. While the yearly maxima increased by about 1.3 °C, the augmentation is higher in spring (1.4 °C–1.6 °C) and summer (1.8 °C–1.9 °C) [29]. At the same time, the number of heavy precipitation events decreased, increasing the number of isolated days with moderate and heavy precipitation [30–32].

Located in the southeastern part of Romania, in the Dobrogea region, Constanța County is a significant contributor to the country's economy, particularly in agriculture and the tourism industry, due to its unique geographical position between the Danube and the Black Sea. The literature focuses on the analysis and forecast of the hydro-meteorological variables; no article has yet investigated the population exposure to the drought effect in the county. Exposure involves socio-economic, demographic, and agricultural dynamics [33], and is reflected in the population's well-being.

Therefore, the main objectives of this article are to analyze the aridity level at different locations in Constanța County and assess the risk of drought. The first goal is achieved by computing the de Martonne index, analyzing its trend, and the drought duration and intensity. The second goal is realized by computing the Drought Risk Index (*DRI*) built considering the climate and the socio-economic aspects. Despite the various studies performed for Romania, such an analysis has not been performed until now. It will shed light on the lesser-known aspects of the impact of drought in the region and could significantly aid authorities in making informed decisions to mitigate the effects of drought.

2. Materials and Methods

2.1. Study Area

Constanța County (Figure 1) is one of the most urbanized counties in Romania. Located in the country's southeastern part, it shares its northern border with Tulcea County. Their conventional border crosses the Casimcea Plateau and the Razim, Zmeica, and Sinoe

Lakes complex. To the east is the Black Sea. On the western side, Constanța County is flanked by the counties of Călărași, Ialomița, Brăila, and the Danube River. The southern neighbor is Bulgaria.

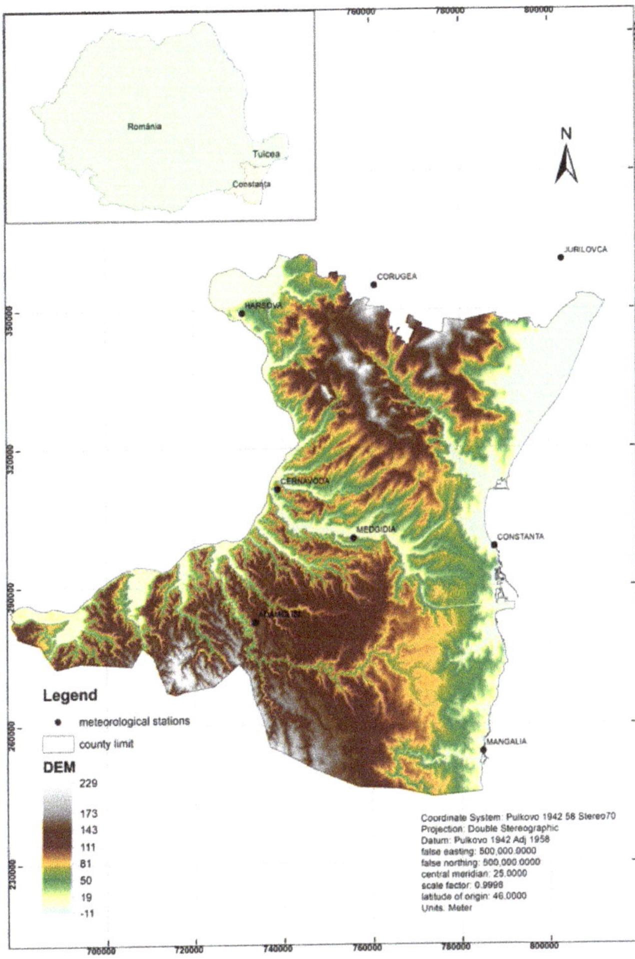

Figure 1. Map of Romania, with Constanța County and the meteorological stations.

Constanța County has predominantly low altitudes (about 200 m). The Măcinului Group, from the northern Massif of Dobrogea, represents the highest form of relief, reaching 467 m at the Pricopan peak. The Casimcea Plateau, with the highest hill of 300–350 m, is the orographic node from which the waters flow to the southwest, south, and southeast.

The Romanian climate is continental temperate, but the Danube Delta, the hydrographical basin Dobrogea (to which Constanța belongs), and the coastal waters give it some specific characteristics. The large water basins, the Black Sea, and the Danube River influence the quantity of precipitation in the area. During the study period, the average multi-year temperature was 11 °C. The precipitation amount is among the lowest in the country. However, over time, the Black Sea has produced exceptional cyclones in Constanța County, which has determined national records of precipitation that still stand today. The Black Sea has substantially impacted the climate, characterized by mild winters. Moderate precipitation and temperature were recorded in autumn and summer.

The Danube River, particularly in the Chiciu–Isaccea sector, the Danube Delta, and the Dobrogea Hydrographic Basin, is a significant water source. The total surface water resources represent approximately 404,136.4 million m^3/year, with the Danube contributing about 12.71% of the total resources. Four important reservoirs, with a volume of about 24.45 million m^3, are found in the Dobrogea Hydrographic Area. The water resources stored in the Dobrogea area are reduced and unevenly distributed in time and space, posing a potential challenge for effective water resource management [34,35]. There are also a few lakes, with salinity ranging from 0.45 g/L at Siutghiol to 75–95 g/L at Techirgiol [36].

2.2. Data Series

The data series used in this study are the monthly precipitation and temperature series recorded during 1965–2018. The National Administration of Meteorology, the Constanța branch, provided the data series recorded at six meteorological stations (Figure 1): Adamclisi, Cernavodă, Constanța, Hârșova, Mangalia, and Medgidia. Unfortunately, the data for the period 2018–2023 are not accessible to the general public.

The average multiannual temperature and precipitation for the investigated period are presented in Table 1. The temperature is higher on the coast (Constanța and Mangalia) than in the rest of the territory, and the precipitation varies between 435 and 500 mm [28,30].

Table 1. Temperature and precipitation- multiannual average during 1965–2018.

Station	Altitude (m)	Temperature (°C)	Precipitation (mm)
Adamclisi	159	11.1	501
Cernavodă	87	11.4	487
Medgidia	72	11.2	456
Hârșova	38	11.2	435
Constanța	14	12.1	453
Mangalia	9	11.7	446

2.3. Methodology

The first research stage was the computation of the annual de Martonne aridity index, I_{DM} [mm/°C], to assess the aridity level at the studied places. The formula used for this aim is [37]

$$I_{DM} = P/(T + 10), \quad (1)$$

where P [mm] is the annual precipitation and T [°C] is the average annual temperature in the region obtained at the previous stage.

The drought intensity is evaluated using the de Martonne index as follows: Hyper-arid if $I_{DM} \in [0, 5)$, Arid (A) if $I_{DM} \in [5, 15)$, Semi-arid (SEA) when $I_{DM} \in [15, 24)$, Moderately arid (MA) for $I_{DM} \in [24, 30)$, Slightly arid (SLA) when $I_{DM} \in [30, 35)$, and Moderately humid for $I_{DM} \in [35, 40)$ [38,39].

Furthermore, we tested the existence of a monotonic trend of IDM against randomness using the Mann–Kendall test [40]. When the randomness hypothesis was rejected, we determined the slope of the linear trend using Sen's slope method [41].

According to the Sendai definition [42], risk includes hazard and vulnerability. Based on this, the methodology proposed aims to estimate the Drought Risk Index (*DRI*) by evaluating the Drought Hazard Index (*DHI*) and Drought Vulnerability Index (*DVI*), in the following stages (Figure 2).

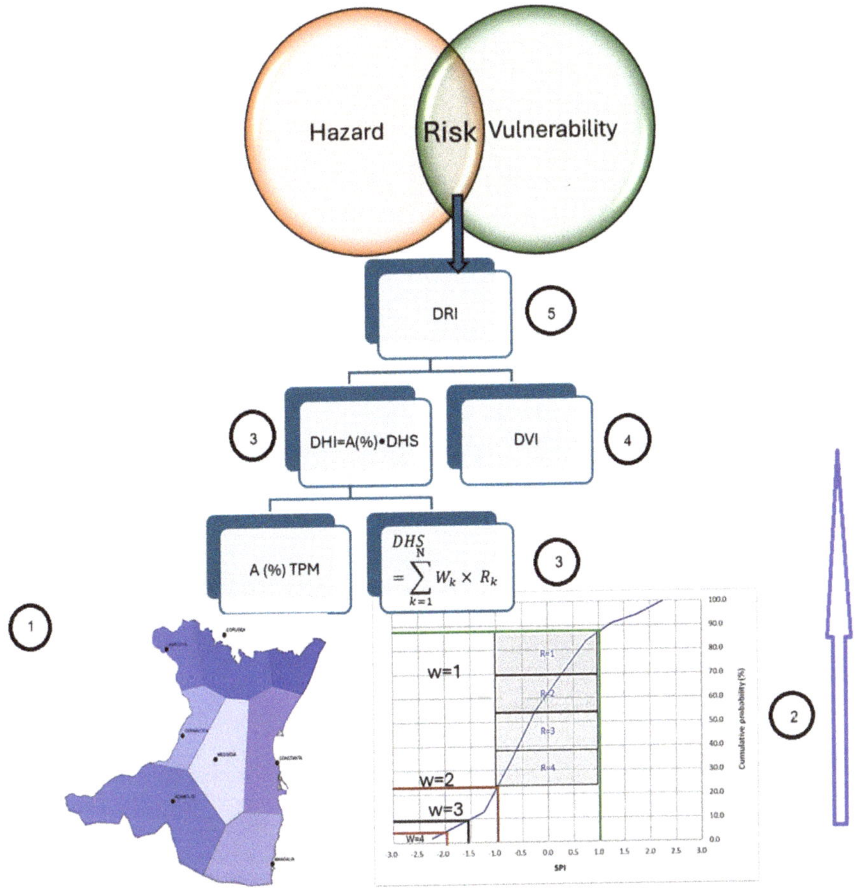

Figure 2. *DRI* index computation—from bottom to top. *A* (%) is the area assigned by the Thiessen Polygon Method. *DHS* is the Drought Hazard Score. *DVI* is the Drought Vulnerability Index.

1. Apply the Thiessen Polygon Method (TPM) to determine the area associated with each station and compute the regional temperature and precipitation.

To this aim, each station is connected by the closest neighbor by lines whose perpendicular bisectors are drawn. The result is a set of polygons, each containing a station. The weight assigned to a station is equal to the ratio between the area of the designated zone and the region's whole area [43,44]. The study area's average temperature (precipitation) is a weighted average of the values recorded at the stations.

2. Compute the *SPI* [45] for a certain period using the precipitation series as input.

First, we fit a gamma distribution to the precipitation series. Then, we determine the cumulative distribution. Then, the cumulative probability, $G(x)$, is determined. Then, the adjustment for the probability of the accumulation zero precipitation is performed by the formula proposed by Edwards and McKee [46], so the cumulative probability will be

$$H(x) = q + (1 - q)G(x), \quad (2)$$

where q is the probability of having null values in the series.

Finally, H is transformed into a standard Gaussian distribution, providing the *SPI* values. According to them, the type of climate can be assessed as follows: Extremely wet for $SPI > 2$, Very wet when $SPI \in [1.5, 2)$, Moderately wet for $SPI \in [1, 1.5)$, Normal drought (ND) for $SPI \in [-1, 1)$, Moderate drought (MD) for $SPI \in [-1.5, -1)$, Severe drought (SD) when $SPI \in [-2, -1.5)$, and Extreme drought (ED) when $SPI < -2$.

Guttman [47] suggested using at least 20 years of monthly records, but the best results are obtained with a series of 50–60 years [48]. Other authors [49,50] appreciate that *SPI* proved its effectiveness in studying long drought or high humidity periods.

The advantages of using this index consist of the following [49]:

- Flexibility: It can be calculated for various time intervals.
- Early warning: The index availability for shorter periods (e.g., one to three months) can help detect drought early and evaluate its severity.
- Cross-location comparison: It allows comparing different locations with varying climates.
- Probabilistic analysis: The index's probabilistic nature enables the analysis of past events, making it suitable for decision-making.

However, there are some drawbacks to using this index:

- Reliance on rainfall records: The index is solely based on rainfall data.
- Lack of soil water ratio component: It does not account for evapotranspiration/potential evapotranspiration (ET/PET) ratios [51].

Since the computation is not easy, different programs were developed, such as the *SPI* [52] and DrinC software [53,54]. We used the last one for this study.

3. Compute the Drought Hazard Index (*DHI*) in the following steps [55]:

- Determine the Drought Hazard Score (*DHS*) for each station, *i*, with the formula

$$DHS = \sum_{k=1}^{N} W_k \times R_k \tag{3}$$

where N is the number of *SPI* values for each time interval, W is the weight, and R is the rating score. The weights are given in correlation to SPI values: $W = 0$ when $SPI > 1$; $W = 1$ for ND, $W = 2$ for MD, $W = 3$ for SD, and $W = 4$ for ED. The ratings (R) are assigned based on the cumulative distribution function (CDF) (Figure 2). Their values are from 1 to 4 in ascending order, based on the quartiles of CDF within each drought category.

- Compute the Drought Hazard Index (*DHI*) by

$$(DHI)_i = A(\%)_i \times (DHS)_i \tag{4}$$

where A (%) is the area assigned by the TPM to each station.

- Normalize the *DHI* using Formula (5), presented below in a general context.

The hazard intensity is evaluated as follows: Reduced when $DHI \in [0, 0.25)$, Moderate for $DHI \in [0.25, 0.50)$, High if $DHI \in [0.50, 0.75)$, and Very high when $DHI \in [0.50, 0.75)$.

4. Compute the Drought Vulnerability Index (*DVI*).

Vulnerability is closely related to a region's socio-economic conditions and is a potential indicator of maximum loss or harm during an event. Given the impact on the population, accurate vulnerability assessments to reflect drought scenarios at the local level are urgently needed in the context of climate change. Therefore, selecting the vulnerability indicators must be relevant to the studied hazard and the regional context [56,57].

In this study, the following socio-economic indicators were utilized to determine *DVI*: the Total Agricultural Land (TAL), the Population Density (PD), the Water Consumption (m^3) per Inhabitant (WA) in each city, and the Built Environment (TC). Each indicator was normalized (in its range), then the average was calculated to obtain the *DVI* value.

The vulnerability intensities and the vulnerability classes are the same as for *DHI*.

In all cases when normalizing was necessary, it was performed using the following formula [58]:

$$X_i = \frac{x_{max} - x_i}{x_{max} - x_{min}}, \tag{5}$$

where x_i is the actual value, X_i is the normalized value of x_i, and $x_{min}(x_{max})$ is the minimum (maximum) values in the set subject to the normalizing procedure.

5. Compute the Drought Risk Index (*DRI*).

DRI is defined as the product between *DHI* and *DVI*. If *DHI* = 0 or *DVI* = 0, there is no risk. The degree of risk is evaluated using the same classes as for *DHI*.

3. Results

The annual de Martonne aridity index computed for the study period for the six stations is presented in Figure 3.

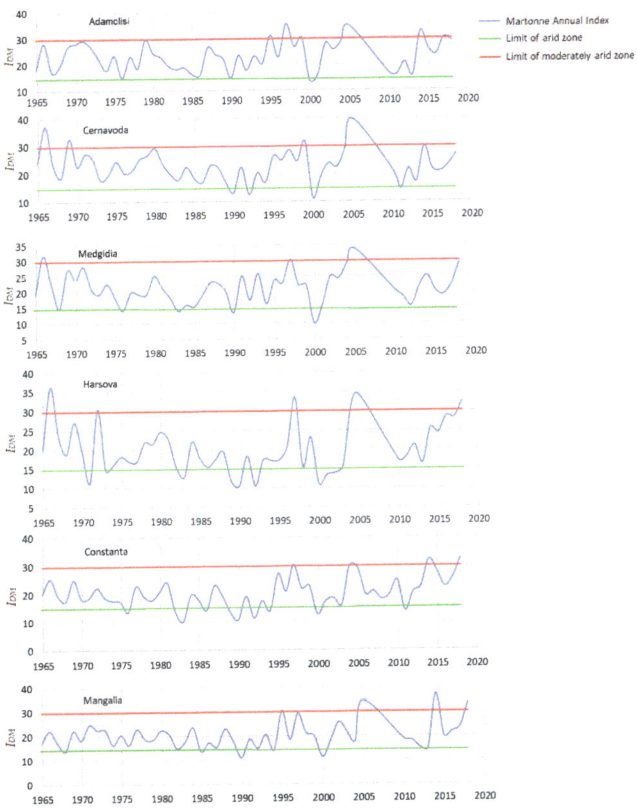

Figure 3. The de Martonne annual aridity index. The red line is the upper limit of the moderately arid zone, the green one is the upper limit of the arid zone, and the blue curve represents the values of the de Martonne annual index.

For Adamclisi, the index values are included (with one exception) in the interval 15–35, indicating a variation between the Slightly arid and Semi-arid climate, with periodic changes from Semi-arid to Arid periods, followed by the reverse behavior. Cernavodă (situated near the Black Sea—Danube Canal) experienced higher variations in the climate, especially after 2000. We notice a passage from an arid period towards a moderately humid one (in 2005), followed by a movement in the opposite direction to the limit of the arid zone (in 2011), and a return inside the "limits" of the semi-arid to moderately arid climate. A similar variation for 2000–2018 is noticed at Medgidia, Hârșova, and Mangalia. The variation of the index, e.g., of the drought episodes, is the lowest at the Black Sea Littoral sites in the Arid–Semi-arid boundaries, at least for the first 35 years covered by this study.

The results of the Mann–Kendall test rejected the hypothesis that there is a monotonic trend for all but the Constanța series, for which an increasing trend with the Sen's slope of 0.09048 was determined (emphasizing increasing aridity at this location).

Figure 4 presents the Thiessen polygons for the precipitation series. The average regional precipitation and temperature computed by TPM are 430.66 mm and 11.3 °C, respectively.

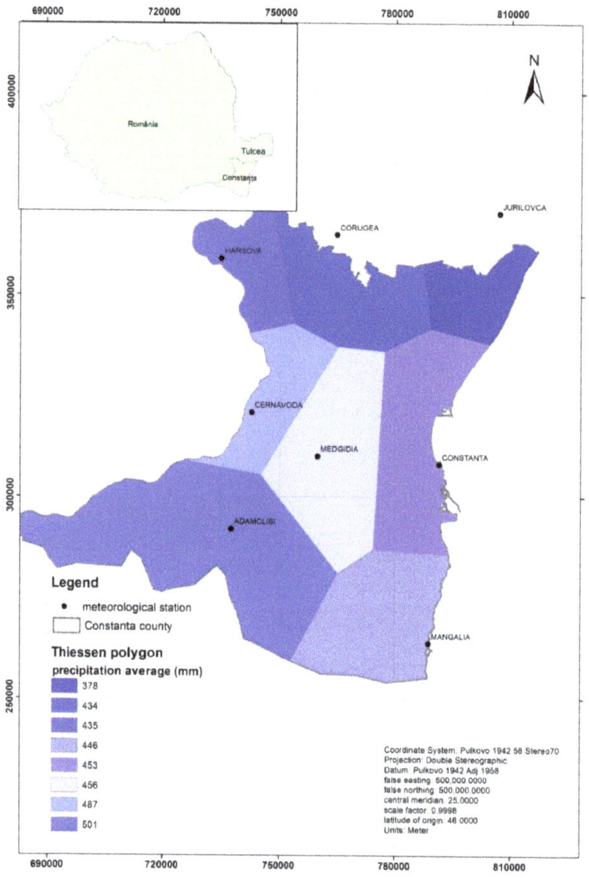

Figure 4. The results of TPM for the precipitation series.

The monthly precipitation series was the input data series for calculating the SPI. DrinC software allows the computation of *SPI* for various intervals (3, 6, 9, 12, 24, or 48 months). Nonetheless, this article focuses on the *DRI* obtained using the *SPI* computed at three months (December, March, June, and September) and 12 months because it is more relevant for monitoring irrigation systems.

The MegaStat Addin in Excel was employed to determine the frequency distribution and cumulative frequencies of the *SPI* values. Figure 5 shows a histogram built using the series Adamclisi for December, and Tables 2 and 3 present the absolute and cumulative frequencies (computed at three months), respectively, together with the classification corresponding to each interval and its weight.

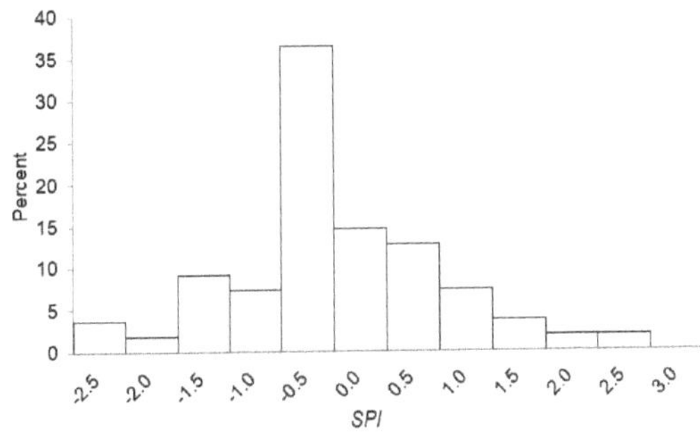

Figure 5. Histogram of SPI for December, at Adamclisi.

Table 2. Absolute frequencies (at 12 months) at five stations.

Interval	Cernavodă	Medgidia	Hârşova	Constanța	Mangalia	Class	*W*
<−2	2	3	1	1	2	Extreme	4
[−2.0, −1.5)	1	2	4	2	1	Severe	3
[−1.5, −1.0)	6	4	3	7	5	Medium	2
[−1.0, 1.0)	39	40	38	35	41	Normal	1
[1.0, 1.5)	4	3	3	5	2		
[1.5, 2.0)	1	1	5	3	1		
>2	2	2	1	2	3		

Table 3. Cumulative frequencies (%) computed at 12 months at five stations.

Interval	Cernavodă	Medgidia	Hârşova	Constanța	Mangalia	Class	*W*
<−2	3.64	5.45	1.82	1.82	3.64	Extreme	4
[−2.0, −1.5)	5.45	9.09	9.09	5.45	5.45	Severe	3
[−1.5, −1.0)	16.36	16.36	14.55	18.18	14.55	Medium	2
[−1.0, 1.0)	87.27	89.09	83.64	81.82	89.09	Normal	1
[1.0, 1.5)	94.55	94.55	89.09	90.91	92.73		
[1.5, 2.0)	96.36	96.36	98.18	96.36	94.55		
>2	3.64	5.45	1.82	1.82	3.64		

DHS was computed using Formula (3), and *DHI* was determined at three months using Equation (4). The surface (*A*) calculated by TPM and percentage surface *A*(%) were used to obtain *DHI* from *DHS*. The values listed in Table 4 were obtained for December. The columns of Table 4 contain, from left to right, the location (column 1), the rating for Normal, Medium, Severe, and Extreme drought (columns 2–5), *DHS* (column 6), the surface and percentage surface (columns 7 and 8), *DHI* computed by (4) (column 9), and *DHI* Normalized computed by (5) (column 10).

Considering the classification presented in the Materials and Methods section, we found that the drought hazard is very high at Cernavodă, Hârşova, and Mangalia. It is reduced at Adamclisi, moderate at Constanța, and high at Medgidia. The results are in concordance with those of the de Martonne aridity index, which indicates the highest extremes (over 35) at Cernavodă, Hârşova, and Mangalia and abrupt variations in time.

Table 4. Computation of *DHS* and *DHI* for December.

Location	Normal	Medium	Severe	Extreme	*DHS*	A	A (%)	*DHI*	*DHI* Normalized	Class
Medgidia	78.20	3.60	3.60	3.60	110.60	1200.00	0.17	18.69	0.51	High
Adamclisi	70.90	7.30	0.00	5.40	107.10	1900.00	0.27	28.66	0.00	Reduced
Cernavodă	70.90	9.10	1.80	3.60	108.90	590.40	0.08	9.05	1.00	Very high
Hârșova	67.30	7.30	5.50	1.80	105.60	871.92	0.12	12.97	0.80	Very high
Constanța	65.50	7.30	3.60	1.80	98.10	1600.00	0.23	22.10	0.33	Moderate
Mangalia	69.10	3.60	5.50	1.80	100.00	938.51	0.13	13.22	0.79	Very high

According to the DVI values are presented in Table 5, column 2, all stations but Constanța are highly or very highly vulnerable to drought. In turn, *DRI* classifies the Adamclisi and Constanța as having a low risk of drought and Medgidia as having a moderate risk of drought. Hârșova and Mangalia experience a high risk of drought, whereas Cernavodă experiences a very high risk (Table 5, last column).

Table 5. *DVI* and *DRI* for December.

Location	*DVI*	*DVI—Class*	*DRI*	*DRI—Class*
Medgidia	0.71	High	0.36	Moderate
Adamclisi	0.50	High	0.00	Low
Cernavodă	0.94	Very high	0.94	Very high
Hârșova	0.74	High	0.59	High
Constanța	0.40	Moderate	0.13	Low
Mangalia	0.86	Very high	0.68	High

The computation performed using *SIP* at three months (March, June, and September) provided almost similar results. Based on them, the drought vulnerability maps and hazard maps were built. Figure 6 displays the *DHI* and *DVI* maps drawn using the *SPI* indices computed at 3 months.

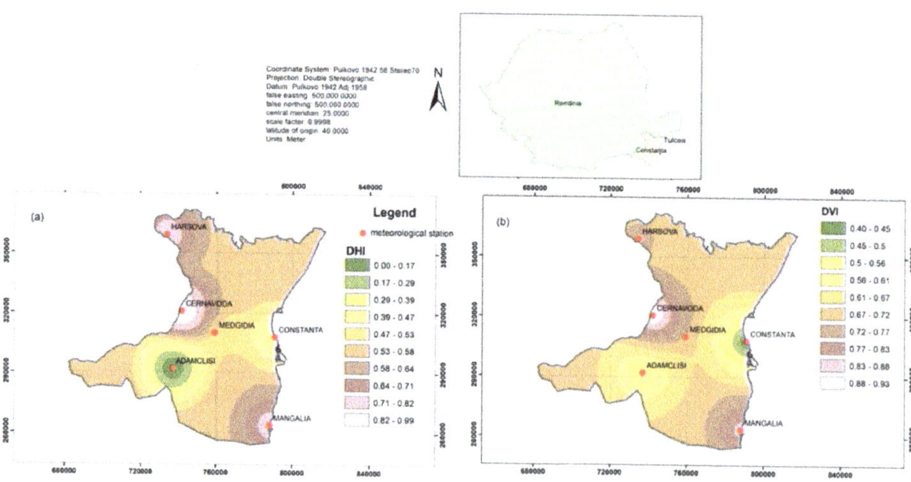

Figure 6. The maps of (**a**) DHI and (**b**) DVI built using the SPI index computed at 3-months.

We notice the uneven distribution of the drought hazard intensity, with the highest intensity appearing in the northwestern and southeastern parts of the region. A similar conclusion is drawn regarding vulnerability to drought.

Figure 7 (left) shows the DRI map built using the SPI computed at 3-months. It indicates the highest risk to drought at Cernavodă and the lowest at Constanța and Adamclisi. The DRI varied between 0.25 and 0.5, which means a medium risk in the rest of the territory. The results are concordant with the actual situation, given the water resources (surface and groundwater) from which both cities benefit.

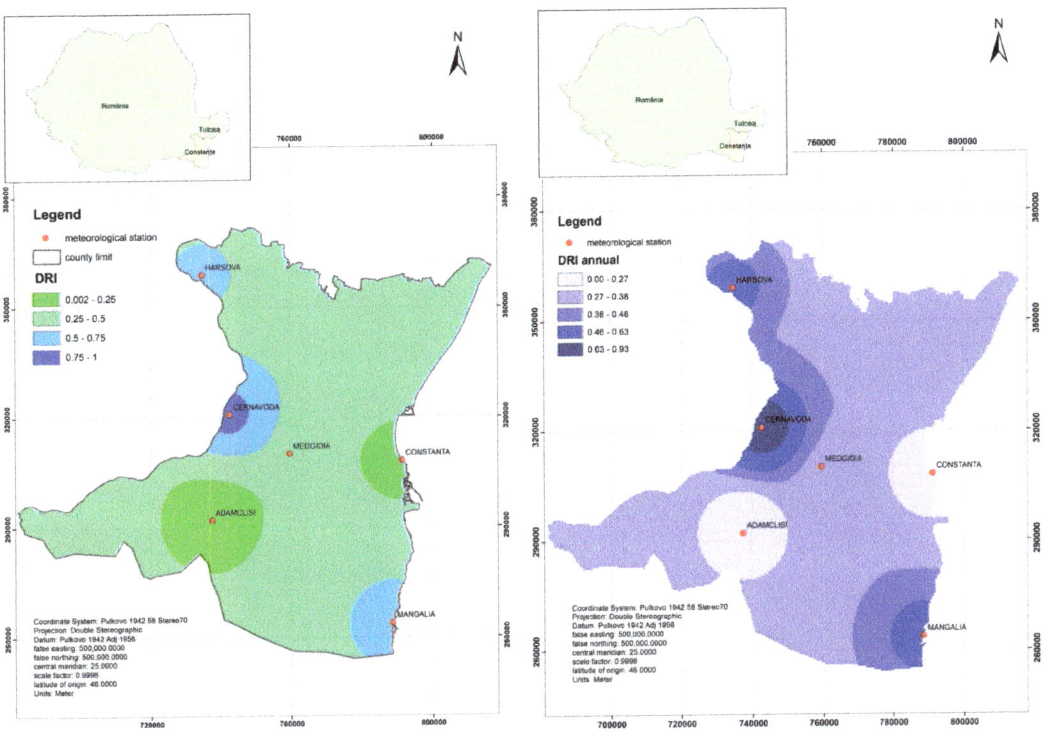

Figure 7. *DRI* map built using the SPI index computed at (**left**) 3-month and (**right**) 12-month.

Applying the same method to compute the SPI for 12-month, we obtained the DRI map presented in Figure 7 (right). A slight attenuation in the maximum DRI values is noticed due to incorporating the seasonal effects (high rainfall, snowmelt) that balance the impact of the water deficit in some measure.

4. Discussion

The SPI computation allows one to determine the duration (DD), severity (DS), and intensity (DI) of drought. DD represents the number of months between the drought's start and its end. DS is the sum of the absolute values of the SPI during the period of drought. DI is obtained by dividing DS by DD [59].

Prolonged drought can be assessed using DD. It is defined as a period where a pattern of precipitation deficiencies persists for more than six months.

The SPI 12-month is presented on the left-hand side of Figures 8 and 9 for Constanța and Mangalia. The right-hand side of the same figures and Table 6 show the values of DD, DS, and DI at all the studied stations. To compute DD, DS, and DI, we used only the drought events for which the SPI values are less than −1.

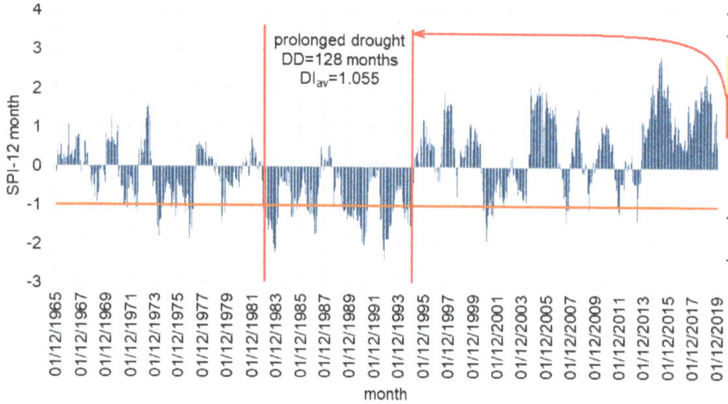

start_date	end_date	DD	DS	DI
1/8/1971	/1/81972	12	8.12	0.677
1/3/1974	1/6/1977	39	35.94	0.922
1/6/1979	1/3/1981	21	11.44	0.545
1/11/1982	1/8/1987	57	57.26	1.005
1/4/1989	1/3/1995	71	78.76	1.109
1/10/2000	1/1/2003	27	22.14	0.820
1/7/2007	1/11/2007	4	4.2	1.050
1/6/2009	1/10/2009	4	2.01	0.503
1/11/2011	1/5/2012	6	4.05	0.675
	max	71	78.76	1.109
	min	4	2.01	0.503

Figure 8. SPI-12 month DD, DS, and DI for Constanța. The yellow-highlighted region contains the periods that belong to the prolonged drought periods. Considering only the period included in the red rectangle, DD = 128 months, and the average drought intensity DI_{av} = 1.055.

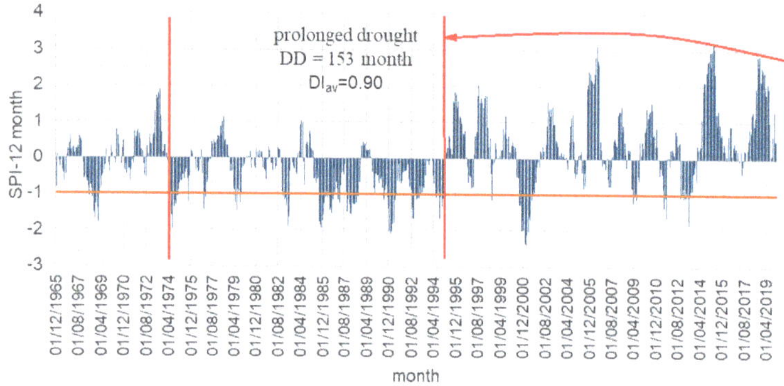

start_date	end_date	DD	DS	DI
1/5/1968	1/8/1969	15	14.95	1.00
1/4/1974	1/11/1975	19	17.84	0.94
1/11/1976	1/7/1977	8	5.15	0.64
1/5/1979	1/3/1980	10	7.64	0.76
1/12/1982	1/1/1984	13	10.27	0.79
1/8/1985	1/11/1988	39	42.92	1.10
1/11/1989	1/3/1995	64	54.64	0.85
1/8/1998	1/11/1998	3	1.42	0.47
1/8/2000	1/3/2002	19	25.58	1.35
1/3/2009	1/9/2009	6	5.41	0.90
1/7/2011	1/10/2011	3	4.09	1.36
10/1/2012	1/1/2014	15	13.2	0.88
	max	64	54.64	1.36
	min	3	1.42	0.47

Figure 9. SPI-12 month, DD, DS, and DI for Mangalia. The yellow-highlighted region contains the periods that are included in the prolonged drought periods.

The number of drought events (NDE) was between 9 at Constanța and Cernavodă and 12 at Mangalia and Medgidia. The drought duration was between 3 months (at Mangalia and Cernavodă) and 71 months (at Constanța). The highest drought severity was 78.76 at Constanța, 65.61 at Medgidia, 54.64 at Mangalia, 50.94 at Adamclisi, 41.25 at Cernavodă, and 37.83 at Hârșova. The drought intensities varied in the intervals 0.503–1.109 at Constanța, 0.473–1.363 at Mangalia, 0.511–1.493 at Adamclisi, 0.438–1.602 at Hârșova, 0.307–1.687 at Medgidia, and 0.463–1.307 at Cernavodă.

The sums of highlighted DD values in the rectangles in Figures 8 and 9 give the prolonged drought duration at Constanța (128 months) and Mangalia (153 months). The corresponding average drought intensities are 1.055 and 0.905. Considering for Constanța the period 1 March 1974 to 1 March 1995, the number of months of prolonged drought will be 188, and the average drought intensity DI_{av} = 0.976.

Table 6 shows, highlighted in yellow, the intervals belonging to prolonged drought periods. Summing up the highlighted values in column 3, for each station, we obtain 106 at Adamclisi, 99 at Hârșova, 109 at Cernavodă, and 27 months at Medgidia. The average drought intensities for these prolonged drought periods were, respectively, 0.902, 1.059, 1.052, and 0.997. Thus, the highest average drought intensity for the prolonged drought periods was recorded at Constanța from 1 November 1982 to 1 March 1995.

Table 6. DD, DS, and DI for (**a**) Adamclisi, (**b**) Hârșova, (**c**) Cernavodă, and (**d**) Medgidia. The yellow-highlighted zones contain the periods included in the prolonged drought periods.

(a)	Start_Date	End_Date	*DD*	*DS*	*DI*
	1 December 1965	1 September 1966	9	4.60	0.511
	1 December 1967	1 May 1969	17	13.04	0.767
	1 March 1974	1 June 1975	15	20.32	1.355
	1 June 1976	1 September 1977	15	20.16	1.344
	1 May 1982	1 March 1984	22	15.81	0.719
	1 August 1984	1 September 1987	37	50.94	1.377
	1 May 1989	1 August 1989	3	1.66	0.553
	1 November 1990	1 March 1992	16	11.56	0.723
	1 October 1992	1 February 1995	28	15.68	0.560
	1 October 2000	1 September 2002	23	34.35	1.493
	1 July 2011	1 June 2014	35	32.30	0.923
		max	37	50.94	1.493
		min	3	1.66	0.511

(b)	Start_Date	End_Date	*DD*	*DS*	*DI*
	1 August 1971	1 September 1972	13	20.83	1.602
	1 November 1973	1 November 1975	24	21.91	0.913
	1 June 1976	1 May 1977	11	6.56	0.596
	1 July 1983	1 April 1984	9	9.61	1.068
	1 June 1986	1 March 1988	21	17.56	0.836
	1 June 1989	1 July 1991	25	37.83	1.513
	1 May 1992	1 March 1995	34	33.92	0.998
	1 July 1995	1 May 1996	10	5.97	0.597
	1 October 2000	1 May 2004	43	47.91	1.114
	1 April 2007	1 August 2007	4	3.78	0.945
	1 March 2009	1 July 2009	4	1.75	0.438
		max	43	47.91	1.602
		min	4	1.75	0.438

(c)	Start_Date	End_Date	*DD*	*DS*	*DI*
	1 January 1969	1 April 1969	3	1.39	0.463
	10 January 1973	1 July 1975	21	21.54	1.026
	3 January 1985	1 March 1988	36	36.91	1.025
	5 January 1989	1 March 1992	34	36.52	1.074
	6 January 1992	1 September 1995	39	41.25	1.058
	9 January 2000	1 August 2002	23	30.06	1.307
	10 January 2006	1 March 2009	29	25.05	0.864
	7 January 2011	1 May 2012	10	10.43	1.043
	5 January 2013	1 August 2014	15	12.99	0.866
		max	39	41.25	1.493
		min	3	1.39	0.511

(d)	Start_Date	End_Date	*DD*	*DS*	*DI*
	1 September 1968	1 July 1969	10	10.27	1.027
	1 March 1974	1 October 1974	7	9.03	1.290
	1 September 1975	1 June 1978	33	25.46	0.772
	1 June 1979	1 July 1980	13	9.03	0.695
	1 May 1982	1 June 1987	61	65.51	1.074
	1 July 1990	1 August 1991	13	17.53	1.348
	1 January 1994	1 October 1995	11	7.55	0.686
	1 August 2000	1 August 2002	24	40.48	1.687
	1 April 2007	1 September 2007	5	5.52	1.104
	1 January 2009	1 September 2009	8	9.03	1.129
	1 July 2012	1 August 2013	13	13.26	1.020
	1 June 2017	1 December 2017	6	1.84	0.307
		max	61	65.51	1.698
		min	5	1.84	0.307

The existence of drought in Dobrogea was less investigated. There are only a few historical references to such periods. Hepites [60] drew up a map of the precipitation regime based on the values recorded between 1884 and 1898, published in the Annales of the Meteorological Institute of Romania in 1900. The precipitation values for the period investigated by Hepites are 406 mm in Mangalia and 412 mm in Constanța. The southern zone recorded annual precipitations between 400 and 500 mm, while it varied between 500 and 600 mm in the inner territory. Hepites indicated that in 1986, an average of 279 mm of precipitation was recorded in Dobrogea. In Mangalia, there was 164 mm, in Hârșova 189 mm, and 261 mm in Constanța.

Otetelișanu and Elefteriu [61] drew a map of the rainfall distribution for 1891–1915. We note that the isohyet of 400 mm passed in the coastal area, the rest of the territory being located between the isohyets of 400 and 500 mm. In the document presented in the Bulletin of the Romanian Royal Society of Geography, pages 209–222, the two researchers noted the low rainfall on the Black Sea coast, considering that there were severe drought periods.

During the period investigated in this study, we determined periods of prolonged drought, but also shorter periods (2007, 2011–2013). They correspond to those determined by other authors [28,62,63]. Importantly, our results are concordant with those of Dobrica [28], who investigated hydrological drought based on the data series covering 1965–2005 (in the Nuntasi Lake basin, in Constanța County), using the Standardized Streamflow Index (SSFI). He found that after 1999, the SSFI had only negative values, indicating hydrological drought.

Different researchers indicated an increase in drought events in Europe. For example, Poljanšek [64] reported the augmentation of the meteorological drought frequency in southern and central Europe since 1950. Stahl et al. [65] and Gudmundsson et al. [66] found hydrological drought during 1950–2015 in the same European zones. In Belarus, drought events became more frequent after 1950 [67]. In a study covering the last 120 years, Ionita and Nagavciuc [68] pointed out after analyzing the SPEI12 index that most Central European and Mediterranean countries experience a significant drying trend. The results are similar to those of Vicente-Serrano et al. [69] based on the SPI12.

The study of Tsakiris and Vangelis [70] identified pan-European drought events in 1950–1952, 1953–1954, 1972–1974, and 2003, confirming some of our findings. The augmentation of average temperatures after the 1990s also increased the severity of drought events in southern Europe, especially in summer [71,72].

5. Conclusions

The present article investigated the intensity of drought and assessed the vulnerability to drought and the drought risk in Constanța County. It analyzed long-term data series collected at six meteorological stations within the county.

The de Martonne index values varied between slightly arid and arid zones at all stations. The results indicated a high and very high vulnerability to drought in most locations and a very high and high drought risk in half of them.

The number of drought events varied between 9 and 12, with durations from 3 to 71 months, and drought severities between 37.83 and 78.76. The prolonged drought duration above 99 months indicates the necessity to better investigate the extent of the drought for taking action to mitigate its effects.

To fully understand the drought's impact, one should compare the SPI for different periods and other drought indicators that emphasize the actual effects on plants and different parts of the economy. The SPI only measures water supply and does not consider evapotranspiration. Therefore, it cannot fully capture how higher temperatures affect moisture availability and demand. Additionally, it does not consider the precipitation intensity and its influence on streamflow, runoff, and water availability. Therefore, the research will be extended using other indices that incorporate evapotranspiration and take into account soil moisture, which is critical in evaluating agricultural drought and water stress. Since we currently do not have access to such data, we shall also use satellite data.

The present study used data from only six meteorological stations. The number of stations will be increased to extend the spatial resolution of the drought assessment and better capture the diverse geographical features of Constanța County. Moreover, more factors (such as agricultural resilience and yield losses, crop-specific vulnerabilities, and infrastructure) should be considered for building the DRI to reflect the drought risk better.

Both natural and human-made factors contribute to the triggering of risks, amplifying dryness and drought in various ways. Understanding the meteorological factors that lead to drought, like atmospheric circulation, is also essential. Therefore, these factors should be considered and analyzed to determine their impact on drought intensity. Based on these findings, forecast models should also be built.

Considering the study's results, we conclude that addressing the complex climatic risks in Dobrogea, such as dryness and drought, urgently requires a collaborative and interdisciplinary approach.

Author Contributions: Conceptualization, C.E.M. and A.B.; methodology, C.E.M., A.B. and A.O.; software, A.O.; validation, C.E.M. and A.B.; formal analysis, C.E.M.; investigation, C.E.M., A.B. and A.O.; resources, C.E.M. and A.B.; data curation, C.E.M.; writing—original draft preparation, A.O. and A.B.; writing—review and editing, A.B.; visualization, A.O.; supervision, A.B.; project administration, C.E.M.; funding acquisition, A.B. All authors have read and agreed to the published version of the manuscript.

Funding: This research received no external funding.

Institutional Review Board Statement: Not applicable.

Informed Consent Statement: Not applicable.

Data Availability Statement: Data will be available on request from the authors.

Acknowledgments: The authors thank the anonymous reviewers and editor for their valuable suggestions that helped improve the manuscript. We especially thank Reviewer 4 for the comments that led us to determine future research directions.

Conflicts of Interest: Author Amela Osman was employed by the company RAJA S.A. The remaining authors declare that the research was conducted in the absence of any commercial or financial relationships that could be construed as a potential conflict of interest.

References

1. Heim, R.R., Jr.; Bathke, D.; Bonsal, B.; Cooper, E.W.T.; Hadwen, T.; Kodama, K.; McEvoy, D.; Muth, M.; Nielsen-Gammon, J.W.; Prendeville, H.R.; et al. A Review of User Perceptions of Drought Indices and Indicators Used in the Diverse Climates of North America. *Atmosphere* **2023**, *14*, 1794. [CrossRef]
2. Safdar, M.; Shahid, M.A.; Zaman, M.; Rasul, F.; Muzammal, H.; Raza, A.; Sabir, R.M.; Zafar, U. Drought Monitoring with Multiple Indices and Management through Various Techniques: A Review. *Eng. Proc.* **2023**, *56*, 307. [CrossRef]
3. Zargar, A.; Sadiq, R.; Naser, B.; Khan, F.I. A Review of Drought Indices. *Environ. Rev.* **2011**, *19*, 333–349. [CrossRef]
4. Zarafshani, K.; Sharafi, L.; Azadi, H.; Van Passel, S. Vulnerability assessment models to drought: Toward a conceptual framework. *Sustainability* **2016**, *8*, 588. [CrossRef]
5. Mishra, A.K.; Singh, V.P. A review of drought concepts. *J. Hydrol.* **2010**, *391*, 202–216. [CrossRef]
6. Araneda-Cabrera, R.J.; Bermúdez, M.; Puertas, J. Benchmarking of drought and climate indices for agricultural drought monitoring in Argentina. *Sci. Total Environ.* **2021**, *790*, 148090. [CrossRef]
7. Wang, L.; Yu, H.; Yang, M.; Yang, R.; Gao, R.; Wang, Y. A Drought Index: The Standardized Precipitation Evapotranspiration Runo Ff Index. *J. Hydrol.* **2019**, *571*, 651–668. [CrossRef]
8. Maftei, C.; Muntean, R.; Vaseashta, A. Monitoring and Modelling of Extreme Hydrometeorological Events. In *Modeling and Monitoring Extreme Hydrometeorological Events*; Maftei, C., Muntean, R., Vaseashta, A., Eds.; IGI Global: Hershey, PA, USA, 2024; Volume 3, pp. 1–21.
9. Communication from the Commission to the European Parliament and the Council—Addressing the Challenge of Water Scarcity and Droughts in the European Union, {SEC(2007) 993}, {SEC(2007) 996}. Available online: https://www.eea.europa.eu/policy-documents/addressing-the-challenge-of-water (accessed on 27 July 2024).
10. Guidelines for Preparation of the Drought Management Plans: Development and Implementation of Risk-Based Drought Management Plans in the Context of the EU Water Framework Directive. Available online: https://www.gwp.org/globalassets/global/gwp-cee_files/idmp-cee/idmp-guidelines-hi4web-final.pdf (accessed on 27 July 2024).
11. Wei, W.; Wang, J.; Ma, L.; Wang, X.; Xie, B.; Zhou, J.; Zhang, H. Global Drought-Wetness Conditions Monitoring Based on Multi-Source Remote Sensing Data. *Land* **2024**, *13*, 95. [CrossRef]
12. World Meteorological Organization—WMO; Global Water Partnership—GWP. Handbook of Drought Indicators and Indices. Available online: https://www.droughtmanagement.info/literature/GWP_Handbook_of_Drought_Indicators_and_Indices_2016.pdf (accessed on 27 July 2024).
13. Haied, N.; Foufou, A.; Khadri, S.; Boussaid, A.; Azlaoui, M.; Bougherira, N. Spatial and Temporal Assessment of Drought Hazard, Vulnerability and Risk in Three Different Climatic Zones in Algeria Using Two Commonly Used Meteorological Indices. *Sustainability* **2023**, *15*, 7803. [CrossRef]
14. Standardized Precipitation Index (SPI), European Drought Observatory. Available online: https://drought.emergency.copernicus.eu/data/factsheets/factsheet_spi.pdf (accessed on 13 September 2024).
15. Espinosa, L.A.; Portela, M.M.; Rodrigues, R. Spatio-temporal variability of droughts over past 80 years in Madeira Island. *J. Hydrol. Reg. Stud.* **2019**, *25*, 100623. [CrossRef]
16. Lorenzo, M.N.; Alvarez, I.; Taboada, J.J. Drought evolution in the NW Iberian Peninsula over a 60 year period (1960–2020). *J. Hydrol.* **2022**, *610*, 127923. [CrossRef]
17. Malakiya, A.D.; Suryanarayana, T.M.V. Assessment of drought using standardized precipitation index (SPI) and reconnaissance drought index (RDI): A case study of Amreli District. *Int. J. Sci. Res.* **2016**, *5*, 1995–2002.
18. Wang, Q.; Zhang, R.; Qi, J.; Zeng, J.; Wu, J.; Shui, W.; Wuu, X.; Li, J. An improved daily standardized precipitation index dataset for mainland China from 1961 to 2018. *Sci. Data* **2022**, *9*, 124. [CrossRef] [PubMed]
19. Chou, J.; Xian, T.; Zhao, R.; Xu, Y.; Yang, F.; Sun, M. Drought Risk Assessment and Estimation in Vulnerable Eco-Regions of China: Under the Background of Climate Change. *Sustainability* **2019**, *11*, 4463. [CrossRef]
20. Brooks, N. Vulnerability, Risk and Adaptation: A Conceptual Framework. Available online: https://www.researchgate.net/publication/200032746_Vulnerability_Risk_and_Adaptation_A_Conceptual_Framework#fullTextFileContent (accessed on 30 July 2024).

21. Disaster Through a Different Lens. Behind Every Effect, There Is a Cause. Available online: https://www.unisdr.org/files/2010 8_mediabook.pdf (accessed on 30 July 2024).
22. Dai, A. Characteristics and trends in various forms of the Palmer Drought Severity Index during 1900–2008. *J. Geophys. Res.* **2011**, *116*, D12115. [CrossRef]
23. Zare, M.; Azam, S.; Sauchyn, D.; Basu, S. Assessment of Meteorological and Agricultural Drought Indices under Climate Change Scenarios in the South Saskatchewan River Basin, Canada. *Sustainability* **2023**, *15*, 5907. [CrossRef]
24. Drought Monitoring in Romania. Available online: https://www.copernicus.eu/ro/node/8534 (accessed on 27 December 2023).
25. Birsan, M.-V.; Micu, D.-M.; Niţă, I.-A.; Mateescu, E.; Szép, R.; Keresztesi, Á. Spatio-temporal changes in annual temperature extremes over Romania (1961–2013). *Rom. J. Phys.* **2019**, *64*, 816.
26. Popescu, A.; Dinu, T.A.; Stoian, E.; Serban, V. Variation of the main agricultural crops yield due to drought in Romania and Dobrogea region in the period 2000–2019. *Sci. Pap. Ser. Manag. Econ. Eng. Agric. Rural Dev.* **2020**, *20*, 20.
27. Bărbulescu, A.; Deguenon, J. About the variations of precipitation and temperature evolution in the Romanian Black Sea Littoral. *Rom. Rep. Phys.* **2015**, *67*, 625–637.
28. Dobrica, G. Assessment of the Hydrological Drought in the Nuntași-Tuzla Lake Watershed, Constanța County. Ph.D. Thesis, Ovidius University of Constanța, Constanța, Romania, 2023. (In Romanian).
29. Mocanu-Vargancsik, C.A.; Barbulescu, A. Study of the Temperature's Evolution Trend on the Black Sea Shore at Constanța, Sea-Conf 2019. *IOP Conf. Ser. J. Phys. Conf. Ser.* **2019**, *1297*, 012010. [CrossRef]
30. Croitoru, A.-E.; Piticar, A.; Burada, D.C. Changes in precipitation extremes in Romania. *Quatern. Int.* **2015**, *415*, 325–335. [CrossRef]
31. Bărbulescu, A.; Maftei, C.E. Evaluating the Probable Maximum Precipitation. Case study from the Dobrogea region, Romania. *Rom. Rep. Phys.* **2023**, *75*, 704. [CrossRef]
32. Bărbulescu, A.; Dumitriu, C.S.; Maftei, C. On the Probable Maximum Precipitation Method. *Rom. J. Phys.* **2022**, *67*, 801.
33. Aadhar, S.; Mishra, V. High-resolution near real-time drought monitoring in south Asia. *Sci. Data* **2017**, *4*, 170145. [CrossRef] [PubMed]
34. The Updated Management Plan of the Danube River, Danube Delta, Hydrographic Space Dobrogea, and Coastal Waters. Available online: https://www.mmediu.ro/app/webroot/uploads/files/PMBH_Actualizat_Text_ABADL.pdf (accessed on 12 August 2024). (In Romanian).
35. National Strategy for Water Management 2023–2035. Available online: https://www.mmediu.ro/categorie/strategia-nationala-pentru-gospodarirea-apelor-2023-2035/444 (accessed on 30 July 2024). (In Romanian).
36. Breier, A. *The Lakes from the Romanian Seaside of the Black Sea—Hydrogeographic*; Editura Academiei Republicii Socialiste Romania: Bucharest, Romania, 1976. (In Romanian)
37. De Martonne, E. Une nouvelle fonction climatologique: L'indice d'aridité. *La Meteorol.* **1926**, *2*, 449–458.
38. Neniu, A.-I.; Vlăduț, A.Ș. The influence of climatic conditions on the forest vegetation within the Getic Subcarpatians—Oltenia sector. *Univ. Craiova Ser. Geogr.* **2020**, *21*, 5–18.
39. McKee, T.B.; Doesken, N.J.; Kleist, J. The relationship of drought frequency and duration to time scale. In Proceedings of the 8th Conference on Applied Climatology, Anaheim, CA, USA, 17–22 January 1993; pp. 179–184.
40. Kendall, M.G. *Rank Correlation Methods*, 4th ed.; Charles Griffin: London, UK, 1975.
41. Sen, P.K. Estimates of the regression coefficient based on Kendall's tau. *J. Am. Stat. Assoc.* **1968**, *63*, 1379–1389. [CrossRef]
42. Hazard Definition and Classification Review: Technical Report. Available online: https://www.undrr.org/publication/hazard-definition-and-classification-review-technical-report (accessed on 13 September 2024).
43. Saliba, Y.; Bărbulescu, A. A comparative evaluation of spatial interpolation techniques for maximum temperature series in the Montreal region. *Rom. Rep. Phys.* **2024**, *76*, 701.
44. Webster, R.; Oliver, M.A. *Geostatistics for Environmental Scientists*; Wiley: Chichester, NH, USA, 2007.
45. Edwards, D.C.; McKee, T.B. Characteristics of 20th Century Drought in the United States at Multiple Times Scales. *Atmos. Sci. Pap.* **1997**, *634*, 1–30.
46. Guttman, N.B. Comparing the Palmer Drought Index and the Standardized Precipitation Index. *JAWRA* **1998**, *34*, 113–121. [CrossRef]
47. Guttman, N.B. Accepting the Standardized Precipitation Index: A Calculation Algorithm. *JAWRA* **1999**, *35*, 311–322. [CrossRef]
48. Wu, H.; Hayes, M.J.; Wilhite, D.A.; Svoboda, M.D. The effect of the length of record on the Standardized Precipitation Index calculation. *Int. J. Clim.* **2005**, *25*, 505–520. [CrossRef]
49. Koudahe, K.; Kayode, A.; Samson, A.; Adebola, A.; Djaman, K. Trend Analysis in Standardized Precipitation Index and Standardized Anomaly Index in the Context of Climate Change in Southern Togo. *Atmos. Clim. Sci.* **2017**, *7*, 401–423. [CrossRef]
50. Ortiz-Gómez, R.; Flowers-Cano, R.S.; Medina-García, G. Sensitivity of the RDI and SPEI Drought Indices to Different Models for Estimating Evapotranspiration Potential in Semiarid Regions. *Water Resour. Manag.* **2022**, *36*, 2471–2492. [CrossRef]
51. Vicente-Serrano, S.M.; Beguería, S.; López-Moreno, J.I. A Multiscalar Drought Index Sensitive to Global Warming: The Standardized Precipitation Evapotranspiration Index. *J. Clim.* **2010**, *23*, 1696–1718. [CrossRef]
52. SPI Program. Available online: https://drought.unl.edu/Monitoring/SPI/SPIProgram.aspx (accessed on 10 July 2024).
53. DrinC (Drought Indices Calculator). Available online: https://drought-software.com/ (accessed on 10 July 2024).

54. Tigkas, D.; Vangelis, H.; Tsakiris, G. The Drought Indices Calculator (DrinC). Available online: https://www.researchgate.net/publication/245542402_The_Drought_Indices_Calculator_DrinC#fullTextFileContent (accessed on 12 July 2024).
55. Dabanli, I. Drought Risk Assessment by Using Drought Hazard and Vulnerability Indexes. Available online: https://www.researchgate.net/publication/325697272_Drought_Risk_Assessment_by_Using_Drought_Hazard_and_Vulnerability_Indexes#fullTextFileContent (accessed on 12 December 2023).
56. Fontaine, M.M.; Steinemann, A.C. Assessing Vulnerability to Natural Hazards: Impact-Based Method and Application to Drought in Washington State. *Nat. Hazard. Rev.* **2009**, *10*, 11. [CrossRef]
57. Jameel, M.; Hameed, S.; Shemal, K.; Al-Ansari, N.; Abed, S. Spatial and Temporal Assessment of Drought in the Northern Prone of Iraq Using Standardized Precipitation Index. *Engineering* **2023**, *15*, 691–708. [CrossRef]
58. Naumann, G.; Barbosa, P.; Garrote, L.; Iglesias, A.; Vogt, J. Exploring drought vulnerability in Africa: An indicator based analysis to be used in early warning systems. *Hydrol. Earth Syst. Sci.* **2014**, *18*, 1591–1604. [CrossRef]
59. Sun, J.; Bi, S.; Bashir, B.; Ge, Z.; Wu, K.; Alsalman, A.; Ayugi, B.O.; Alsafadi, K. Historical Trends and Characteristics of Meteorological Drought Based on Standardized Precipitation Index and Standardized Precipitation Evapotranspiration Index over the Past 70 Years in China (1951–2020). *Sustainability* **2023**, *15*, 10875. [CrossRef]
60. Hepites, Ş.C. Distribution of rain by districts and basins in Romania. *Bull. Polytech. Soc.* **1896**, 350–353. (In Romanian: Repartițiunea ploaiei pe districte și pe basinuri în România. *Buletinul Societații Politehnice*)
61. Otetelișanu, E.; Elefteriu, G.D. Map of atmospheric precipitation in Romania—Average annual distribution from 1891–1915, Bull. *Royal Soc. Rom.* **1920**, 209–222. (In Romanian: Harta precipitațiunilor atmosferice din România—Repartițiunea mijlocie anuală din perioada 1891–1915, *Buletinul Societății Regale din România*)
62. Angearu, C.-V.; Ontel, I.; Boldeanu, G.; Mihailescu, D.; Nertan, A.; Craciunescu, V.; Catana, S.; Irimescu, A. Multi-Temporal Analysis and Trends of the Drought Based on MODIS Data in Agricultural Areas, Romania. *Remote Sens.* **2020**, *12*, 3940. [CrossRef]
63. Maftei, C.; Dobrica, G.; Cerneaga, C.; Buzgaru, N. Drought Land Degradation and Desertification—Case Study of Nuntasi-Tuzla Lake in Romania. In *Water Safety, Security and Sustainability: Threat Detection and Mitigation*; Vaseashta, A., Maftei, C., Eds.; Springer International Publishing: Cham, Switzerland, 2021; pp. 583–597.
64. Poljanšek, K.; Marín Ferrer, V.; De Groeve, N.; Clark, I. Science for Disaster Risk Management. In *Knowing Better and Losing Less*; Publications Office of the European: Reims, Luxembourg, 2017.
65. Stahl, K.; Tallaksen, L.M.; Hannaford, J.; van Lanen, H.A.J. Filling the White Space on Maps of European Runoff Trends: Estimates from a Multi-Model Ensemble. *Hydrol. Earth Syst. Sci.* **2012**, *16*, 2035–2047. [CrossRef]
66. Gudmundsson, L.; Seneviratne, S.I. Observation-Based Gridded Runoff Estimates for Europe (E-RUN Version 1.1). *Earth Syst. Sci. Data* **2016**, *8*, 279–295. [CrossRef]
67. Danilovich, I.S.; Loginov, V.F.; Groisman, P.Y. Changes of Hydrological Extremes in the Center of Eastern Europe and Their Plausible Causes. *Water* **2023**, *15*, 2992. [CrossRef]
68. Ionita, M.; Nagavciuc, V. Changes in drought features at the European level over the last 120 years. *Nat. Hazards Earth Syst. Sci.* **2021**, *21*, 1685–1701. [CrossRef]
69. Vicente-Serrano, S.M.; Domínguez-Castro, F.; Murphy, C.; Hannaford, J.; Reig, F.; Peña-Angulo, D.; Tramblay, Y.; Trigo, R.M.; Mac Donald, N.; Luna, M.Y.; et al. Long-term variability and trends in meteorological droughts in Western Europe (1851–2018). *Int. J. Climatol.* **2021**, *41*, E690–E717. [CrossRef]
70. Tsakiris, G.; Vangelis, H. Establishing a drought index incorporating evapotranspiration. *Eur. Water* **2005**, *9*, 3–11.
71. Vicente-Serrano, S.M.; Lopez-Moreno, J.I.; Beguería, S.; Lorenzo-Lacruz, J.; Sanchez-Lorenzo, A.; García-Ruiz, J.M.; Azorin-Molina, C.; Morán-Tejeda, E.; Revuelto, J.; Trigo, R.; et al. Evidence of increasing drought severity caused by temperature rise in southern Europe. *Environ. Res. Lett.* **2014**, *9*, 044001. [CrossRef]
72. Hänsel, S.; Hoy, A.; Brendel, C.; Maugeri, M. Record summers in Europe: Variations in drought and heavy precipitation during 1901–2018. *Int. J. Clim.* **2022**, *42*, 6235–6257. [CrossRef]

atmosphere

MDPI

Article

Temperature and Precipitation Extremes in the Brazilian Legal Amazon: A Summary of Climatological Patterns and Detected Trends

Wanderson Luiz-Silva [1,*], Anna Carolina Fernandes Bazzanela [1,2], Claudine Pereira Dereczynski [1], Antonio Carlos Oscar-Júnior [3] and Igor Pinheiro Raupp [2]

[1] Department of Meteorology, Federal University of Rio de Janeiro, Rio de Janeiro 21941-909, Brazil; bazzanelaannacarolina@ufrj.br (A.C.F.B.); claudine.dereczynski@igeo.ufrj.br (C.P.D.)
[2] Brazilian Electric Energy Research Center (CEPEL), Rio de Janeiro 21941-911, Brazil; raupp@cepel.br
[3] Department of Geography, Rio de Janeiro State University, Rio de Janeiro 20550-900, Brazil; antonio.junior@uerj.br
* Correspondence: wanderson@igeo.ufrj.br

Abstract: The continuous understanding of extreme weather events in the Amazon is fundamental due to the importance of this biome for the regional and planetary climate system. Climate characterization and the identification of changes in the current climate can be key findings for adaptation and mitigation measures. This study examined climatology and trends in 20 climate extreme indices associated with air temperature and precipitation in the Brazilian Legal Amazon (BLA). Daily observed data, interpolated at grid points, were analyzed from 1961 to 2020. Statistical tests were employed to determine the trend's significance and magnitude. The results indicate that prolonged heat, hot days, and annual temperature records have become increasingly frequent in practically all of BLA over the last decades. Warm days and nights are increasing at approximately +11 days/decade. Heat waves have gone from 10 to 20 consecutive days on average in the 1960s to around 30–40 days in recent years. Indices associated with the intensity and frequency of extreme precipitation show a reduction, especially in the rainiest portion of the BLA, the western sector. In the east/south region of BLA, where consecutive dry days reach 100 days/year, they continue to increase at a rate of +1.5 days/decade, a fact related to the delay at the beginning of the rainy season. These aspects deserve attention since they impact local circulation, reducing the convergence of humidity not only over the BLA but also in central-southern region of Brazil.

Keywords: Amazon; climatology; climate change; climate extremes; precipitation; temperature

1. Introduction

The general circulation of the atmosphere, as well as precipitation in South America and even in more distant places in the Northern Hemisphere, suffer a direct effect of the components of the hydrological cycle on the Amazon, especially condensation and evaporation [1,2]. In addition, the Amazon Forest biome is home to about 15% of global biodiversity, thus being biologically the wealthiest region on Earth, contributing to the biogeochemical functioning of the Earth system [3]. According to [4], the large-scale degradation of the Amazon Forest would leave a lasting negative legacy for the functioning and diversity arising from the atmosphere–biosphere interaction.

The Amazon of the coming decades may be even more vulnerable than climate models assume, since deforestation, forest fragmentation, and changes in land cover and land use have negative ecological consequences [5,6]. This fact raises important questions about how deforestation and forest fragmentation influence regional and global climate. Several studies have investigated regional climate change associated with deforestation of the Amazon rainforest [7–9]. There is evidence that a local reduction in rainfall and an increase in air temperature occur due to deforestation of the Amazon rainforest. However, the conclusions of such studies depend on a series of parameterized processes, such as convection schemes in coupled climate models [10]. In the long term, global warming and consequent climate change receive a considerable contribution from releasing carbon into the atmosphere through deforestation and forest degradation, which are significant sources of greenhouse gas emissions into the atmosphere [11].

According to the Intergovernmental Panel on Climate Change (IPCC) [12], more frequent and severe droughts in the coming decades may affect large areas of the Amazon basin and other tropical forest formations due to the accumulation of greenhouse gases. These disturbances are caused by an unevenness in the planet's energy balance and the hydrological cycle [13]. This trend can interact synergistically with the variability of sea surface temperature anomalies, generating even more frequent and intense extreme precipitation and drought events driven by deforestation [14], as observed in 2005 [15].

It is known that the most critical mode of interannual climate variability globally, the El Niño—Southern Oscillation (ENSO) phenomenon, has a marked effect on the spatial and temporal variability of rainfall and hydrology in the Amazon [16–19]. El Niño events in 2005, 2010, 2015, and 2023, with episodes of heat waves co-occurring, caused significant impacts on the hydrological cycle of the Amazon [20–24]. Global climate change can intensify extreme episodes of drought or flooding derived from ENSO [25].

According to [26], the high precipitation deficit observed in 2015, which affected 80% of the Amazon basin—especially in the eastern portion—outlined the exceptional character of the drought episode that year. On the other hand, [27] highlights the influence of La Niña episodes on floods in the basin. The local population of the Amazon uses the rivers of the region for their subsistence, which also serves as transportation. The area's existing and planned hydroelectric reservoirs are also essential for Brazil's energy demand [28]. These activities are affected by both medium-term climate variability and long-term climate change. The intensity and extent of floods and droughts in the basin are directly related to the significant seasonal and interannual variations and the changes observed in these patterns over the decades [29,30].

Most studies on climate change in the Amazon focus on verification of trends. However, prior knowledge of the spatial characterization of climate extremes in recent decades is also essential. Research shows that, for the entire Amazon region, annual minimum, maximum, and average temperatures show upward trends, while some areas show significant decreasing trends in the region's dry season rainfall volumes [16,31–35]. According to [22], warming may intensify the severity of droughts in the future due to the reduction in precipitation and the increase in potential evapotranspiration. There is evidence that the length of the dry season in the Amazon has been increasing, thus affecting local biodiversity and river flow [16,36–38]. For research on local and regional impacts, it is fundamental that such modifications over the years are made available in detail for each region of the study area.

In this context, this work aims to structure a compendium of climatology and observed trends associated with climate extreme indices related to air temperature and precipitation in the Brazilian Legal Amazon (BLA). The BLA is a political definition attributed by the

Brazilian government to the area of the Amazon rainforest belonging to Brazil (Figure 1). It is composed of nine Brazilian states: Acre (AC), Amapá (AP), Amazonas (AM), Mato Grosso (MT), Pará (PA), Tocantins (TO), Rondônia (RO), Roraima (RR), and the part of the state of Maranhão (MA) west of the 44°W meridian. Management strategies that consider the probability of an increase in extreme weather events in line with climate variability, especially the continuous warming trend, are fundamental for the future sustainability of the Amazon rainforest and the ecosystem services it offers.

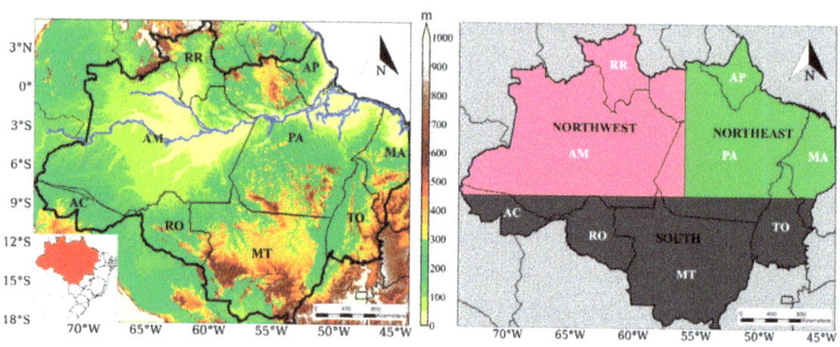

Figure 1. Brazilian Legal Amazon (BLA) region (**left**) with topography (m) and the three subdivisions of the study area (**right**): northwest (NW), northeast (NE), and south (S). The Brazilian states that compose the BLA are Acre (AC), Amapá (AP), Amazonas (AM), Maranhão (MA), Mato Grosso (MT), Pará (PA), Rondônia (RO), Roraima (RR), and Tocantins (TO).

2. Methodology

2.1. Study Area

The BLA (Figure 1) includes the hydrographic basins of the Tocantins–Araguaia and Amazon rivers and encompasses three different biomes (Amazon, Cerrado, and Pantanal). According to data from the Brazilian Institute of Geography and Statistics (IBGE), the BLA occupies an area of approximately 5 million square kilometers, accounting for 70% of the total spatial extension of the world's tropical forests. BLA has Brazil's most significant hydroelectric potential [39], which is currently close to 110 thousand MW.

The region is mainly covered by the Amazon Rainforest, the largest tropical forest in the world and home to significant biodiversity. BLA has a predominantly equatorial (or humid tropical) climate, with substantial annual rainfall accumulations, no or short dry periods, and high temperatures throughout the year [40]. Annual rainfall volumes range from 2000 to 3000 mm [19], and flooding usually occurs between December and May [41].

2.2. Observed Data

The characterization of the climate of a given region depends primarily on three factors: the quantity and quality of data available per area and the size of the time series of the variable analyzed. For a country with continental dimensions like Brazil, the information available regionally is small. This scarcity of data can be observed mainly in the country's north, northeast, and midwest regions, making it an obstacle to research aimed at the spatial detection of trends in climate extremes.

This work includes daily data on minimum and maximum air temperatures and precipitation from the Brazilian Daily Weather Gridded Data (BR-DWGD) [42]. This database was developed from the combination of information from the National Institute of Meteorology (INMET) surface meteorological stations and rainfall stations of the National Water and Basic Sanitation Agency (ANA). The data have a strong temporal coverage, covering the period from January 1961 to July 2020 (2022, for precipitation) with a spatial resolution of 0.1° latitude × 0.1° longitude. The interpolation method used is the Inverse Distance Weighted (IDW) [43]. Unfortunately for the BLA, the data density is the lowest compared to the other Brazilian regions, with one rainfall station every 250 thousand km^2 until 2003 and one every 150 thousand km^2 in recent years. Extensive areas, such as the south of Pará, lack weather stations.

In 2021, the World Meteorological Organization (WMO) updated the climate normals to the 1991–2020 baseline period to provide the most recent data for climate information. Therefore, in this study, the period considered for air temperature climatology is 1991–2019, while for precipitation, it is 1991–2020. For trends, the entire available period is considered, that is, 1961–2019 for air temperature and 1961–2020 for precipitation.

2.3. Climate Extreme Indices

The Expert Team on Climate Change Detection and Indices (ETCCDI) of the Climatology Commission (CCl) of the World Meteorological Organization (WMO) elaborated the climate extremes indices used in this research. Twenty indices based on daily data of minimum temperature, maximum temperature, and precipitation were used (Table 1). The indicators consider extreme climate events' duration, frequency, and intensity [44,45]. Several studies in Brazil and around the world have been using these indicators of climate extremes to detect changes in the present climate [32,33,46–51] and to project changes in the future climate [32,52–55]. The *xclim* library from Python was employed to compute the extreme climate indices from the observational data. We also used Python 3.9.7 tools to structure all the figures presented in this work.

Table 1. List of climate extremes indices evaluated in this research.

Index	Definition	Unit
TMAXmean	Mean maximum temperature (TX)	°C
TMINmean	Mean minimum temperature (TN)	°C
DTR	Daily temperature range	°C
TXx	Maximum value of daily maximum temperature	°C
TXn	Minimum value of daily maximum temperature	°C
TNx	Maximum value of daily minimum temperature	°C
TNn	Minimum value of daily minimum temperature	°C
TX90p	Number of days in which TX > 90th percentile	days
TX10p	Number of days in which TX < 10th percentile	days
TN90p	Number of days TN > 90th percentile	days
TN10p	Number of days in which TN < 10th percentile	days
WSDI	Annual count of days with at least 6 consecutive days in which TX > 90th percentile	days
SU35	Annual count of days in which TX > 35 °C	days
PRCPTOT	Annual total precipitation (PRCP) on wet days (PRCP > 1 mm)	mm
R95p	Annual total precipitation when PRCP > 95th percentile	mm
RX5day	Maximum rainfall on 5 consecutive days	mm
SDII	The ratio between PRCPTOT and the number of wet days (PRCP > 1 mm)	mm/day
R30mm	Annual count of days with PRCP >= 30 mm	days
CDD	Maximum number of consecutive days with PRCP < 1 mm	days
CWD	Maximum number of consecutive days with PRCP > 1 mm	days

2.4. Statistical Trend Tests

The Mann–Kendall nonparametric statistical test analyzes the significance of indicator trends [56,57]. The null hypothesis of this test is that there is no increasing or decreasing trend in the time series. The test assumes that the succession of values occurs independently and that the probability distribution remains the same (simple random series); the data need not belong to a particular distribution. This test is an appropriate method to assess the significance of possible changes in climate series [58]. Another advantage is that its results are less affected by outlier values than other tests, as its calculation is based on the sign of differences and not directly on the values of the variable. The formula for calculating the Mann–Kendall statistical test is:

$$Z = \frac{S + u}{[VAR(S)]^{\frac{1}{2}}}, \quad (1)$$

where VAR(S) is the variance and $u = -1$ if $S > 0$; $u = 0$ if $S = 0$; and $u = +1$ if $S < 0$; thus:

$$S = \sum_{k=1}^{n-1} \sum_{j=k+1}^{n} signal(x_j - x_k), \quad (2)$$

where:

$$signal(x_j - x_k) = 1 \text{ if } x_j - x_k > 0;$$
$$signal(x_j - x_k) = 0 \text{ if } x_j - x_k = 0; \quad (3)$$
$$signal(x_j - x_k) = -1 \text{ if } x_j - x_k < 0$$

and:

$$VAR(S) = \frac{1}{18}\left[n\cdot(n-1)\cdot(2n+5) - \sum_{p=1}^{g} t_p\cdot(t_p-1)\cdot(2t_p+5)\right]. \quad (4)$$

In Equation (2), x corresponds to the time series, ranging from k ($j = k + 1$) to n. The number of repeated value groups is given by g (a null difference between the compared values). The number of repeated values in each group r (to g) is represented by v in Equation (4). The trend is indicated by the Z value in Equation (1), where a positive (negative) Z means an increasing (decreasing) trend by the normal distribution. Suppose the probability (p) of the Mann–Kendall test is equal to or less than the significance level α in a bilateral test. Hence, a statistically significant trend exists, whereas a p-value greater than α confirms a nonsignificant trend. The significance level α adopted for this study was 0.05 or 5%; that is, a confidence level β of 0.95 or 95%.

Compared to other parametric techniques, the Mann–Kendall test is very robust concerning normality biases and non-stationary values in the time series. For this reason, it is widely used in long-term trend surveys [28,33,35,59,60].

The Sen's Slope test assesses the magnitude of trends in indices series. The Sen's Slope test is a nonparametric method (based upon an assumption of linear trend) used to estimate the magnitude of trends [61]. To calculate Sen's Slope, the curvatures of all value pairs in a time series are computed as:

$$SEN = median\left(\frac{x_j - x_i}{j - i}\right), \forall j > i \quad (5)$$

where xj is the indicator's value in a specific year, and xi is the value in the previous year. As Sen's Slope test is insensitive to outlier values and missing data, this test is more rigorous than the usual regression slope. It provides a more realistic measure of time series trends.

3. Results and Discussions

3.1. Climatology and Trends in Extreme Events

3.1.1. Air Temperature Extremes

Figure 2 shows the annual climatologies (1991–2019) and trends (1961–2019) of the indices associated with air temperature in the BLA. TMAXmean and TMINmean represent the average annual field of maximum and minimum temperatures, respectively, while DTR represents the average daily temperature range. Table 2 presents the trends in air temperature extremes in the NW, NE, and S regions.

Due to the high levels of solar radiation that fall on the surface of the BLA throughout the year, the air temperature is relatively high in practically its entire extension [62]. TMAXmean exhibits a very homogeneous behavior in almost all of the BLA, ranging from 32 to 33 °C, except for some stretches in the eastern sector, where it reaches 34 °C (Figure 2a). The eastern portion of the BLA has a typical tropical climate, with higher temperatures during the driest time of the year (in the austral winter). In the north of Pará and the west of Amapá, TMAXmean is lower, not exceeding 31 °C, probably due to the higher precipitation frequency in this sector. The trends observed in TMAXmean show that the eastern portion of the BLA has magnitudes of up to +0.4 °C/decade, while the western sector exhibits higher magnitudes, reaching up to +0.6 °C/decade (Figure 2b). From 1980 to 2013, [31] also observed an increased rate of +0.4 °C/decade in maximum temperature across the BLA. It is noted that the most accentuated warming in the BLA began in the 1980s (Figure 2c), with average rates of up to +0.3 °C/decade, mainly in the NW region, but also in the NE and S regions, with statistically significant rates of +0.2 °C/decade (Table 2). According to [63], suggestions for the cause of the observed changes since 1980 come from patterns in tropical Atlantic and Pacific Sea surface temperatures.

The TMINmean field presents more significant spatial heterogeneity, with a clear division between the northern and southern sectors of the study area (Figure 2d). In the south–central portion, which covers the states of Acre, Mato Grosso, and Rondônia, the TMINmean does not exceed 21 °C in most of the region, impacted significantly by the occasional cold events that occur in the region during the austral autumn and winter [64]. In the south of Mato Grosso, the mean TMINmean is below 18 °C, values that are also more influenced by the incursion of cold air masses, especially between April and September [65]. In the north–central portion of the BLA, TMINmean exceeds 22 °C, with the highest values observed in the eastern sector of Amazonas and Amapá, in the north of Pará, and Roraima. As for TMAXmean, through the analysis of the verified TMINmean trends, it is evident that there is a statistically significant warming trend in almost the entire extension of the BLA (Figure 2e). Despite the difference in TMINmean between the three BLA regions, with values about 3 °C higher in the NW and NE regions than in the S region, a clear upward trend in this index has been observed in the three regions (NW, NE, and S) over the last few years (Figure 2f). The magnitude of the TMINmean trend is greater than that of TMAXmean, reaching average rates of up to +0.4 °C/decade, especially in the NE region of BLA (Table 2). In the eastern sector of the BLA, the magnitude of the TMINmean trend is more significant than in the other areas, with some points exceeding +0.8 °C/decade. In the BLA's western and northern sectors, the TMINmean trend's magnitude is more homogeneous, around +0.4 °C/decade. As seen earlier, the sharpest upward trend of TMAXmean is observed in the western and northwestern sectors of BLA, which is the opposite signal to that observed for TMINmean.

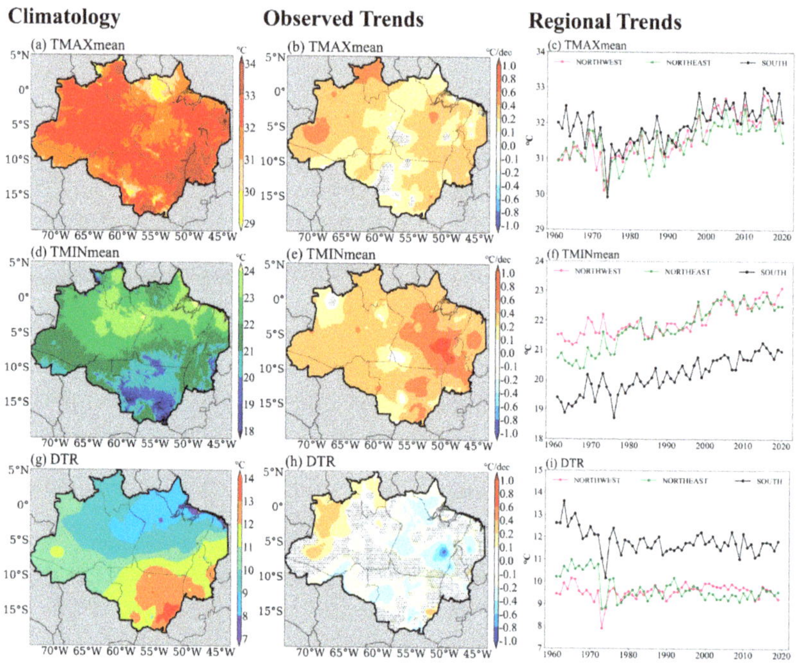

Figure 2. Climatology (**left**) for the period 1991–2019, trends (**center**), and time series in the NW, NE, and S regions of ALB (**right**) for the period 1961–2019 of the indices of climate extremes associated with air temperature: (**a**–**c**) TMAXmean, (**d**–**f**) TMINmean, and (**g**–**i**) DTR. In the observed trends, areas without dotting represent statistically significant trends at the 95% confidence level.

Table 2. Magnitudes of trends in the climate extremes indices associated with air temperature in 1961–2019 in the ALB's NW, NE, and S regions. Bold values denote statistically significant trends at the 95% confidence level.

Trends in Temperature Extremes Indices			
Indices	NW	NE	S
TMAXmean (°C/dec.)	**+0.3**	**+0.2**	**+0.2**
TMINmean (°C/dec.)	**+0.3**	**+0.4**	**+0.3**
DTR (°C/dec.)	0.0	**−0.2**	**−0.1**
TXx (°C/dec.)	**+0.3**	**+0.3**	**+0.4**
TXn (°C/dec.)	**+0.2**	**+0.1**	+0.1
TNx (°C/dec.)	**+0.3**	**+0.4**	**+0.3**
TNn (°C/dec.)	**+0.3**	**+0.5**	**+0.4**
TX90p (days/dec.)	**+10.8**	**+10.5**	**+8.1**
TN90p (days/dec.)	**+14.0**	**+11.9**	**+13.0**
TX10p (days/dec.)	**−6.8**	**−9.4**	**−3.7**
TN10p (days/dec.)	**−14.6**	**−34.0**	**−16.9**
SU35 (days/dec.)	**+4.3**	**+3.6**	**+8.0**
WSDI (days/dec.)	**+2.3**	**+2.2**	**+1.7**

In the southern and eastern sectors of the BLA, the highest values of TMAXmean and the lowest values of TMINmean are observed, resulting in high values of DTR (Figure 2g). In the north of the BLA, TMAXmean and TMINmean present their lowest and highest values due to the greater cloud cover, thus reducing the DTR value. Hence, the DTR field exhibits a north–south gradient, with lower values in the north, approximately between 8 and 9 °C, and higher values in the south, between 13 and 14 °C. In the central portion of the BLA, the magnitude of the DTR trends is smaller, in addition to presenting a few areas without statistical significance. A significant DTR reduction is observed in the central and eastern portions over −0.2 °C/decade (Figure 2h). This DTR pattern is in line with what was observed in the TMAXmean and TMINmean trends since TMINmean presents a higher upward trend than TMAXmean, thus resulting in a reduction in the mean daily temperature amplitude (Figure 2i), which reaches mean rates of up to −0.2 °C/decade in the NE region of BLA (Table 2). The only sign of a significant increase in DTR is observed in the northwestern sector of BLA, an area where the magnitudes of TMINmean trends are lower than those of TMAXmean. This increasing trend of DTR in the western Amazon had previously been verified by [36] for the period from 1980 to 2013; however, it had a slightly lower rate.

Regarding the intensity of extreme maximum temperature events (Figure 3), as observed in the climatological fields of TMAXmean, the highest values of TXx (highest annual maximum temperature) are noted in the southern and eastern sectors of the BLA. At the same time, the lowest is established in the northern and western portions (Figure 3a). In the south of Mato Grosso, TXx exceeds 38 °C, conditions that are more common between the end of winter and the beginning of the austral spring in the face of atmospheric blockages [66]. Regarding TXx trends, statistically significant warming is again observed in most of the BLA (Figure 3b). In the S region of the BLA, the warm extremes have already approached 40 °C over the last decade (Figure 3c), with an average rate of increase of +0.4 °C/decade (Table 2). A trend of a more significant rise in TXx in the BLA's west, east, and southeast sectors, reaching close to +0.8 °C/decade in some points, evidently indicating a growing trend in the daytime hot extremes. When combined with periods of drought, these extreme episodes of high maximum temperature can increase the risk of forest fires directly affecting the region's population, flora, fauna, lakes, and rivers [67].

In the same region where the highest TXx values occur (south of the BLA), the lowest TXn values (lowest maximum annual temperature) are also recorded, which are below 20 °C (Figure 3d). In the case of TXn, the lowest values are found in the southern and western portion of the study area, while the highest (>26 °C) are found in the central, northern, and eastern sectors. The low values of TXn in the south of the BLA result from the entry of masses of polar origin throughout the austral winter, which reach lower latitudes through the continent's interior, as previously mentioned. The signal is not so homogeneous regarding TXn trends, which show a decreasing trend in some regions (Figure 3e). In the state of Acre and some isolated areas of BLA, it is possible to observe a statistically significant trend of TXn reduction from −0.2 to −0.4 °C/decade. Such a reduction in TXn may be associated with increased cloudiness and precipitation in these areas. The TXn-mediated time series in the three BLA regions indicate that the annual variability is more significant in the NW and S regions than in the NE region (Figure 3f). Another point to highlight is that the trend signal does not have statistical significance in most of the BLA, except for the states of Rondônia, Roraima, and the center–north of Amazonas and Mato Grosso. The latter has the highest warming magnitude in TXn (+0.6 °C/decade).

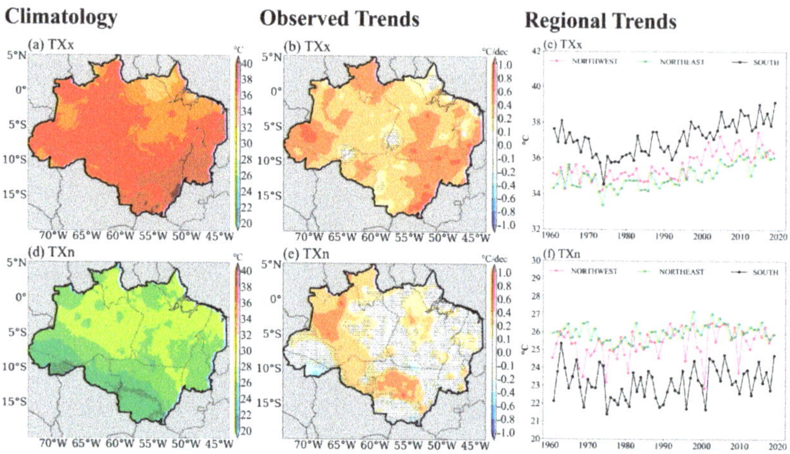

Figure 3. Climatology (**left**) for the period 1991–2019, trends (**center**), and time series in the NW, NE, and S regions of ALB (**right**) for the period 1961–2019 of the indices of climate extremes associated with air temperature: (**a**–**c**) TXx and (**d**–**f**) TXn. In the observed trends, areas without dotting represent statistically significant trends at the 95% confidence level.

Among the indices associated with the absolute extreme values of minimum temperature (Figure 4), TNx (highest annual minimum temperature) is the one that shows a more spatially uniform behavior in its climatology (Figure 4a). In most of the BLA, the values are around 24 to 26 °C. The eastern sector of the BLA (especially along the Amazon River), Roraima, northern Pará, and the central portion of Tocantins stand out, where TNx values are above 26 °C. Regarding TNx trends, there is a statistically significant warming trend in most of the BLA (Figure 4b). The warming magnitude in the BLA's west, north, and south portions is approximately +0.4 °C/decade. In contrast, in the eastern portion, this warming is more pronounced, exceeding +0.8 °C/decade in some locations, such as in the central region of Tocantins, in the southeast of Pará, and in the northeast of Mato Grosso, which encompasses the Xingu River basin, a trend also found by [34]. Such an increase in TNx over the last decades is observed in the time series shown in Figure 4c.

On the other hand, regarding climatology, TNn (lowest minimum annual temperature) is the absolute extreme index that presents the most remarkable spatial heterogeneity, with marked differences between the northern and southern sectors of the BLA (Figure 4d). In the north, the highest TNn values (>20 °C) are recorded due to its more equatorial location, while in the south, the values of this index are below 12 °C. In the extreme south of the BLA, TNn is below 10 °C in extreme cold events. The warming trend for TNn is statistically significant across the east–central BLA (Figure 4e). In addition, a trend above +1.0 °C/decade covers a portion of Mato Grosso, southeast of Pará and northwest of Tocantins. The NE region of BLA has the highest average growth rate of TNn over the last few years, with a magnitude of +0.5 °C/decade (Table 2). In this area, on average, TNn went from 17 °C in the 1960s to about 20 °C in the 2010s (Figure 4f).

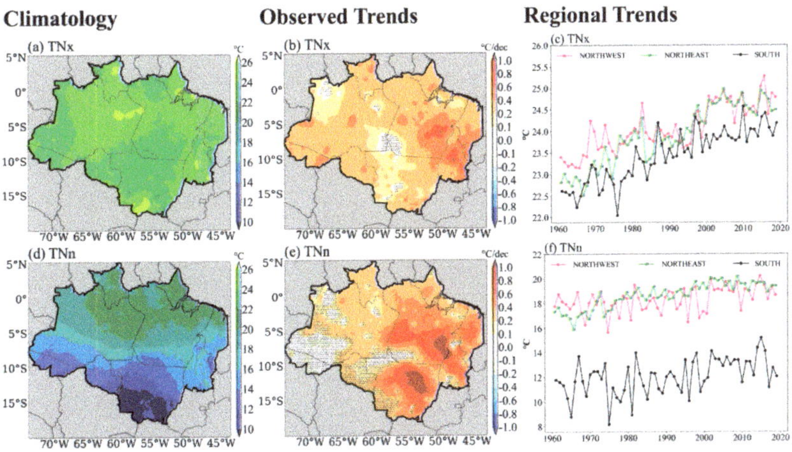

Figure 4. Climatology (**left**) for the period 1991–2019, trends (**center**), and time series in the NW, NE, and S regions of the ALB (**right**) for the period 1961–2019 of the indices of climate extremes associated with air temperature: (**a–c**) TNx and (**d–f**) TNn. In the observed trends, areas without dotting represent statistically significant trends at the 95% confidence level.

Figure 5 shows the detected trends in the frequency of extreme temperature events in the BLA. The indices associated with percentiles exhibit an upward trend in warm days and nights and a significant reduction in cold days and nights. About hot days (TX90p), an increase of more than +15 days/decade was observed in the extreme north of the study area and the central portion of Pará (Figure 5a). In recent years, TX90p has exceeded 100 days a year on average in the three BLA regions (Figure 5b). There is also a significant upward trend in the other areas, although to a lesser extent. For warm nights (TN90p), there is a more pronounced upward trend throughout the BLA, especially in the midwest and far north, exceeding the +20 days/decade mark in some regions (Figure 5c). In the last two decades, the BLA has recorded an average of more than 120 warm nights per year (Figure 5d). Among the three BLA regions, the NW region has the highest average rate of increase in the frequency of hot nights over the last few years, with a magnitude of up to +14 days/decade (Table 2). According to [68], a raised frequency of extreme temperatures can curtail human productivity and boost all-cause human mortality among disenfranchised people in the Amazon. Increased temperatures may also affect livestock and diminish crop yields, aggravated by deforestation-induced decrement in precipitation [69].

As for cold days (TX10p), there was a reduction in this index throughout BLA (Figure 5e), with the most significant decreases occurring in the central portion of the state of Pará and the northwest sector of the study area as a whole, exceeding −15 days/decade. In the three BLA regions, cold days went from 50 to 75 days per year to about 25 to 50 days per year on average in 60 years (Figure 5f). The TN10p index also stands out significantly, revealing a substantial statistically significant reduction in cold nights throughout the eastern and northern sectors of the BLA (Figure 5g). The signal magnitude in these regions is over −25 days/decade, while in the southern and western portions, the magnitude of the trend is considerably lower, staying below −15 days/decade. Cold nights in the year have decreased from approximately 150 to less than 50 nights per year on average in recent decades in the three BLA regions (Figure 5h).

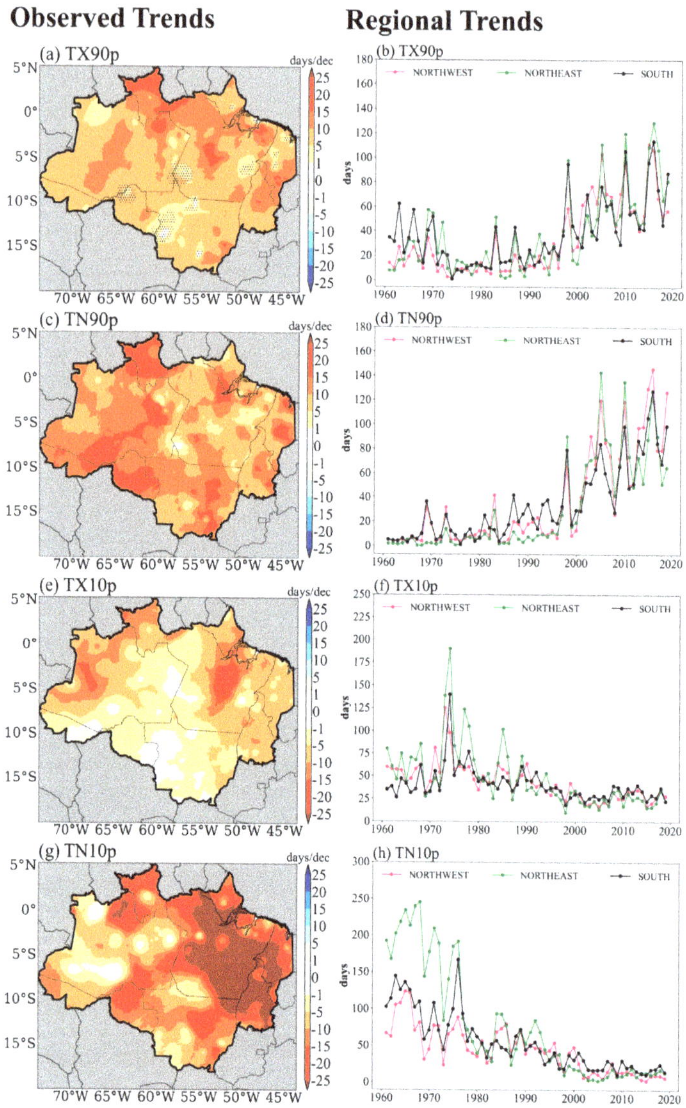

Figure 5. Trends (**left**) and time series in the NW, NE, and S regions of ALB (**right**) for the period 1961–2019 of the indices of climate extremes associated with air temperature: (**a**,**b**) TX90p, (**c**,**d**) TN90p, (**e**,**f**) TX10p, and (**g**,**h**) TN10p. In the observed trends, areas without dotting represent statistically significant trends at the 95% confidence level.

Figure 6 shows the climatology and trends of the SU35 (days with TX above 35 °C) and WSDI (duration of heat waves) indices. In the case of the SU35, the values are higher in the southeast and east sectors of the BLA (Figure 6a). Most of these hot days are recorded without precipitation, with a higher incidence of solar radiation. About a third of the year (SU35 > 100 days) registers maximum temperatures above 35 °C in the extreme south of Mato Grosso and in the state of Tocantins, which predominates during the driest time of the year (June to September). In the central–west of the BLA, the mean SU35 values do not exceed 30 days. In this region, maximum temperatures are milder due to the predominance of cloudiness and the high frequency of rainy days [19]. Figure 6b shows an upward trend in SU35 at several points in the BLA. The magnitude of this trend is over +10 days/decade throughout the eastern sector of the study area, in the state of Rondônia, and the east part of the state of Mato Grosso, indicating a higher frequency of extremely hot days. Therefore, the S region of the BLA presents the most evident increase in SU35 (Figure 6c), reaching more than 60 days a year in recent years in this condition, at an average growth rate of +8 days/decade (Table 2).

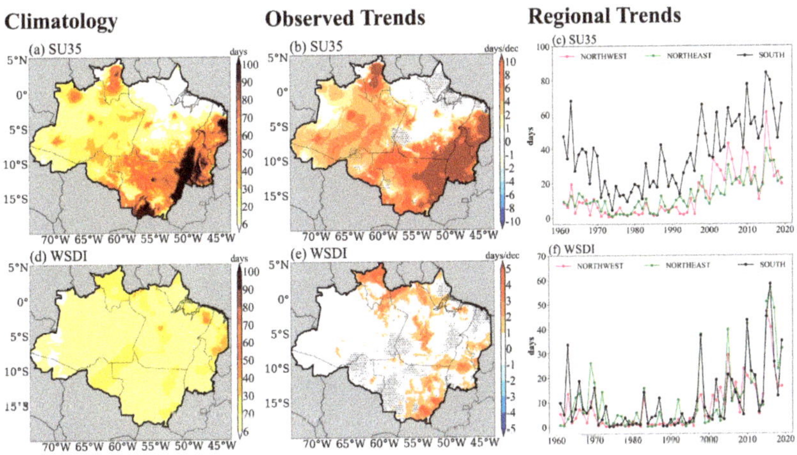

Figure 6. Climatology (**left**) for the period 1991-2019, trends (**center**), and time series in the NW, NE, and S regions of the ALB (**right**) for the period 1961-2019 of the climate extremes indices associated with air temperature: (**a–c**) SU35,and (**d–f**) WSDI. In the observed trends, areas without dotting represent statistically significant trends at the 95% confidence level.

As for WSDI (duration of heat waves), the eastern portion of the BLA experiences, on average, more than 30 days per year with very high maximum temperatures, in this case, above the 90th percentile (Figure 6d). On the other hand, in the southern, central, and northern portions of the BLA, the average WSDI values are less than 20 days. In some parts in the far west of Acre and Amazonas, the index values are less than 6 days; there are no heat waves. This variation in the index is again associated with the pattern of a higher frequency of rainy days in the western sector of the region. Similarly to the other trends, WSDI shows a statistically significant upward trend of +3 days/decade in several areas of the BLA, with special emphasis on the southeast of Mato Grosso and north of Roraima (Figure 6e). The increase in the duration of heat waves becomes more evident from the 2000s onwards in the three BLA regions (Figure 6f). [70] identified that the accretion of the northerly South Atlantic Subtropical Anticyclone (SASA) wind circulation weakens the influx of moisture to the southeastern Amazon, coinciding with the most intense heat waves recorded in this region, and ratifying the trends found in SU35 and WSDI (Figure 6), especially after the 2000s.

3.1.2. Precipitation

Figure 7 presents the annual climatologies (1991–2020) and the trends (1961–2020) of the indices of extremes related to precipitation intensity in the BLA. Regarding the PRCPTOT, it is observed that the highest annual rainfall totals are recorded in the northwest of the BLA, exceeding 3100 mm/year, while in the eastern and southern portions, the accumulated rainfall is less than 1600 mm/year (Figure 7a). This spatial difference is highlighted by the seasonality of precipitation, physiographic aspects, and atmospheric systems acting in the region. The accentuated precipitation in the northwestern part of the study area is associated with the Intertropical Convergence Zone (ITCZ) [71], where the southeast trade winds undergo an orographic uplift over the Andes [72]. In addition, the South American Monsoon System (SAMS) begins in the northwest of the continent, involving the northwest of the BLA, with the first convective activities between the end of winter and the beginning of the austral spring [73,74]. In addition, in the north of Pará and Amapá, rainfall volumes are also significant due to the influence of the ITCZ during the late summer and early austral autumn [75,76], and their associated atmospheric disturbances, especially coastal instability lines [77].

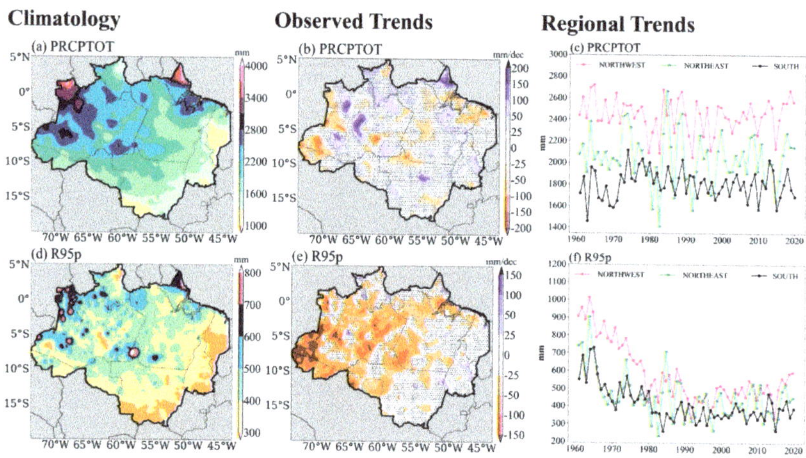

Figure 7. Climatology (**left**) for the period 1991–2020, trends (**center**), and time series in the NW, NE, and S regions of the ALB (**right**) for the period 1961–2020 of the indices of precipitation-related climate extremes: (**a–c**) PRCPTOT and (**d–f**) R95p. In the observed trends, areas without dotting represent statistically significant trends at the 95% confidence level.

Unlike temperature, the behavior of the trend is not uniform in the precipitation indices, presenting very heterogeneous trends in the BLA. In the case of PRCPTOT, isolated signs of both increase and reduction are observed over the entire length of the BLA (Figure 7b). In the state of Acre and the extreme west of the state of Amazonas, there is a reduction of more than −100 mm/decade, while in the central portion of Amazonas, there is a trend of increase in PRCPTOT greater than +100 mm/decade. This increase in the index is also significant in the north of Amapá, where the magnitude of the trend exceeds +200 mm/decade. However, when analyzing the time series in the three regions of the BLA (NW, NE, and S), it is noted that there are no statistically significant trends in PRCPTOT (Figure 7c and Table 3).

In climatological terms, the R95p index shows that the most intense precipitation events are observed at specific points in the BLA's central, western, and northern sectors (Figure 7d). The SACZ often causes heavy rain in the central portion of the study area, another characteristic system of SAMS [78,79]. In some parts of these regions, the accumulated rainfall of these events, added to the local convection, is over 800 mm/year. In the north/northeast of the BLA, the instability lines that form along the coast during the afternoon due to sea breeze circulation are responsible for the expressive volumes of rain in this area [80].

Regarding the trends of R95p, their direction is more spatially uniform than the trends of the PRCPTOT index, with a predominance of reduction in the study region (Figure 7e). The magnitude of this trend is more evident in the central–western portion of the BLA, with some areas exceeding −150 mm/decade. In the eastern and southeastern sectors of the BLA, it is observed that some localities show a subtle upward trend, less than +75 mm/decade. The temporal evolution of R95p mediated in the three BLA regions shows a sharp decrease over the last decades (Figure 7f), especially in the NW region, where the mean rate of decline is −70.2 mm/decade (Table 3). The high frequency of El Niño events since the 1990s may be related to the reduction in heavy rainfall in this part of the Amazon [81]. In the eastern and southeastern sectors of the BLA, it is observed that some localities show a subtle upward trend, less than +25 mm/decade.

Table 3. Magnitudes of trends in the indices of climate extremes related to precipitation in 1961–2020 for the ALB's NW, NE, and S regions. Bold values represent statistically significant trends at the 95% confidence level.

Trends in Precipitation Extreme Indices			
Indices	NW	NE	S
PRCPTOT (mm/dec.)	−3.4	−11.8	−5.5
R95p (mm/dec.)	**−70.2**	**−27.3**	**−30.4**
RX5day (mm/dec.)	**−3.1**	−1.8	**−2.6**
SDII (mm/day/dec.)	**−0.4**	−0.1	**−0.1**
R30mm (days/dec.)	**−1.3**	**−0.5**	**−0.5**
CDD (days/dec.)	−0.4	+0.3	**+1.5**
CWD (days/dec.)	**+3.2**	**+2.0**	**+1.0**

In the RX5day index, associated with the maximum precipitation over five consecutive days, there is a predominance of accumulated rainfall over 120 mm/5 days in most of the BLA (Figure 8a). The only significant difference is in the north of Amapá, where this index exceeds 200 mm/5 days due to the performance of the ITCZ. Similar to the R95p index, a significant downward trend in RX5day is also evident in the western sector of BLA (Figure 8b). The magnitude of the signal is more apparent in Acre, in the southwestern sector of Amazonas, and in the extreme south of Rondônia, where it exceeds −15 mm/decade. The trend signal is quite spatially heterogeneous in the central and eastern portions of the BLA. Thus, the trend measured in the NE region is not statistically significant (Figure 8c and Table 3).

The climatology of daily precipitation intensity, represented by the SDII index (Figure 8d), presents a spatial distribution similar to that of the R95p index (Figure 7d), with higher values in the central and eastern portions of the BLA and the extreme north. In these regions, the average SDII value is around 9 to 11 mm/day, slightly higher in the north of Amapá (>14 mm/day). In general, the values of both indices are higher in the northeast and northwest edges of the study area, regions identified as hotspot areas [82,83], that is, critical points of significant biodiversity that present significant vulnerability to robust changes in climate [84]. SDII also exhibits a significant downward trend in the central–western portion of BLA, exceeding −1 mm/day per decade in some locations (Figure 8e). The analysis of the water balance in the Amazon and the simulation of deforestation and land surface processes in the BLA shows that these elements play an essential role in reducing SDII in the west of the region [85]. In the NE region of the BLA, the SDII signal is quite varied. The temporal evolution has no statistical significance (Figure 8f), with a mean magnitude of only −0.1 mm/day per decade (Table 3).

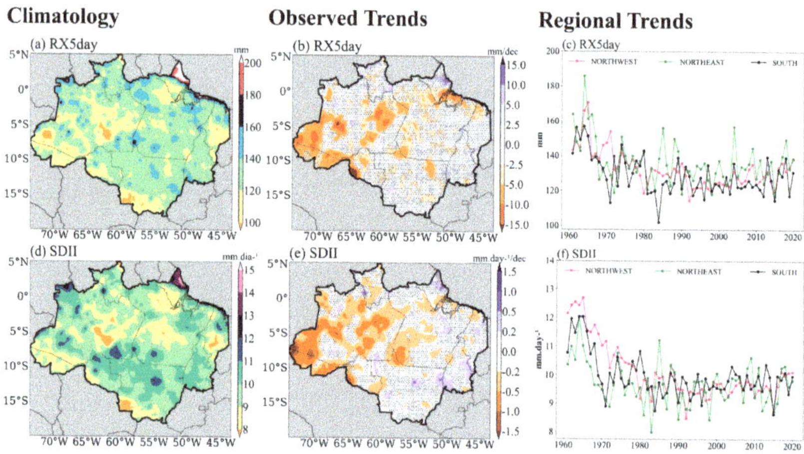

Figure 8. Climatology (**left**) for the period 1991–2020, trends (**center**), and time series in the NW, NE, and S regions of ALB (**right**) for the period 1961–2020 of the indices of precipitation-related climate extremes: (**a–c**) RX5day and (**d–f**) SDII. In the observed trends, areas without dotting represent statistically significant trends at the 95% confidence level.

Figure 9 shows the annual climatologies and trends of indices related to the frequency and duration of precipitation extremes in the BLA. The highest values of the R30mm index (days with rain above 30 mm) are found in some points of the central and northwestern portions of the BLA, as well as in the north of Amapá, exceeding 18 days a year (Figure 9a), mainly due to the ITCZ and coastal instability lines, as previously mentioned. In the eastern and southeastern sectors of the study area, the R30mm values are less than 10 days. Such frequencies of extreme rainfall are relatively low since precipitation is usually well-distributed throughout the year in the BLA. The trend of the R30mm index shows a notable negative trend in the central–west sector of BLA (Figure 9b). The magnitude of this reduction is established around −3 days/decade, being highly pronounced in all areas of BLA (Figure 9c). The local heterogeneity of trends probably happens because the forest plays a vital role in maintaining local and regional rainfall, contributing to the hydrological cycle through recycling and transporting moisture within and outside the region, depending on the land use in each area [86].

The annual climatology of the indices of consecutive dry days (CDD) and consecutive wet days (CWD) in the BLA is also shown in Figure 9. Figure 9d shows that the south and east of the BLA are the driest regions, where CDD values are highest. On average, the eastern sector of Mato Grosso and the south–central portion of Tocantins experience more than 100 consecutive days without rain in the year, indicating that the dry season is longer in this area, usually throughout the austral winter [40]. In the north and west of the study area, the average CDD is less than 20 days, where local thermodynamics contribute to more frequent convective precipitation [19]. Regarding CDD trends, there is a clear division between the east–southeast and west–northwest sectors of the BLA (Figure 9e). In the east–southeast sector, there is a significant upward trend in CDD, especially in the state of Tocantins and in the eastern sector of Mato Grosso, where the magnitude of the trend is over +4 days/decade. In the southern region of the BLA, CDD increased from the 1980s onwards, from 60 days in the 1970s to 70 days in the 2010s (Figure 9f). This region's average

CDD growth rate is +1.5 days/decade (Table 3). This growth in CDD is probably associated with the increase in the dry period in the Amazon and the consequent late onset of the monsoons [87,88]. Thus, recent analyses of the tree cover distribution in this area show a change in climatic conditions that could cause critical transitions from a forest ecosystem to a savannah ecosystem [89]. In isolated areas of the west–northwest sector, there is a reduction trend in the CDD index, although to a lesser extent, with a decrease of less than −2 days/decade.

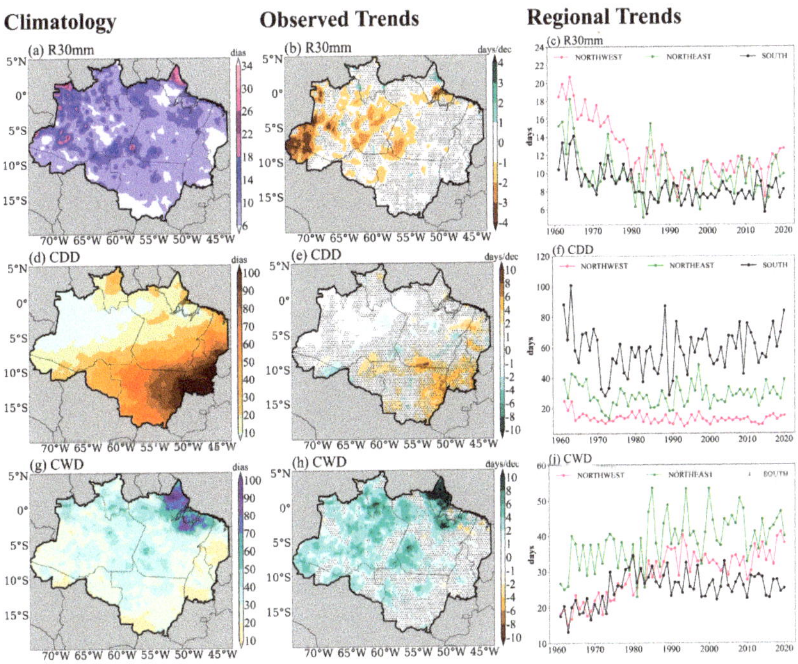

Figure 9. Climatology (**left**) for the period 1991–2020, trends (**center**), and time series in the NW, NE, and S regions (**right**) for the period 1961–2020 of the indices of precipitation-related climate extremes: (**a–c**) R30mm, (**d–f**) CDD, and (**g–i**) CWD. In the observed trends, areas without dotting represent statistically significant trends at the 95% confidence level.

CWD ratifies this striking characteristic of the dry period in the southeast of the BLA, since the average is a maximum of 20 consecutive days with rain per year, while in the central–northern portion, the index values are greater than 40 days (Figure 9g). In the north of Pará and the state of Amapá, the average value of the CWD is greater than 80 days during periods close to the ITCZ, which is in line with the results of other indices for this same region. In the western sector of the BLA, it is possible to observe an increase in CWD, with a magnitude greater than +2 days/decade (Figure 9h). The highest magnitudes of this index are observed in the state of Amapá, where the values exceed +10 days/decade. If consecutive wet days increase in the NW region of the BLA (Figure 9i), even with the reduction in the intensity of the precipitation extremes indices (R95p, RX5day, and SDII), this may be an indication that rainfall is better distributed, occurring with less intensity in the region.

The main trends found in the BLA over the last few decades are summarized in Figure 10. Despite the unchanged annual rainfall accumulation, there has been an accentuated decrease in the frequency and intensity of extreme precipitation events in the western portion of the BLA. On the other hand, there is an expressive growth in the duration of dry periods in the southeastern region of the Amazon, which indicates an irregular distribution of annual rainfall volumes. In addition, heat waves and extremely warm days continue to increase significantly throughout practically the entire BLA, especially in the western, eastern, and southern sectors, which is in line with warming trends in most of Brazil [32,33,90].

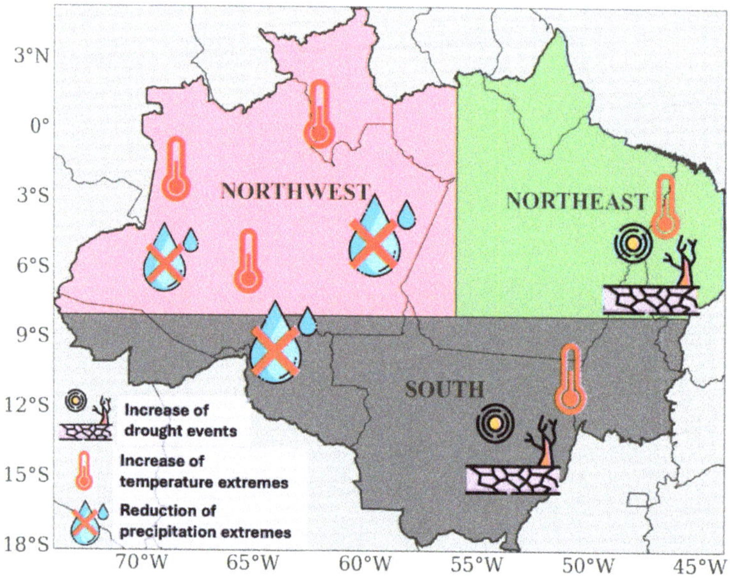

Figure 10. Summary of trends in the behavior of climate extremes related to temperature and precipitation identified over the last few years in the BLA.

4. Conclusions and Final Considerations

Detailed knowledge about extreme weather events in tropical forest regions such as the Amazon is essential due to their impacts on local ecosystems, economic activities, and the region's population. The climatological characterization of temperature and precipitation extremes presented here showed spatial and temporal patterns directly affecting such environmental and socioeconomic segments. The concentrations of greenhouse gases, such as carbon dioxide and methane, are the main contributors to the increase in average temperatures at the global and regional levels. Thus, the regional climate of the Amazon can be impacted by the rise in the concentrations of these gases, reducing the return of moisture to the atmosphere by evapotranspiration. This study compiled the climatology and detected trends in climate extremes in the BLA.

Climate extremes indices related to air temperature show a statistically significant increase in the duration, frequency, and intensity of extremely hot events in practically all of the BLA over the last few years. The frequency of days with very high maximum temperatures and annual maximum temperature records shows a sharp increase, especially in the eastern and southern portions of the BLA. In some areas, the number of days with maximum temperatures above 35 °C doubled from the end of the 20th century to the beginning of the 21st century, and today, TXx reaches 40 °C or more frequently. This warming is likely increasing evaporation and evapotranspiration in the BLA. Development in the Amazon region has gone beyond agricultural boundaries, resulting in widespread changes in land cover, especially in the southeastern portion of the BLA. In addition, studies show that the intensity of heatwaves has a greater impact than their duration concerning the health of the population in the Amazon.

The climatology of precipitation (PRCPTOT) over the BLA shows maximums in the NW region (greater than 2500 mm/year) and lower values in the eastern and southern sectors (less than 1600 mm/year). In the wetter region of the BLA, the extremes related to intensity (R95p, RX5day, SDII) and frequency of precipitation (R30mm and CWD) show a negative trend, which implies a better rainfall distribution in this area. On the other hand, in the less rainy region, east and south of the BLA, where CDD is over 100 days/year, there is a statistically significant increase, around 4 days/decade. Such a substantial increase in dry periods usually occurs between late autumn and early austral spring. In fact, since the beginning of the twenty-first century, southern BLA has experienced a prolonged dry season related to a late onset of SAMS. Research also shows that anthropogenic forcings, such as the conversion of forests to agricultural land and deforestation, affect the length of the dry period in the region. In addition, deforestation and the increase in greenhouse gases associated with human-caused fires lead to more aerosols, which, combined with urban pollution, can modify the hydrological cycle in BLA.

The compendium of trends in climate extremes in the BLA, as carried out in this research, highlights that the rise of air masses over the Amazon can probably be inhibited by deforestation in the region, thus reducing precipitation due to lower moisture recycling. These elements impact the local circulation, reducing the convergence on the SACZ, an essential component in the delay of the rainy season in the southeast of the Amazon and the central portion of Brazil. Finally, BLA warming may further increase the severity of droughts due to increased potential evapotranspiration. Future work that considers a detailed investigation of trends in sensible and latent heat fluxes, and their relationships with temperature and precipitation extremes, may improve the quantification of climate change impacts in the Amazon.

Author Contributions: Conceptualization, W.L.-S.; Methodology, W.L.-S. and A.C.F.B.; Software, A.C.F.B.; Validation, A.C.F.B.; Formal analysis, W.L.-S.; Data curation, W.L.-S., A.C.F.B. and C.P.D.; Writing—original draft, W.L.-S.; Writing—review & editing, C.P.D. and A.C.O.-J.; Visualization, W.L.-S. and A.C.F.B.; Supervision, C.P.D. and A.C.O.-J.; Project administration, I.P.R. All authors have read and agreed to the published version of the manuscript.

Funding: This research received no external funding.

Institutional Review Board Statement: Not applicable.

Informed Consent Statement: Not applicable.

Data Availability Statement: All dataset used in this research is available in public domain: https://sites.google.com/site/alexandrecandidoxavierufes/brazilian-daily-weather-gridded-data (accessed on 10 February 2025)

Conflicts of Interest: The authors declare no conflict of interest.

References

1. Malhi, Y.; Roberts, J.T.; Betts, R.A.; Killeen, T.J.; Li, W.; Nobre, C.A. Climate change, deforestation, and the fate of the Amazon. *Science* **2008**, *319*, 169–172. [CrossRef] [PubMed]
2. Marengo, J.A.; Espinoza, J.C. Extreme seasonal droughts and floods in Amazonia: Causes, trends and impacts. *Int. J. Climatol.* **2016**, *36*, 1033–1050. [CrossRef]
3. Hubbell, S.P.; He, F.; Condit, R.; Borda-de-Água, L.; Kellner, J.; ter Steege, H. Colloquium paper: How many tree species are there in the Amazon and how many of them will go extinct? *Proc. Natl. Acad. Sci. USA* **2008**, *105*, 11498–11504. [CrossRef] [PubMed]

4. Malhi, Y.; Aragão, L.E.O.C.; GBLAraith, D.; Huntingford, C.; Fisher, R.; Zelazowski, P.; Sitch, S.; Meir, P. Exploring the likelihood and mechanism of climate-change-induced dieback of the Amazon rainforest. *Proc. Natl. Acad. Sci. USA* **2009**, *106*, 20610–20615. [CrossRef]
5. Perz, S.; Brilhante, S.; Brown, F.; Caldas, M.; Ikeda, S.; Mendoza, E.; Overdevest, C.; Reis, V.; Reyes, J.F.; Rojas, D.; et al. Road building, land use and climate change: Prospects for environmental Governance in the Amazon. *Philos. Trans. R. Soc. B* **2008**, *363*, 1889–1895. [CrossRef]
6. Silvério, D.V.; Brando, P.M.; Macedo, M.N.; Beck, P.S.A.; Bustamante, M.; Coe, M.T. Agricultural expansion dominates climate changes in southeastern Amazonia: The overlooked non-GHG forcing. *Environ. Res. Lett.* **2015**, *10*, 104015. [CrossRef]
7. Alves, L.M.; Marengo, J.A.; Fu, R.; Bombardi, R.J. Sensitivity of Amazon regional climate to deforestation. *Am. J. Clim. Change* **2017**, *6*, 75–98. [CrossRef]
8. Guimberteau, M.; Ciais, P.; Ducharne, A.; Boisier, J.P.; Aguiar, A.P.D.; Biemans, H.; De Deurwaerder, H.; Galbraith, D.; Kruijt, B.; Langerwisch, F.; et al. Impacts of future deforestation and climate change on the hydrology of the Amazon Basin: A multi-model analysis with a new set of land-cover change scenarios. *Hydrol. Earth Syst. Sci.* **2017**, *21*, 1455–1475. [CrossRef]
9. Oliveira, B.F.A.; Bottino, M.J.; Nobre, P.; Nobre, C.A. Deforestation and climate change are projected to increase heat stress risk in the Brazilian Amazon. *Commun. Earth Environ.* **2021**, *2*, 207. [CrossRef]
10. Nobre, P.; Malagutti, M.; Urbano, D.F.; Almeida, R.A.F.; Giarolla, E. Amazon deforestation and climate change in a coupled model simulation. *J. Clim.* **2009**, *22*, 5686–5697. [CrossRef]
11. Saatchi, S.S.; Harris, N.L.; Brown, S.; Lefsky, M.; Mitchard, E.T.A.; Salas, W.; Zutta, B.R.; Buermann, W.; Lewis, S.L.; Hagen, S.; et al. Benchmark map of forest carbon stocks in tropical regions across three continents. *Proc. Natl. Acad. Sci. USA* **2011**, *108*, 9899–9904. [CrossRef] [PubMed]
12. IPCC. Climate Change 2021: The Physical Science Basis. In *Contribution of Working Group I to the Sixth Assessment Report of the Intergovernmental Panel on Climate Change*; Masson-Delmotte, V.P., Zhai, A., Pirani, S.L., Connors, C., Péan, S., Berger, N., Caud, Y., Chen, L., Goldfarb, M.I., Gomis, M., Eds.; Cambridge University Press: Cambridge, UK; New York, NY, USA, 2021.
13. Betts, R.A.; Cox, P.M.; Collins, M.; Harris, P.P.; Huntingford, C.; Jones, C.D. The role of ecosystem-atmosphere interactions in simulated Amazonian precipitation decrease and forest dieback under global climate warming. *Theor. Appl. Climatol.* **2004**, *78*, 157–175. [CrossRef]
14. Brando, P.M.; Goetz, S.J.; Baccini, A.; Nepstad, D.C.; Beck, P.S.A.; Christman, M.C. Seasonal and interannual variability of climate and vegetation indices across the Amazon. *Proc. Natl. Acad. Sci. USA* **2010**, *107*, 14685–14690. [CrossRef] [PubMed]
15. Marengo, J.A.; Nobre, C.A.; Tomasella, J.; Oyama, M.D.; Oliveira, G.S.; Oliveira, R.; Camargo, H.; Alves, L.M.; Brown, I.F. The drought of Amazonia in 2005. *J. Clim.* **2008**, *21*, 495–516. [CrossRef]
16. Marengo, J.A.; Souza, C.M., Jr.; Thonicke, K.; Burton, C.; Halladay, K.; Betts, R.A.; Alves, L.M.; Soares, W.R. Changes in climate and land use over the Amazon region: Current and future variability and trends. *Front. Earth Sci.* **2018**, *6*, 228. [CrossRef]
17. Barbosa, M.L.F.; Delgado, R.C.; Andrade, C.F.; Teodoro, P.E.; Silva Junior, C.A.; Wanderley, H.S.; Capristo-Silva, G.F. Recent trends in the fire dynamics in Brazilian Legal Amazon: Interaction between the ENSO phenomenon, climate and land use. *Environ. Dev.* **2021**, *39*, 100648. [CrossRef]
18. Jimenez, J.C.; Marengo, J.A.; Alves, L.M.; Sulca, J.C.; Takahashi, K.; Ferrett, S.; Collins, M. The role of ENSO flavours and TNA on recent droughts over Amazon forests and the Northeast Brazil region. *Int. J. Climatol.* **2021**, *41*, 3761–3780. [CrossRef]
19. Luiz-Silva, W.; Oscar-Júnior, A.C.; Cavalcanti, I.F.A.; Treistman, F. An overview of precipitation Climatology in Brazil: Space-time variability of Frequency and intensity associated with atmospheric systems. *Hydrol. Sci. J.* **2021**, *66*, 289–308. [CrossRef]
20. Zeng, N.; Yoon, J.-H.; Marengo, J.A.; Subramaniam, A.; Nobre, C.A.; Mariotti, A.; Neelin, J.D. Causes and impacts of the 2005 Amazon drought. *Environ. Res. Lett.* **2008**, *3*, 014002. [CrossRef]
21. Lewis, S.L.; Brando, P.M.; Phillips, O.L.; Van Der Heijden, G.M.F.; Nepstad, D. The 2010 Amazon drought. *Science* **2011**, *331*, 554. [CrossRef]
22. Jiménez-Muñoz, J.C.; Mattar, C.; Barichivich, J.; Santamaría-Artigas, A.; Takahashi, K.; Malhi, Y.; Sobrino, J.A.; van der Schrier, G. Record-breaking warming and extreme drought in the Amazon rainforest during the course of El Niño 2015–2016. *Sci. Rep.* **2016**, *6*, 33130. [CrossRef] [PubMed]
23. Espinoza, J.C.; Jimenez, J.C.; Marengo, J.A.; Schongart, J.; Ronchail, J.; Lavado-Casimiro, W.; Ribeiro, J.V.M. The new record of drought and warmth in the Amazon in 2023 related to regional and global climatic features. *Sci. Rep.* **2024**, *14*, 8107. [CrossRef] [PubMed]
24. Lima, L.S.; Oliveira e Silva, F.E.; Anastácio, P.R.D.; Kolanski, M.M.P.; Pereira, A.C.P.; Menezes, M.S.R.; Cunha, E.L.T.P.; Macedo, M.N. Severe droughts reduce river navigability and isolate communities in the Brazilian Amazon. *Commun. Earth Environ.* **2024**, *5*, 370. [CrossRef]

25. Cai, W.; Santoso, A.; Wang, G.; Yeh, S.-W.; An, S.-I.; Cobb, K.M.; Collins, M.; Guilyardi, E.; Jin, F.-F.; Kug, J.-S.; et al. ENSO and greenhouse warming. *Nat. Clim. Change* **2015**, *5*, 849–859. [CrossRef]
26. Panisset, J.S.; Libonati, R.; Gouveia, C.M.P.; Machado-Silva, F.; França, D.A.; França, J.R.A.; Peres, L.F. Contrasting patterns of the extreme drought episodes of 2005, 2010 and 2015 in the Amazon Basin. *Int. J. Climatol.* **2018**, *38*, 1096–1104. [CrossRef]
27. Towner, J.; Ficchí, A.; Cloke, H.L.; Bazo, J.; Perez, E.C.; Stephens, E.M. Influence of ENSO and tropical Atlantic climate variability on flood characteristics in the Amazon basin. *Hydroloy Earth Syst. Sci.* **2021**, *25*, 3875–3895. [CrossRef]
28. Luiz-Silva, W.; Xavier, L.N.R.; Maceira, M.E.P.; Rotunno, O.C. Climatological and hydrological patterns and verified trends in precipitation and streamflow in the basins of Brazilian hydroelectric plants. *Theor. Appl. Climatol.* **2019**, *137*, 353–371. [CrossRef]
29. Davidson, E.A.; Araújo, A.C.; Artaxo, P.; Balch, J.K.; Brown, I.F.; Bustamante, M.M.C.; Coe, M.T.; DeFries, R.S.; Keller, M.; Longo, M.; et al. The Amazon basin in transition. *Nature* **2012**, *481*, 321–328. [CrossRef]
30. Sorribas, M.V.; Paiva, R.C.D.; Melack, J.M.; Bravo, J.M.; Jones, J.M.; Carvalho, L.; Beighley, E.; Forsberg, B.; Costa, M.H. Projections of climate change effects on discharge and inundation in the Amazon basin. *Clim. Change* **2016**, *136*, 555–570. [CrossRef]
31. Almeida, C.T.; Oliveira-Júnior, J.F.; Delgado, R.C.; Cubo, P.; Ramos, M.C. Spatiotemporal rainfall and temperature trends throughout the Brazilian Legal Amazon, 1973–2013. *Int. J. Climatol.* **2017**, *37*, 2013–2026. [CrossRef]
32. Avila-Diaz, A.; Benezoli, V.; Justino, F.; Torres, R.; Wilson, A. Assessing current and future trends of climate extremes across Brazil based on reanalyses and earth system model projections. *Clim. Dyn.* **2020**, *55*, 1403–1426. [CrossRef]
33. Regoto, P.; Dereczynski, C.P.; Chou, S.C.; Bazzanela, A.C. Observed changes in air temperature and precipitation extremes over Brazil. *Int. J. Climatol.* **2021**, *41*, 5125–5142. [CrossRef]
34. Luiz-Silva, W.; Regoto, P.; Vasconcellos, C.F.; Garcia, K.C.; Guimarães, F.B.F. A comprehensive analysis of observed and projected climate extremes of temperarure and precipitation in Belo Monte hydropower plant—Eastern Amazon, Brazil. *Int. J. Climatol.* **2022**, *42*, 9710–9729. [CrossRef]
35. Terassi, P.M.B.; Galvani, E.; Gobo, J.P.A.; Oscar-Júnior, A.C.S.; Luiz-Silva, W.; Sobral, B.S.; Gois, G.; Biffi, V.H.R. Exploring climate extremes in Brazil's Legal Amazon. *Stoch. Environ. Res. Risk Assess.* **2024**, *38*, 1403–1422. [CrossRef]
36. Silva, P.E.; Santos e Silva, C.M.; Spyrides, M.H.C.; Andrade, L.M.B. Precipitation and air temperature extremes in the Amazon and northeast Brazil. *Int. J. Climatol.* **2019**, *39*, 579–595. [CrossRef]
37. Agudelo, J.; Espinoza, J.C.; Junquas, C.; Arias, P.A.; Sierra, J.P.; Olmo, M.E. Future projections of low-level atmospheric circulation patterns over south tropical South America: Impacts on precipitation and Amazon dry season length. *J. Geophys. Res. Atmos.* **2023**, *128*, e2023JD038658. [CrossRef]
38. Sierra, J.P.; Espinoza, J.C.; Junquas, C.; Wongchuig, S.; Polcher, J.; Moron, V.; Fita, L.; Arias, P.A.; Schrapffer, A.; Pennel, R. Impacts of land-surface heterogeneities and Amazonian deforestation on the wet season onset in southern Amazon. *Clim. Dyn.* **2023**, *61*, 4867–4898. [CrossRef]
39. Soito, J.L.S.; Freitas, M.A.V. Amazon and the expansion of hydropower in Brazil: Vulnerability, impacts and possibilities for adaptation to global climate change. *Renew. Sustain. Energy Rev.* **2011**, *15*, 3165–3177. [CrossRef]
40. Alvares, C.A.; Stape, J.L.; Sentelhas, P.C.; Gonçalves, J.L.M.; Sparovek, G. Köppen's climate classification map for Brazil. *Meteorol. Z.* **2013**, *22*, 711–728. [CrossRef]
41. Latrubesse, E.M.; Arima, E.Y.; Dunne, T.; Park, E.; Baker, V.R.; d'Horta, F.M.; Wight, C.; Wittmann, F.; Zuanon, J.; Baker, P.A.; et al. Damming the rivers of the Amazon basin. *Nature* **2017**, *546*, 363–369. [CrossRef]
42. Xavier, A.C.; Scanlon, B.R.; King, C.W.; Alves, A.I. New improved Brazilian daily weather gridded data (1961–2020). *Int. J. Climatol.* **2022**, *42*, 8390–8404. [CrossRef]
43. Borges, P.A.; Franke, J.; Anunciação, Y.M.T.; Weiss, H.; Bernhofer, C. Comparison of spatial interpolation methods for the estimation of precipitation distribution in Distrito Federal, Brazil. *Theor. Appl. Climatol.* **2016**, *123*, 335–348. [CrossRef]
44. Karl, T.R.; Nicholls, N.; Ghazi, A. CLIVAR/GCOS/WMO workshop on indices and indicators for climate extremes: Workshop summary. *Clim. Change* **1999**, *42*, 3–7. [CrossRef]
45. Frich, P.; Alexander, L.V.; Della-Marta, P.; Gleason, B.; Haylock, M.; Klein Tank, A.M.G.; Peterson, T. Observed coherent changes in climatic extremes during the 2nd half of the 20th century. *Clim. Res.* **2002**, *19*, 193–212. [CrossRef]
46. Dereczynski, C.P.; Chou, S.C.; Lyra, A.; Sondermann, M.; Regoto, P.; Tavares, P.; Chagas, D.; Gomes, J.L.; Rodrigues, D.C.; Skansi, M.M. Downscaling of climate extremes over South America—Part I: Model evaluation in the reference climate. *Weather Clim. Extrem.* **2020**, *29*, 100273. [CrossRef]
47. Dunn, R.J.H.; Alexander, L.V.; Donat, M.G.; Zhang, X.; Bador, M.; Herold, N.; Lippmann, T.; Allan, R.; Aguilar, E.; Barry, A.A.; et al. Development of an updated global land in situ-based data det of temperature and precipitation extremes: HadEX3. *JGR Atmos.* **2020**, *125*, e2019JD032263. [CrossRef]

48. Luiz-Silva, W.; Silva, F.P.; Dereczynski, C.P.; França, J.R.A. Towards a link between climate extremes and thermodynamic patterns in the city of Rio de Janeiro-Brazil: Climatological aspects and identified changes. *J. Geosci. Environ. Prot.* **2023**, *11*, 131–160. [CrossRef]
49. Kara, Y.; Yavuz, V.; Temiz, C.; Lupo, A.R. Exploring spatio-temporal precipitation variations in Istanbul: Trends and patterns from five stations across two continents. *Atmosphere* **2024**, *15*, 539. [CrossRef]
50. Santos, A.L.M.; Gonçalves, W.A.; Andrade, L.M.B.; Rodrigues, D.T.; Batista, F.F.; Lima, G.C.; Silva, C.M.S. Space-time characterization ox extreme precipitation indices for the semiarid region of Brazil. *Climate* **2024**, *12*, 43. [CrossRef]
51. Yang, Y.; Chang, W. Analysis of spatial and temporal distribution and changes in extreme climate events in Northwest China from 1960 to 2021: A case study of Xinjiang. *Sustainability* **2024**, *16*, 4960. [CrossRef]
52. Li, C.; Zwiers, F.; Zhang, X.; Li, G.; Sun, Y.; Wehner, M. Changes in annual extremes of daily temperature and precipitation in CMIP6 models. *J. Clim.* **2021**, *34*, 3441–3460. [CrossRef]
53. Medeiros, J.F.; Oliveira, C.P.; Avila-Diaz, A. Evaluation of extreme precipitation climate indices and their projected changes for Brazil: From CMIP3 to CMIP6. *Weather Clim. Extrem.* **2022**, *38*, 100511. [CrossRef]
54. Reboita, M.S.; Kuki, C.A.C.; Marrafon, V.H.; Souza, C.A.; Ferreira, G.W.S.; Teodoro, T.; Lima, J.W.M. South America climate change revealed through climate indices projected by GCMs and Eta-RCM ensembles. *Clim. Dyn.* **2022**, *58*, 459–485. [CrossRef]
55. Veiga, S.F.; Nobre, P.; Giarolla, E.; Capistrano, V.B.; Silva, M.B., Jr.; Casagrande, F.; Soares, H.C.; Kubota, P.Y.; Figueroa, S.N.; Bottino, M.J.; et al. Climate change over South America simulated by the Brazilian Earth system model under RCP4.5 and RCP8.5 scenarios. *J. S. Am. Earth Sci.* **2023**, *131*, 104598. [CrossRef]
56. Mann, H.B. Non-parametric tests against trend. *Econometrica* **1945**, *13*, 245–259. [CrossRef]
57. Kendall, M.A.; Stuart, A. *The Advanced Theory of Statistics*; Charles Griffin: London, UK, 1967.
58. Sneyers, R. On the statistical analysis of series of observations. *World Meteorol. Organ.* **1990**, *415*, 192.
59. Zilli, M.T.; Carvalho, L.M.V.; Liebmann, B.; Silva Dias, M.A. A comprehensive analysis of trends in extreme precipitation over southeastern coast of Brazil. *Int. J. Climatol.* **2017**, *37*, 2269–2279. [CrossRef]
60. Costa, R.L.; Baptista, G.M.M.; Gomes, H.B.; Silva, F.D.S.; Rocha Júnior, R.L.; Salvador, M.A.; Herdies, D.L. Analysis of climate extremes indices over northeast Brazil from 1961 to 2014. *Weather Clim. Extrem.* **2020**, *28*, 100254. [CrossRef]
61. Sen, P.K. Estimates of the regression coefficient based on Kendall's tau. *J. Am. Stat. Assoc.* **1968**, *63*, 1379–1389. [CrossRef]
62. Marengo, J.A. On the hydrological cycle of the Amazon basin: A historical review and current state-of-the-art. *Rev. Bras. Meteorol.* **2006**, *21*, 1–19.
63. Gloor, M.; Barichivich, J.; Ziv, G.; Brienen, R.; Schöngart, J.; Peylin, P.; Cintra, B.B.L.; Feldpausch, T.; Phillips, O.; Baker, J. Recent Amazon climate as background for possible ongoing and future changes of Amazon humid forests. *Glob. Biogeochem. Cycles* **2015**, *29*, 1384–1399. [CrossRef]
64. Ricarte, R.M.D.; Herdies, D.L.; Barbosa, T.F. Patterns of atmospheric circulation associated with cold outbreaks in southern Amazonia. *Meteorol. Appl.* **2015**, *22*, 129–140. [CrossRef]
65. Escobar, G.C.J.; Vaz, J.C.M.; Reboita, M.S. Surface atmospheric circulation associated with "friagens" in central-west Brazil. *Anuário Do Inst. De Geociências—UFRJ* **2019**, *42*, 241–254. [CrossRef]
66. Marengo, J.A.; Ambrizzi, T.; Barreto, N.; Cunha, A.P.; Ramos, A.M.; Skansi, M.; Carpio, J.M.; Salinas, R. The heat wave of October 2020 in central South America. *Int. J. Climatol.* **2022**, *42*, 2281–2298. [CrossRef]
67. Doughty, C.E.; Metcalfe, D.B.; Girardin, C.A.J.; Amézquita, F.F.; Cabrera, D.G.; Huasco, W.H.; Silva-Espejo, J.E.; Araujo-Murakami, A.; Costa, M.C.; Rocha, W.; et al. Drought impact on forest carbon dynamics and fluxes in Amazonia. *Nature* **2015**, *519*, 78–82. [CrossRef]
68. Butt, E.W.; Baker, J.C.A.; Bezerra, F.G.S.; von Randow, C.; Aguiar, A.P.D.; Spracklen, D.V. Amazon deforestation causes strong regional warming. *Proc. Natl. Acad. Sci. USA* **2023**, *120*, e2309123120. [CrossRef] [PubMed]
69. Flach, R.; Abrahão, G.; Bryant, B.; Scarabello, M.; Soterroni, A.C.; Ramos, F.M.; Valin, H.; Obersteiner, M.; Cohn, A.S. Conserving the Cerrado and Amazon biomes of Brazil protects the soy economy from damaging warming. *World Dev.* **2021**, *146*, 105582. [CrossRef]
70. Costa, D.F.; Gomes, H.B.; Silva, M.C.L.; Zhou, L. The most extreme heat waves in Amazonia happened under extreme dryness. *Clim. Dyn.* **2022**, *59*, 281–295. [CrossRef]
71. Schneider, T.; Bischoff, T.; Haug, G.H. Migrations and dynamics of the intertropical convergence zone. *Nature* **2014**, *513*, 45–53. [CrossRef]
72. Insel, N.; Poulsen, C.J.; Ehlers, T.A. Influence of the Andes Mountains on South American moisture transport, convection, and precipitation. *Clim. Dyn.* **2010**, *35*, 1477–1492. [CrossRef]
73. Gan, M.A.; Kousky, V.E.; Ropelewski, C.F. The South America monsoon circulation and its relationship to rainfall over west-central Brazil. *J. Clim.* **2004**, *17*, 47–66. [CrossRef]

74. Marengo, J.A.; Liebmann, B.; Grimm, A.M.; Misra, V.; Silva Dias, P.L.; Cavalcanti, I.F.A.; Carvalho, L.M.V.; Berbery, E.H.; Ambrizzi, T.; Vera, C.S.; et al. Recent developments on the South American monsoon system. *Int. J. Climatol.* **2012**, *32*, 1–21. [CrossRef]
75. Fu, R.; Dickinson, R.E.; Chen, M.; Wang, H. How do tropical sea surface temperatures influence the seasonal distribution of precipitation in the Equatorial Amazon? *J. Clim.* **2001**, *14*, 4003–4026. [CrossRef]
76. Souza, E.B.; Kayano, M.T.; Ambrizzi, T. Intraseasonal and submonthly variability over the Eastern Amazon and Northeast Brazil during the autumn rainy season. *Theor. Appl. Climatol.* **2005**, *81*, 177–191. [CrossRef]
77. Sousa, A.C.; Candido, L.A.; Satyamurty, P. Convective cloud clusters and squall lines along the coastal Amazon. *Mon. Weather Rev.* **2021**, *149*, 3589–3608. [CrossRef]
78. Robertson, A.W.; Mechoso, C.R. Interannual and interdecadal variability of the South Atlantic Convergence Zone. *Mon. Weather Rev.* **2000**, *128*, 2947–2957. [CrossRef]
79. Carvalho, L.M.V.; Jones, C.; Liebmann, B. The South Atlantic Convergence Zone: Intensity, form, persistence, and relationships with intraseasonal to interannual activity and extreme rainfall. *J. Clim.* **2004**, *17*, 88–108. [CrossRef]
80. Alcântara, C.R.; Silva Dias, M.A.F.; Souza, E.P.; Cohen, J.C.P. Verification of the role of the low level jets in Amazon squall lines. *Atmos. Res.* **2011**, *100*, 36–44. [CrossRef]
81. Shimizu, M.H.; Ambrizzi, T.; Liebmann, B. Extreme precipitation events and their relationship with ENSO and MJO phases over northern South America. *Int. J. Climatol.* **2017**, *37*, 2977–2989. [CrossRef]
82. Espinoza, J.C.; Chavez, S.; Ronchail, J.; Junquas, C.; Takahashi, K.; Lavado-Casimiro, W. Rainfall hotspots over the southern tropical Andes: Spatial distribution, rainfall intensity, and relations with large-scale atmospheric circulation. *Water Resour. Res.* **2015**, *51*, 3459–3475. [CrossRef]
83. Chavez, S.P.; Takahashi, K. Orographic rainfall hot spots in the Andes-Amazon transition according to the TRMM precipitation radar and in situ data. *JGR Atmos.* **2017**, *122*, 5870–5882. [CrossRef]
84. Sherbinin, A. Climate change hotspots mapping: What have we learned? *Clim. Change* **2014**, *123*, 23–37. [CrossRef]
85. Llopart, M.; Reboita, M.S.; Coppola, E.; Giorgi, F.; Rocha, R.P.; Souza, D.O. Land use change over the Amazon Forest and its impacts on the local climate. *Water* **2018**, *10*, 149. [CrossRef]
86. Aguiar, A.P.D.; Vieira, I.C.G.; Assis, T.O.; Dalla-Nora, E.L.; Toledo, P.M.; Santos-Junior, R.A.O.; Batistella, M.; Coelho, A.S.; Savaget, E.K.; Aragão, L.E.O.C.; et al. Land use change emission scenarios: Anticipating a forest transition process in the Brazilian Amazon. *Glob. Change Biol.* **2016**, *22*, 1821–1840. [CrossRef] [PubMed]
87. Pascale, S.; Carvalho, L.M.V.; Adams, D.K.; Castro, C.L.; Cavalcanti, I.F.A. Current and future variations of the monsoons of the Americas in a warming climate. *Curr. Clim. Change Rep.* **2019**, *5*, 125–144. [CrossRef]
88. Bombardi, R.J.; Moron, V.; Goodnight, J.S. Detection, variability, and predictability of monsoon onset and withdrawal dates: A review. *Int. J. Climatol.* **2020**, *40*, 641–667. [CrossRef]
89. Staal, A.; Dekker, S.C.; Hirota, M.; van Nes, E.H. Synergistic effects of drougth and deforestation on the resilience of the south-eastern Amazon rainforest. *Ecol. Complex.* **2015**, *22*, 65–75. [CrossRef]
90. Curado, L.F.A.; Paulo, S.R.; Paulo, I.J.C.; Maionchi, D.O.; Silva, H.J.A.; Costa, R.O.; Silva, I.M.C.B.; Marques, J.B.; Lima, A.M.S.; Rodrigues, T.R. Trends and patterns of daily maximum, minimum and mean temperature in Brazil from 2000 to 2020. *Climate* **2023**, *11*, 168. [CrossRef]

atmosphere

MDPI

Article

Characteristics and Driving Mechanisms of Heatwaves in China During July and August

Jinping Liu [1,2,3,*] and Mingzhe Li [1]

1 College of Surveying and Geo-Informatics, North China University of Water Resources and Electric Power, Zhengzhou 450046, China; mingzhe.li@stu.ncwu.edu.cn
2 Key Laboratory of Mine Spatio-Temporal Information and Ecological Restoration, Ministry of Natural Resources, Jiaozuo 454003, China
3 Hydraulics and Geotechnics Section, KU Leuven, Kasteelpark Arenberg 40, BE-3001 Leuven, Belgium
* Correspondence: liujp@radi.ac.cn

Abstract: Against the backdrop of global warming, heatwaves in China have become more frequent, posing serious risks to public health and socio-economic stability. However, existing identification methods lack precision, and the driving mechanisms of heatwaves remain unclear. This study applies the Excess Heat Factor (EHF) to characterize heatwaves across China from 2013 to 2023, analyzing their spatiotemporal patterns and exploring key drivers such as atmospheric circulation and soil moisture. Key findings reveal significant regional differences: (1) Frequency and Duration—The southeastern coastal regions (e.g., the Yangtze River Delta) experience higher annual heatwave frequencies (1.75–3.5 events) but shorter durations (6.5–8.5 days). In contrast, the arid northwest has both frequent (1.5–3.5 events per year) and prolonged (8.5–14.5 days) heatwaves, while the Tibetan Plateau sees weaker and shorter events. (2) Driving Factors—Heatwaves in the Yangtze River Delta are primarily driven by an intensified subtropical high, leading to subsidence and clear-sky conditions. In Fujian, anomalous low-level winds enhance heat accumulation, while coastal areas show strong soil moisture–temperature coupling, where drier soils intensify warming. Conversely, soil moisture has a weaker influence on the Tibetan Plateau, suggesting a dominant atmospheric control. It is important to note that the EHF index used in this study does not directly account for humidity, which may limit its applicability in humid regions. Additionally, the ERA5 and ERA5-Land reanalysis data were not systematically validated against ground observations, introducing potential uncertainties.

Keywords: heatwave; driving mechanisms; regional variability; atmospheric circulation; soil moisture; excessive heat factor

1. Introduction

Global warming is a key focus of contemporary climate science research. According to the Sixth Assessment Report of the IPCC (2023), the global average temperature has risen by approximately 1.1 °C over the past century [1]. This warming has led to a significant increase in the frequency, duration, and intensity of heatwaves [2–4]. Notable events include the European heatwaves of 2003, 2006 [5–7], and 2015 [8–10]; the extreme heatwave that swept across Central and Eastern Europe and western Russia in 2010 [11–13]; and the severe heat event in India in 2015 [14]. It is expected that such extreme events are likely to increase in the future [8,15–17]. China has been affected by heatwave events multiple times throughout its history, with regions such as North China and East China being particularly affected [18,19]. Recent studies indicate that heatwave events in China, especially those with spatiotemporal correlations, have significantly increased in terms of impact range, intensity, duration, and shifting [20–23]. This trend underscores the urgency and importance of accurately identifying and analyzing heatwave events in China.

Currently, heatwave identification primarily relies on absolute temperature thresholds (e.g., daily maximum temperature $\geq$ 35 °C) or relative thresholds (e.g., the 90th percentile) [24–28]. The absolute threshold method is simple and intuitive but fails to account for regional climate variations [29]. For instance, a temperature of 35 °C may have vastly different impacts in northern and southern Chinese cities. The relative threshold method is more flexible and adaptable to regional differences, making it widely used in heatwave identification and definition [30,31]. However, it does not fully consider the cumulative effects of nighttime temperatures or human adaptation factors [32,33].

To overcome these limitations, this study employs the Excess Heat Factor (EHF) [32,34] to identify heatwave events and quantify their intensity. EHF not only accounts for extreme daytime temperatures but also incorporates the cumulative effects of nighttime heat and human adaptation to high-temperature environments [33,35–37], enabling a more comprehensive and accurate assessment of heatwave intensity [32,38]. In addition to heatwave identification, investigating the driving factors behind heatwaves is crucial. Understanding key meteorological drivers enhances heatwave forecasting and warnings [39].

Current research on heatwave drivers mainly focuses on large-scale climate modes [40,41], sea surface temperature (SST) variations [42–44], and urbanization-related heatwave effects [45–48]. Large-scale climate patterns significantly influence heatwave characteristics: ENSO events markedly increase the frequency, duration, and spatial extent of heatwaves [49], while the positive phase of the AMO has intensified heatwave occurrences over the past two decades through warming effects and enhanced internal atmospheric variability [40]. Rising SSTs are positively correlated with heatwave activity, with moderate heatwaves showing higher sensitivity to SST changes [43]. Urbanization exacerbates heatwave risks through the urban heat island effect; daytime intensification is mainly due to differences in surface evapotranspiration, while nighttime effects are largely driven by increased anthropogenic heat emissions and strengthened warm advection [50]. Despite these advances, the specific roles of atmospheric circulation and soil moisture in heatwave formation and evolution remain insufficiently explored [51–54].

In summary, this study aims to identify heatwave events in China during July–August from 2013 to 2023 using the EHF, and to analyze their spatial patterns and driving mechanisms. The specific objectives are as follows: (1) to quantify heatwave intensity with EHF, addressing limitations of traditional threshold methods by incorporating heat accumulation and nighttime temperature; (2) to reveal spatial distribution, frequency, duration, and intensity patterns of heatwaves, highlighting regional disparities; and (3) to investigate the roles of atmospheric circulation anomalies and soil moisture variations in driving heatwave events.

2. Materials and Methods

2.1. Study Area

China is located in East Asia, bordered by the Pacific Ocean to the east, with a complex and diverse topography (Figure 1). The terrain is characterized by a high western region and a low eastern region, forming a step-like distribution [55–57]. Approximately 67% of the country consists of plateaus, mountains, and hills, including the Tibetan Plateau and the Loess Plateau, while the remaining 33% is mainly composed of basins and plains, such as the Sichuan Basin and the North China Plain [58]. China's complex terrain features (including vast mountain ranges and plateaus) interact with global climate change, amplifying the frequency and intensity of heatwaves in recent years. This dual effect has created substantial challenges across multiple sectors: reduced crop yields in agriculture, exacerbated urban heat island effects, and elevated health risks particularly for vulnerable populations [59]. With its vast territory and long-term accumulation of climate observation data, China provides unique and rich data resources for in-depth studies on the driving mechanisms and impacts of heatwaves. Since the 1960s, China's mean annual air temperature (MAT) has shown a significant upward trend, with an accelerated warming rate observed after 1980 [60]. Over the past 50 years, the average air temperature in most cities has increased by 0.5 °C, resulting in a total MAT rise of 1.44 °C across China. The warming rate generally increases from south to north, except for the Tibetan Plateau [61]. Multi-model ensemble projections based on the highest-ranked models (BMME) indicate that by the end of the 21st century, the annual growth rates of air temperature and precipitation in China will continue to rise. The projected increase under the SSP5-8.5 scenario is higher than that under the SSP2-4.5 scenario, with particularly pronounced warming in high-latitude and high-altitude regions [62].

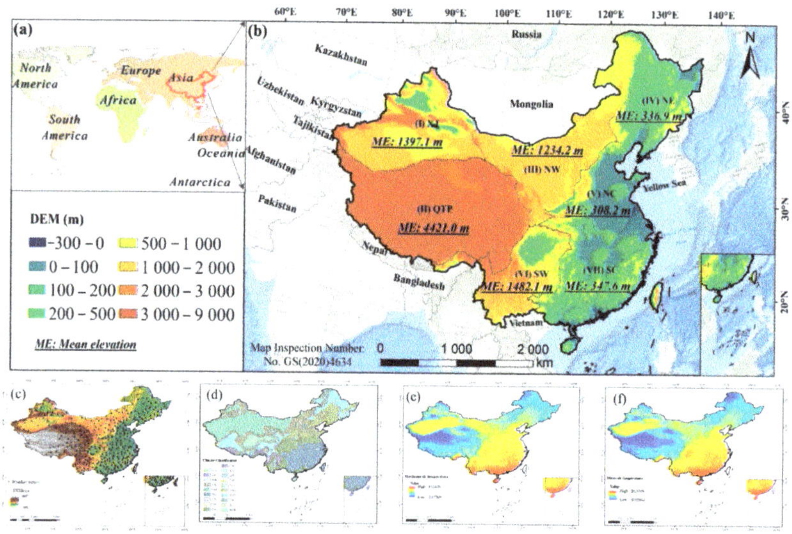

Figure 1. Figures (**a**,**b**) illustrate China's geographical location within the global context and its neighboring countries in Asia (cited from [63]). Figure (**c**) presents the distribution of meteorological stations used in this study. Figure (**d**) displays China's Köppen–Geiger climate classification. Figures (**e**,**f**) show the multi-year average temperature and multi-year average maximum temperature of China, respectively.

2.2. Data Collection

The air temperature data used in this study were obtained from the National Centers for Environmental Information (NCEI) under the U.S. National Oceanic and Atmospheric Administration (NOAA). The dataset includes daily meteorological observations from global stations spanning from 1929 to the present. This study selected daily maximum, minimum, and average air temperature data in China from 2013 to 2023, covering over 400 meteorological stations nationwide. We screened over 400 meteorological stations and selected those with continuous temperature observations for July–August from 2013 to 2023. After quality control, we removed records with missing values, standardized temperature units, and calculated the daily average temperatures for each station. The processed data were then converted into standard NetCDF format, yielding a dataset of 388 qualified stations for analysis. The choice of July–August as the study period is based on the climate zoning standards issued by the China Meteorological Administration (GB/T 35221-2017) [64], which indicate that heatwave events in most regions of China primarily occur from mid-July to mid-August. This study utilizes NOAA meteorological station data to calculate EHF. To explore the relationship between large-scale atmospheric circulation and heatwaves, this study utilized the fifth-generation reanalysis dataset (ERA5) provided by the European Centre for Medium-Range Weather Forecasts (ECMWF). Daily data on geopotential height and wind fields were extracted from the ERA5 dataset, with a spatial resolution of 0.25° × 0.25°. In addition, soil moisture data were sourced from the ERA5-Land reanalysis dataset provided by ECMWF, with daily surface soil moisture data (0–10 cm below the surface) from 2013 to 2023 to support the study of the impact of soil moisture on heatwaves. The wind fields data from ERA5 and the soil moisture data from ERA5-Land show good consistency with ground-based observations [65–67]. Terrain data were obtained from the 2022 release of the Global Bathymetric Chart of the Oceans (GEBCO), which includes a global digital elevation model (DEM) covering land elevation and ocean depth. The spatial resolution is in arc-seconds, with units in meters, and the coordinate projection system is GCS_WGS_1984.

2.3. *Methodology*

2.3.1. Calculation Method of the EHF

In previous studies, various methods have been proposed to define the concept of heatwaves [68,69]. However, a unified standard for the exact definition of heatwaves has not yet been established [70–72]. Existing methods primarily use absolute or relative air temperature thresholds combined with duration criteria. The absolute threshold method applies fixed air temperature standards (e.g., 35 °C), but overlooks regional variability. The relative threshold method, based on local climate percentiles (e.g., 90th or 95th), is more adaptable but has limitations: it neglects cumulative heat exposure, lacks a comprehensive intensity measure, and ignores physiological adaptation. Additionally, heatwave duration definitions vary widely (1–7 days) [72], leading to inconsistencies. Given that most studies currently focus on the health impacts of heatwaves, and existing research suggests that daily mean air temperature is often a better predictor of health outcomes than daily maximum air temperature [73], the EHF method—calculated based on daily mean air temperature—offers several advantages. It accounts for the cumulative effects of heat exposure, incorporates human adaptation factors, and provides a more comprehensive representation of the actual impact mechanisms of extreme heat on human health. Therefore, this study adopts the EHF method to define and identify heatwave events. EHF has been widely used in heat event research in regions such as Australia and Europe [74–76]. The EHF was proposed by Nairn et al., and its core concept is based on the notions of "excess heat" and "heat stress" [36,69].The EHF not only considers the abnormal air temperature rise over three days relative to the previous month, but also incorporates the 95th percentile threshold in climatology [77]. This allows it to comprehensively reflect the key characteristics of heatwaves, including abnormal air temperature increases, exceedance of specific climatic thresholds, and the accumulation of heat prior to the onset of the heatwave. Although the EHF calculation does not directly incorporate humidity parameters, it indirectly reflects the influence of relative humidity by using the average daily air temperature as the simple mean of the daily maximum and minimum air temperature. This is because changes in relative humidity directly affect the diurnal air temperature range, which is reflected in the daily average air temperature values [32,75].

The calculation of EHF for each meteorological station is based on two excess heat indices, namely EHI_{sig} and EHI_{accl}. Before computing these indices, it is necessary to determine the extreme heat threshold and daily average temperature for each station.

In this study, the period from 2013 to 2023 was set as the reference period, and the extreme heat threshold for each station was determined using the following method: First, the daily average temperature series from 2013 to 2023 for each station was arranged in ascending order. Then, the 95th percentile value was taken as the extreme heat threshold for that station. The method for calculating the daily average temperature at each station is shown in Equation (1).

$$T_i = \frac{T_{\max} + T_{mim}}{2} \quad (1)$$

After obtaining the daily average temperature and the extreme heat threshold for each station, EHI_{sig} and EHI_{accl} are calculated accordingly. Their calculation methods are shown in Equations (2) and (3), respectively.

$$EHI_{sig} = \frac{(T_i + T_{i-1} + T_{i-2})}{3} - T_{95} \quad (2)$$

In the equations, T_i represents the daily average temperature on day i, and T_{95} denotes the extreme heat threshold for each station. The choice of a three-day period for calculating the heatwave index is based on studies of the delayed effects of high air temperature on human health. Research has shown that the lag period varies by region: approximately one day in Melbourne [78] and three days in Adelaide [19]. Due to higher summer air temperature in Adelaide, the residents have a greater tolerance for heat, consistent with observations in Barcelona [79] and London [80]. Nairn and Fawcett [36], based on data from Langlois et al. [34], confirmed that consecutive extreme high-air temperature days significantly increase mortality rates.

A positive EHI_{sig} indicates that the three-day average air temperature is warmer than the local annual climate anomaly; if it is negative or zero, the air temperature does not meet the threshold for a heatwave event.

$$EHI_{accl} = \frac{(T_i + T_{i-1} + T_{i-2})}{3} - \frac{(T_{i-3} + \cdots + T_{i-32})}{30} \quad (3)$$

In the equation, T_i is defined as above. EHI_{accl} can be referred to as the adaptation index, which evaluates heat adaptation by comparing the three-day daily average air temperature with the recent 30-day daily average air temperature. Human physiological adaptation to high air temperature typically takes 2 to 6 weeks [36]. This study adopts a 30-day period as the reference for recent air temperature based on related research [32]. Based on the calculated EHI_{sig} and EHI_{accl}, the EHF is defined as follows:

$$EHF = \max(1, EHI_{accl}) \times EHI_{sig} \quad (4)$$

If $EHI_{accl} \geq 1$, it is directly used as a multiplier to emphasize the lack of human adaptation to the heatwave. If $EHI_{accl} < 1$, its value is set to 1, in which case $EHF = EHI_{sig}$, reflecting only the absolute temperature anomaly. A positive EHF value defines the heatwave conditions on day i. The occurrence of a heatwave requires the EHF to be positive for at least three consecutive days (i.e., EHF for three consecutive days).

2.3.2. Definition of the Characteristics of Heatwaves

Based on the EHF, we have defined the key characteristics of heatwave events, including Heatwave Number (HWN), Heatwave Frequency (HWF), Heatwave Duration (HWD), and Heatwave Intensity (HWI), with specific definitions provided in Table 1. Among these, HWN represents the total number of heatwave events during the study period; HWF refers to the total number of days identified as heatwave days within a season (i.e., the sum of days within periods where EHF is positive for at least three consecutive days), used to reflect the overall duration of the heatwave's impact on the season; HWD is calculated by determining the duration of the longest heatwave event in each season, revealing the potential sustained threat of heatwaves; HWI records the peak intensity of heatwave events in a season, reflecting the extreme strength of the heatwaves. It is important to note that HWF, as a measure of the total number of days involved in heatwave events, is influenced by both HWN and HWD, indicating that the frequency and duration of heatwaves are equally important in assessing their overall impact.

Table 1. Definition of Heatwave Characteristics Based on EHF.

Indicators	Definition
HWN	Number of heatwaves
HWF	Number of days with EHF > 0 in heatwave events
HWD	Longest consecutive days with EHF > 0 in heatwave events
HWI	Highest EHF recorded in heatwave events

3. Results

3.1. Extreme Heat Thresholds

Figure 2 shows the extreme heat thresholds for meteorological stations in China from July to August during 2013 to 2023. This threshold is calculated by arranging the daily average temperatures from meteorological stations for 2013–2023 in ascending order and determining the 95th percentile value, which represents the extreme high-temperature threshold for each station. These thresholds are influenced by multiple factors such as terrain, atmospheric circulation, and soil moisture, resulting in significant regional variations. In the eastern regions, particularly in the Yangtze River Basin, the eastern coastal areas, and the southern regions, the extreme heat thresholds are generally higher, ranging between 25 °C and 35 °C, with some areas exceeding 35 °C. These regions experience high summer air temperatures with prolonged durations, and the heat thresholds in the Yangtze River Basin and surrounding areas are particularly notable. The eastern coastal areas, influenced by the maritime climate, are hot and humid during the summer, and the extreme heat thresholds are also at a relatively high level. The southern regions, especially in South China near the equator, experience frequent high air temperature year-round, with high extreme heat thresholds, indicating that these areas are consistently and widely affected by heatwave events [25]. In the northwestern regions, such as the Tarim Basin, Hexi Corridor, and Junggar Basin, the extreme heat thresholds are similarly high. These areas are characterized by low rainfall, high evaporation, and rapid surface air temperature rise. Combined with dry soil and sparse vegetation cover, extreme heat events are particularly severe [81,82]. The extreme heat thresholds on the Tibetan Plateau are relatively low, but in low-altitude river valleys such as the Qaidam Basin and the Yarlung Tsangpo Valley, the extreme heat thresholds are relatively higher. These areas are flat, surrounded by high mountains, and are prone to the formation of local heat low-pressure systems, leading to significant air temperature increases [83]. Additionally, sparse vegetation cover and intense solar radiation further exacerbate the frequency and intensity of extreme heat events [84]. Overall, the extreme heat thresholds in China from 2013 to 2023 exhibited a distinct east–high, west–low spatial pattern. The highest thresholds were observed in the Yangtze River Basin, eastern coastal areas, and southern China, followed by the arid regions in the northwest. In contrast, most areas of the Tibetan Plateau had much lower thresholds, with higher values only in certain river valleys. This indicates that eastern and northwestern China are key regions at high risk of extreme heat.

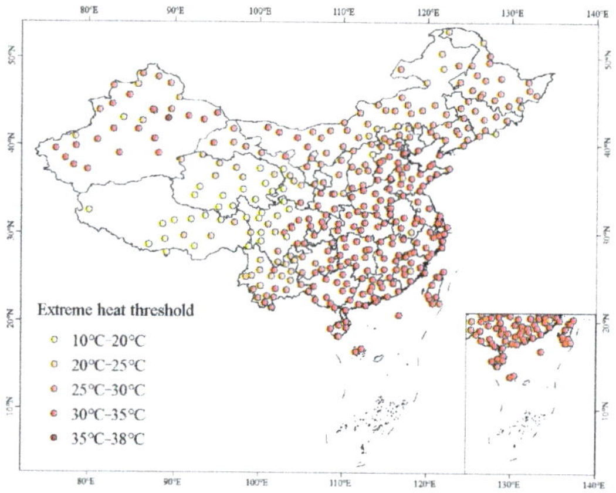

Figure 2. Extreme heat thresholds at meteorological stations in China from July to August 2013–2023.

3.2. *Heatwave and Its Characterization Analysis*

Based on the EHF, we obtained key characteristics of heatwave events, including the HWN, HWF, HWD, and HWI. Their spatial distribution is shown in Figures 3–6. Figure 3 shows the spatial distribution of the number of HWN. From the figure, it can be observed that the regions with higher HWN are in eastern China, including North-Central China, East China, Hainan Province, and the northwestern Xinjiang region. In these areas, the number of heatwave events typically ranges from 2 to 3.5 per year. In contrast, the number of heatwave events is lower in the Qinghai Province and the eastern Tibetan Plateau, where it occurs only 0.5 to 1 time per year. This indicates that heatwaves occur more frequently in eastern and northwestern monsoon regions, while the western plateau and mountainous areas are less frequently affected by heatwaves. In most areas nationwide, the number of heatwave events ranges from 1 to 2 per year. Figure 4 presents the distribution of HWF. The data show that the regions with the highest average HWF per year are still concentrated in the eastern regions, particularly East China and Hainan Province, where the annual average days range from 12.5 to 14.5 days. This indicates that these regions not only experience a high HWF but also have prolonged durations. In comparison, the southwest region has a lower HWF per year, typically ranging from 6.75 to 9 days. The eastern part of the northwest region generally experiences heatwaves for 6.75 to 11.25 days per year, while the Tibetan Plateau has a wider range of HWF, fluctuating between 9 and 13.5 days per year. Notably, regions experiencing two or more heatwave events annually generally have an average HWF exceeding 13.5, such as Zhejiang Province and Hainan Province, where the average HWF can even reach 15.75, reflecting the persistence and frequency of extreme heat events in these regions. Analysis of HWN and HWF indicates that eastern China, northern central China, Hainan, and Xinjiang are high-frequency heatwave regions, experiencing 2 to 3.5 events per year. In contrast, the western plateau regions show the lowest frequency (0.5–1 event per year). HWF in eastern China and Hainan reaches 12.5–15.75 days per year, significantly higher than in other regions. Notably, although the HWN over the Tibetan Plateau is low, certain areas still exhibit HWF values of 9–13.5 days per year, suggesting that individual heatwave events there may last longer.

From Figure 5, the spatial distribution of HWD can be observed. The Tibetan Plateau and southern China, including Yunnan, Guangdong, and Guangxi, generally have shorter average HWD, with most areas being below 8.5 days, indicating that these regions experience fewer prolonged heatwave events. In stark contrast, the Tarim Basin in northwest China and the Sichuan and Chongqing regions in the west have generally longer heatwave durations, with most areas ranging between 10.5 and 14.5 days. Notably, the spatial distribution of HWF and HWD shows certain similarities: low-value regions are concentrated in Yunnan and parts of North China, while high-value areas are mainly found in the eastern monsoon regions, particularly along the southeastern coast and in the Sichuan and Chongqing areas in the west. Figure 6 displays the spatial distribution of HWI, which shows a clear latitudinal variation. Specifically, the HWI is relatively low in the southern regions, indicating that heatwave events in these areas are milder. In contrast, the HWI is higher in the northern regions, reflecting that heatwaves in these areas are more intense.

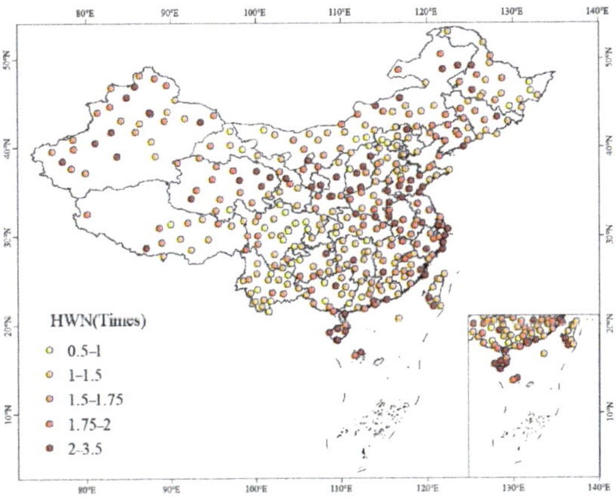

Figure 3. Annual average number of heatwave events (HWN) at meteorological stations in China from July to August 2013–2023.

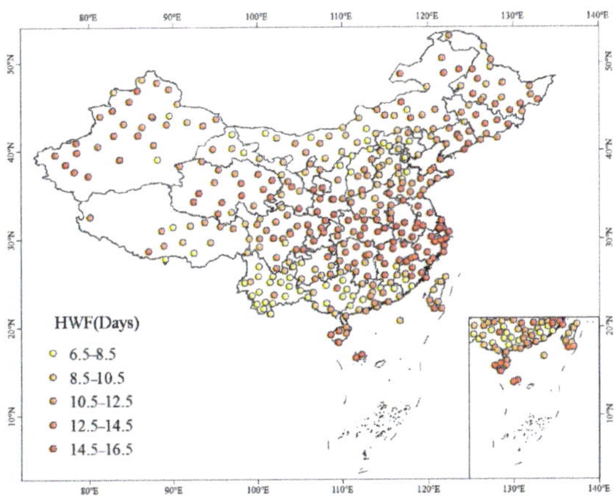

Figure 4. Annual average frequency of heatwave events (HWF) at meteorological stations in China from July to August 2013–2023.

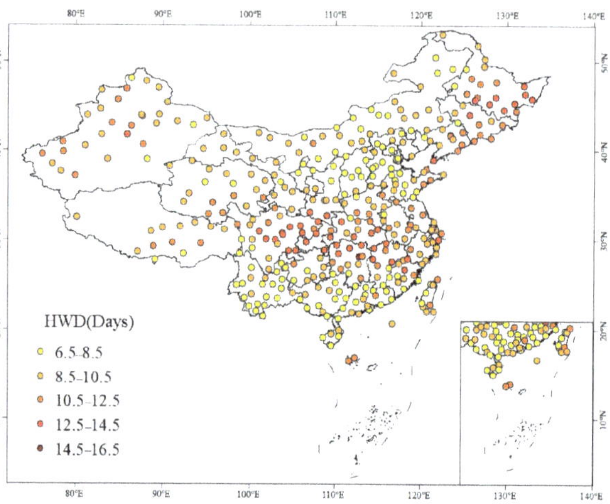

Figure 5. Annual average duration of heatwave events (HWD) at meteorological stations in China from July to August 2013–2023.

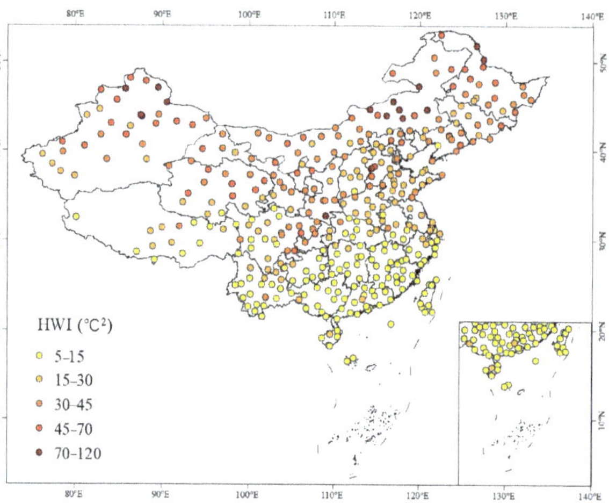

Figure 6. Annual average intensity of heatwave events (HWI) at meteorological stations in China from July to August 2013–2023.

Combined analysis of HWF, HWD, HWN, and HWI reveals significant spatial heterogeneity in heatwave events across China, with distinct regional patterns: (1) Eastern China, northern Central China, and Hainan show the highest heatwave frequency (HWN: 2–3.5 events/year) and total days (HWF: 12.5–15.75 days/year), along with relatively long durations (HWD: 10.5–14.5 days), forming "high-frequency–long-duration" risk zones. These may be linked to monsoon stagnation and intensified urban heat island effects. (2) Northwestern arid regions (e.g., the Tarim Basin in Xinjiang) and the Sichuan Basin are characterized by strong heatwaves (high HWI) and extended durations (HWD: 10.5–14.5 days), with moderate to high numbers (HWN: 2–3.5 events/year), displaying a "long-duration–high-intensity" pattern. This may result from the combined effects of arid climates and basin topography. (3) The Tibetan Plateau experiences the weakest heatwave activity overall (HWN: 0.5–1 event/year, HWD < 8.5 days), yet some valley areas (e.g., the

Yarlung Tsangpo River valley) can accumulate up to 9–13.5 heatwave days per year. This suggests that strong solar radiation and local topography may prolong individual heatwave events. (4) Southern coastal regions (e.g., Guangdong, Guangxi, Yunnan) exhibit a "high-frequency–short-duration–low-intensity" pattern, with higher event numbers (HWN: 1–2 events/year), short durations (HWD < 8.5 days), and low intensity, largely influenced by maritime regulation.

Additionally, heatwave intensity in northern China shows a marked latitudinal gradient, generally higher than in the south, likely due to larger diurnal temperature ranges and stronger radiation under a continental climate. Overall, China's heatwave risk types demonstrate clear east–west divergence (frequency and duration dominate in the east; intensity in the northwest) and north–south contrast (higher intensity in the north; lower in the south).

3.3. Relationship Between Heatwaves, Atmospheric Circulation, and Soil Moisture

This study also selected several typical regions as key research areas, including the Yangtze River Delta, Fujian, Guangdong, Guangxi, Hainan, and the Tibetan Plateau (see Figure 7).

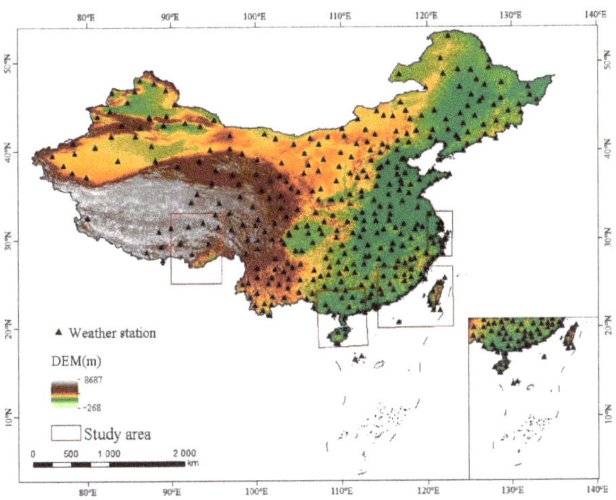

Figure 7. Selected typical research areas for studying the driving factors of heatwaves.

3.3.1. Atmospheric Circulation and Heatwaves

Atmospheric circulation significantly affects heatwaves in the Yangtze River Delta. As shown in Figure 8, during 18–25 July 2017, fluctuations in the 580 hPa level led to surface warming, enhancing boundary layer instability, which in turn triggered a strengthened circulation system. From July 22 to 25, the increase in surface air temperature amplified the upward movement of the lower atmosphere, promoting the formation and development of convective clouds, which affected the water vapor cycle and energy transfer in the atmosphere, creating localized circulation patterns. Additionally, the rise in surface air temperature also influenced the wind field structure in the atmosphere, driving horizontal atmospheric movement and reshaping the atmospheric circulation. During heatwaves, surface air temperatures exhibit a significant increasing trend, and these areas are often marked in red or dark red, representing extremely high air temperature. The formation of these high-air temperature areas is usually the result of intense solar radiation, the characteristics of surface cover, and local climatic conditions working together.

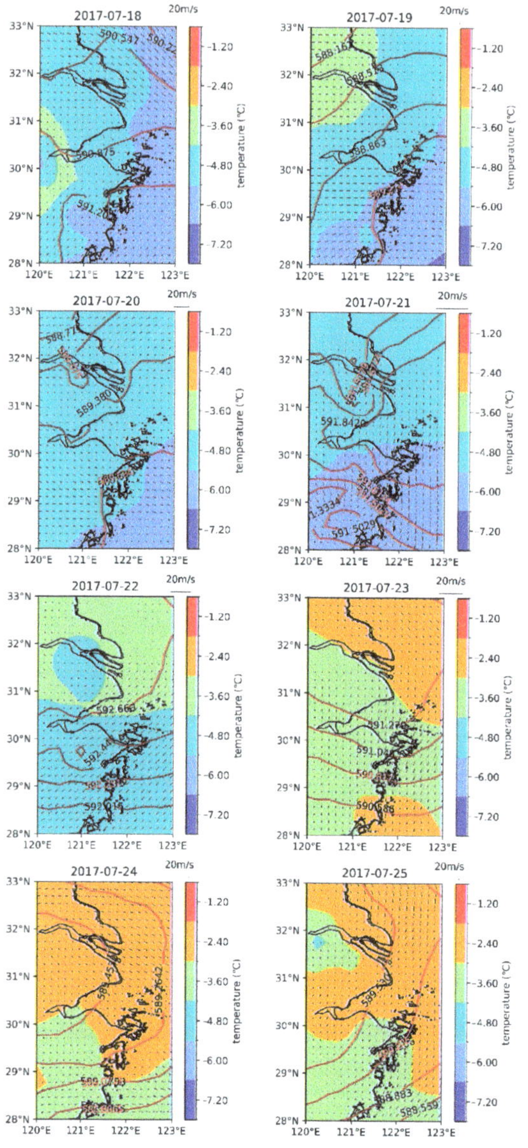

Figure 8. Schematic of atmospheric circulation in the Yangtze River Delta from 18 to 25 July 2017.

Anomalous changes in atmospheric circulation play a key role in sustaining and intensifying heatwaves, especially the flow patterns of the lower atmosphere. As shown in Figure 9, during the heatwave in Fujian from 22 to 29 July 2022, the increase in surface air temperature caused the lower atmosphere to expand due to heating, generating upward movement (labeled as "Upward Airflow"). This upward airflow facilitated the formation and development of convective clouds, further influencing the water vapor cycle and energy transfer in the atmosphere. At the same time, the wind field structure in the atmosphere also changed, and the horizontal pressure gradient force drove the atmosphere to produce horizontal movement (labeled as "Horizontal Airflow"), forming a specific circulation pattern that influenced the formation of sustained high air temperature in the region. In the case illustrated by Figure 8, fluctuations at the 580 hPa level triggered localized atmospheric circulation anomalies, significantly enhancing boundary layer instability. This process promoted surface temperature rise and the development of localized extreme heat events. In contrast, the case in Figure 9 reflects a different mechanism: intense surface heating induced low-level upward motion and horizontal airflow changes, which subsequently affected convective cloud formation and moisture cycling, thereby sustaining and intensifying the heatwave. The first case highlights how circulation anomalies can initiate heatwaves through boundary layer processes (580 hPa fluctuation → enhanced turbulence → increased sensible heat flux), while the second case demonstrates a surface–atmosphere positive feedback mechanism that maintains the heatwave (high temperature → suppressed convection → increased radiative heating).

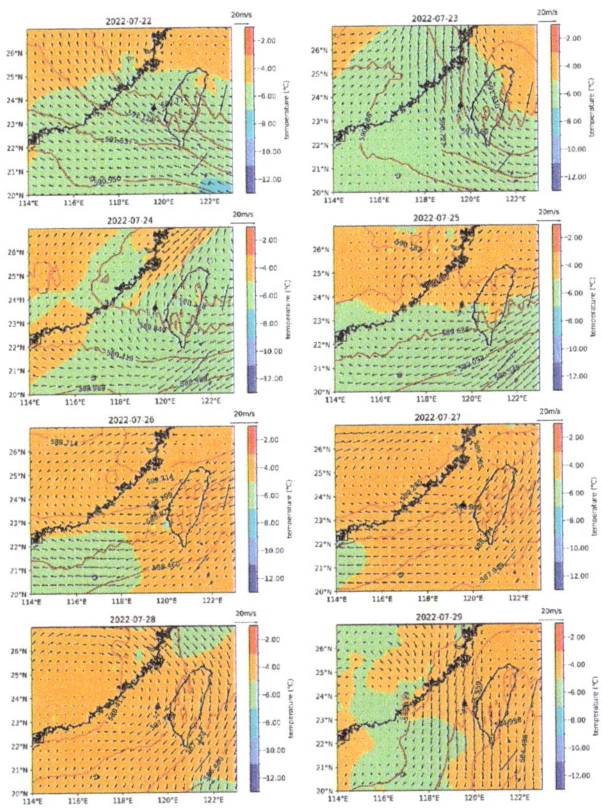

Figure 9. Schematic of atmospheric circulation in Fujian from 22 to 29 July 2022.

The area shown in Figure 10 is located at the junction of Guangdong, Guangxi, and Hainan, with a focus on the distribution and variation of the subtropical high-pressure system. The subtropical high-pressure system is one of the main weather systems affecting many regions during the summer. During the heatwave in this region from 24 to 31 July 2023, the subtropical high-pressure system anomalously intensified and remained in control of the area for a long period. This anomalous intensification led to clear and mostly cloudless weather, allowing solar radiation to reach the surface directly, which further exacerbated the rise in surface air temperature. Meanwhile, the strengthening of the subtropical high-pressure system also suppressed the upward movement of the lower atmosphere, reducing cloud cover and water vapor condensation, thereby intensifying the strength and duration of the heatwave.

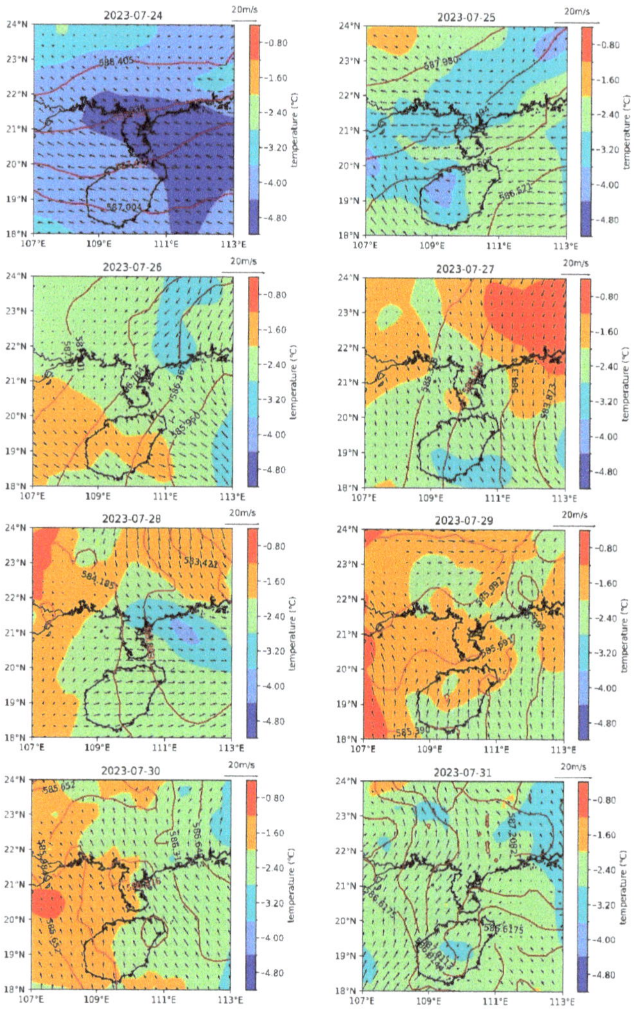

Figure 10. Schematic of atmospheric circulation in Guangdong, Guangxi, and Hainan from 24 to 31 July 2023.

From 10 to 18 August 2022, the Tibetan Plateau experienced an unusual heatwave event (Figure 11). While atmospheric circulation is a key factor in climate systems and affects regional air temperature distribution through the transport of warm and moist air, its role in this particular heatwave event was not significant. Instead, this heatwave event was closely tied to the geographic location and topographic characteristics of the Tibetan Plateau. The high elevation and diverse terrain of the Tibetan Plateau create a unique and complex climate system, where surface energy balance, topography, and local circulation interact to influence air temperature changes. This interplay causes energy to accumulate in localized areas, leading to the formation of localized heat islands and local circulation, which, to some extent, affects the air temperature distribution. The balance between surface radiation absorption and heat release directly influences air temperature variation. Moreover, the undulating topography and diverse landforms regulate air movement patterns, thereby affecting the formation and intensity of local circulations. Therefore, to fully understand this heatwave event, it is crucial to consider not only atmospheric circulation but also other factors such as surface energy balance, topographical features, and local circulation dynamics.

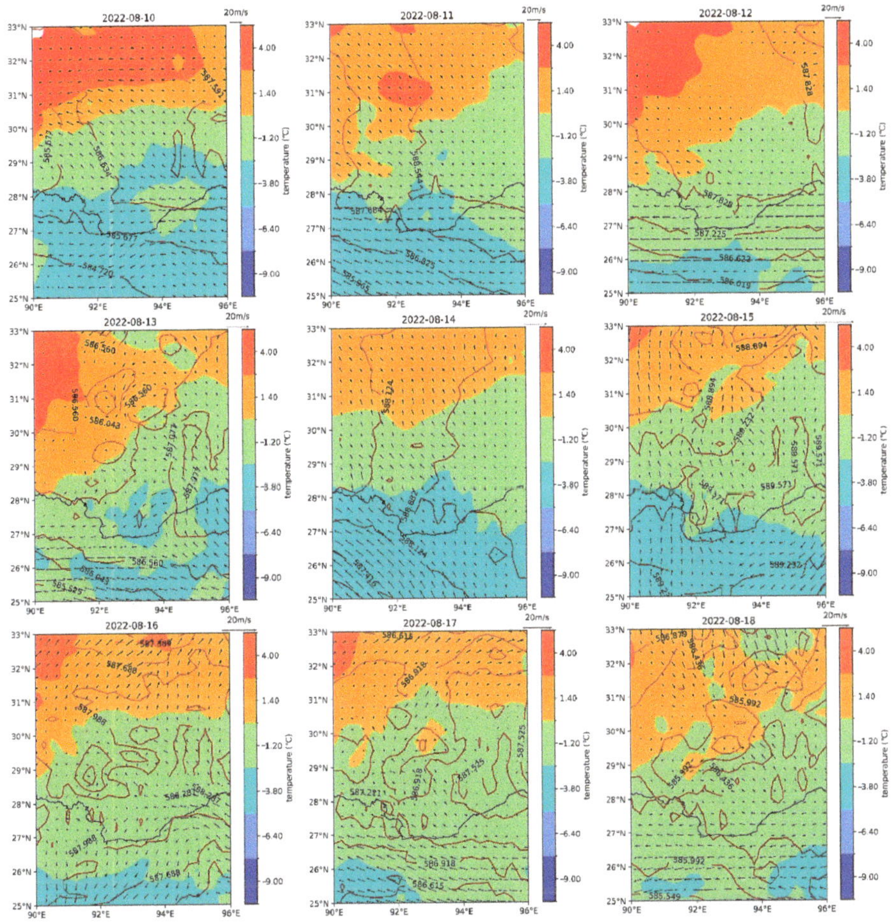

Figure 11. Schematic of atmospheric circulation in the Tibetan Plateau from 10 to 18 August 2022.

In the analysis of atmospheric circulation, we particularly focus on the subtropical high-pressure belt near 30° N. This high-pressure system plays a crucial role in the global climate, and when it anomalously strengthens and advances northward, its influence expands to more regions. When such an enhanced high-pressure system continuously controls a region, the area experiences prolonged subsiding airflow, causing the air to undergo adiabatic warming. This downward airflow continuously transports heat from the upper atmosphere to the surface, causing rapid air temperature increases. The extent and speed of this air temperature rise are often closely linked to the intensity and duration of the heatwave. Particularly when the abnormal high-pressure system combines with specific circulation patterns, it can form a "heat dome". In this climatic pattern, the high-pressure system acts as a lid, tightly trapping hot air in the high-air temperature area and preventing the entry of cooler air. This closed state causes air temperature within the heat dome to continue rising, creating an extreme heat zone. Therefore, when predicting and responding to heatwave events, it is essential to closely monitor changes in the subtropical high-pressure belt in atmospheric circulation maps and its interactions with other circulation systems. This approach will enable more accurate predictions of heatwave occurrence and evolution.

3.3.2. The Role of Soil Moisture in Driving Heatwaves

Soil moisture is a key factor in regulating energy and water exchange between the atmosphere and the land surface. While it has a noticeable but relatively limited influence on the driving mechanisms of heatwaves in certain regions, its impact remains secondary compared to atmospheric circulation. During heatwave periods, normal soil moisture (represented in red) helps stabilize surface air temperature, whereas abnormally low soil moisture (represented in blue) can indicate a rapid increase in land surface air temperature. As shown in Figure 12, in Guangdong, Guangxi, and Hainan, soil moisture exhibited a declining trend from 24 to 28 July 2023 due to intense evaporation caused by the heatwave and insufficient precipitation. This reduction in soil moisture may have further intensified surface air temperature increases, creating favorable conditions for heatwave formation. A similar trend was observed in the Yangtze River Delta (Figure 13) and Fujian (Figure 14) regions, where soil moisture also decreased during the heatwave, driven by mechanisms comparable to those in Guangdong, Guangxi, and Hainan.

Although soil moisture varies during heatwave events, comprehensive analysis suggests that it primarily acts as a secondary factor. The dominant driver of heatwaves is changes in atmospheric circulation patterns, particularly the intensification and persistence of the subtropical high. These circulation patterns enhance subsidence, leading to clear, dry, and hot weather conditions. Soil moisture influences heatwaves indirectly by affecting the surface energy balance and hydrological cycle; low soil moisture reduces vegetation transpiration, exacerbating surface warming. However, compared to atmospheric circulation, its impact is relatively minor and is further constrained by factors such as vegetation cover and soil type. Additionally, soil moisture variations influence atmospheric moisture content: high humidity promotes water vapor evaporation, releasing latent heat that further warms the atmosphere, whereas low humidity increases boundary-layer instability, promoting convective cloud development, which indirectly affects heatwave intensity and spread. In summary, soil moisture plays an auxiliary role in influencing surface air temperature, water vapor dynamics, and circulation patterns in coastal heatwave events, while atmospheric circulation remains the dominant factor.

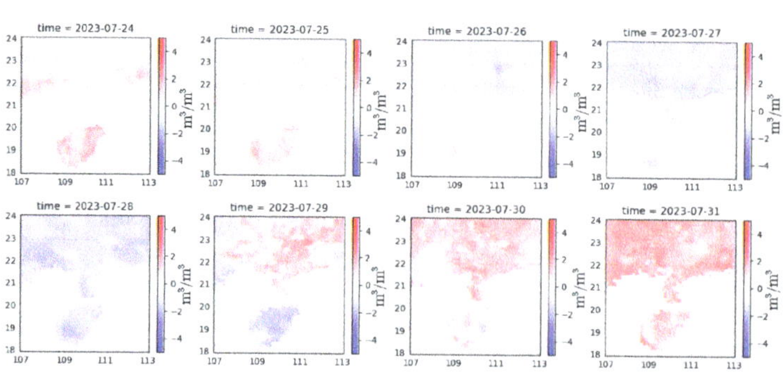

Figure 12. Anomalies in soil moisture in Guangdong, Guangxi, and Hainan during 24–31 July 2023.

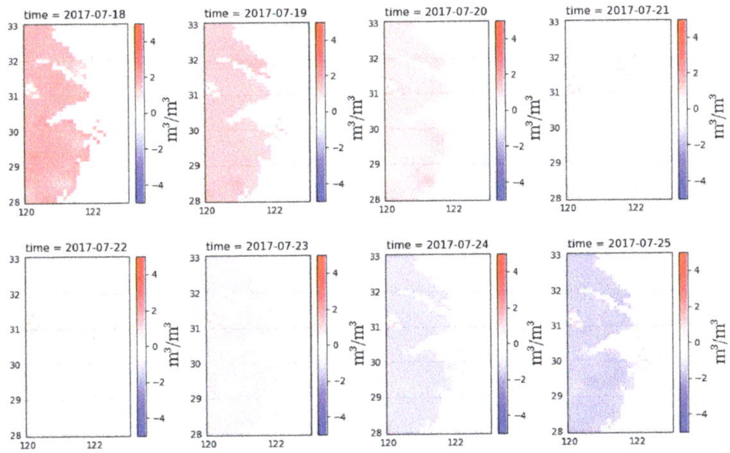

Figure 13. Anomalies in soil moisture in the Yangtze River Delta during 18–25 July 2017.

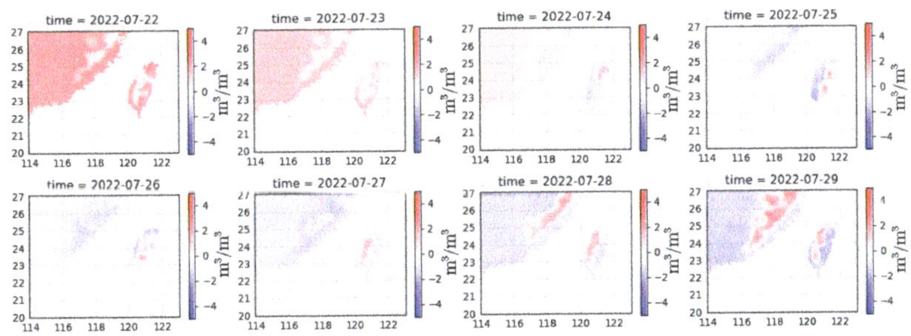

Figure 14. Anomalies in soil moisture in Fujian during 22–29 July 2022.

A detailed analysis of soil moisture data from the Tibetan Plateau during 10–18 August 2022 (Figure 15) requires consideration of its unique high-altitude and low-pressure climate. First, the high elevation of the plateau results in generally lower air temperatures, even during summer, making it significantly cooler than coastal regions. These cooler conditions slow down soil moisture evaporation, helping to maintain relatively stable soil moisture levels. Additionally, August is the peak of the rainy season, providing an abundant source of moisture to the soil and further sustaining its stability. However, frequent heatwaves can accelerate soil moisture loss, potentially leading to a decline in soil moisture. Nevertheless, due to the region's relatively lower air temperature, soil moisture evaporation does not occur as rapidly as in coastal areas. As a result, despite fluctuations, soil moisture levels in the Tibetan Plateau remain more stable compared to those in Guangdong, Guangxi, Hainan, the Yangtze River Delta, and Fujian, where intense solar radiation and higher air temperature cause greater soil moisture variability.

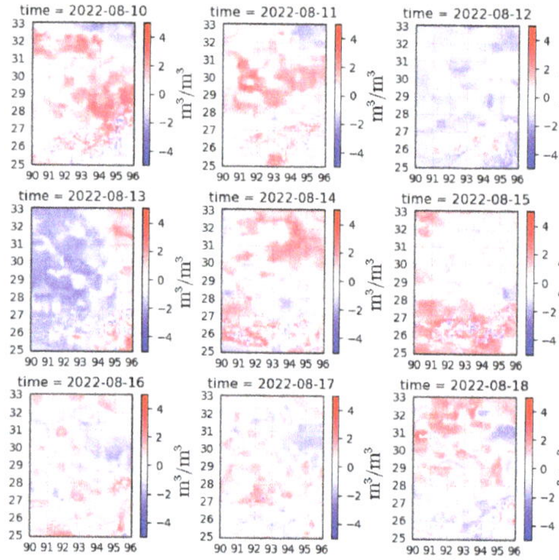

Figure 15. Anomalies in soil moisture in the Tibetan Plateau during 10–18 August 2022.

4. Discussion

Through a systematic analysis of the spatial and temporal distribution characteristics and driving mechanisms of heatwaves in China, this study elucidates the interaction between regional climate responses and extreme events in the context of global warming [85,86]. The study finds that the regional differences in heatwave distribution across China are closely related to the heterogeneous responses of regional climate systems, a pattern that aligns with the broader impacts of climate change [27]. Ji et al. [27] reported a higher frequency of heatwaves in the Yangtze River Basin and longer durations of heatwaves in Xinjiang from 1961 to 2020, and they noted that this trend is consistent with the conclusions drawn in this study. Similarly, Wu et al., 2023 [87] pointed out that the frequency of heatwaves was higher in the northwest, eastern, and central regions of China during the period of 1990 to 2019, which aligns with our findings. However, discrepancies exist in the Tibetan Plateau. Our study shows that heatwave events are relatively rare in this region, while Wu et al., 2023 [87] observed a very high frequency of heatwaves in the southeastern Tibetan region. This may be due to differences in definitions or data resolution. Specifically, these discrepancies may arise from differences in the heatwave identification method and definitions used in our study.

In terms of regional differences in heatwave occurrence, the eastern coastal areas and the Yangtze River Basin experience a higher frequency of heatwaves, primarily related to the anomalous strengthening of the subtropical high-pressure system. This is consistent with earlier studies linking climate circulation anomalies to extreme heat events [88–90]. In contrast, the arid northwest regions, such as Xinjiang, experience longer-lasting heatwaves, which may be associated with uneven surface energy distribution due to low soil moisture. This region appears to be trapped in a vicious cycle of 'dry soil $\rightarrow$ rising temperatures $\rightarrow$ increased evaporation $\rightarrow$ further soil drying', thereby intensifying the positive feedback mechanism between the soil and atmosphere [91,92]. These spatial differences suggest that heatwave events are not driven by a single factor but are the result of interactions among multiple meteorological driving factors [93].

The westward extension and strengthening of the subtropical high-pressure system, along with the subsidence process, have become the primary driving factors for the frequent heatwaves in eastern China. Key mechanisms include the suppression of convection and the enhancement of surface net radiation, which in turn increases temperatures [94]. This mechanism aligns with the circulation-dominated heatwave theory framework [95] and has been validated by the 2022 heatwave event in southern China. Studies show that the anomalous westward extension of the Northwest Pacific subtropical high led to widespread warming [90]. In contrast, the regulating effect of soil moisture on heatwaves varies by region: in the humid southern areas, evaporative cooling mitigates the high-temperature phenomenon [96], while in arid regions, the enhanced energy exchange between the surface and atmosphere exacerbates the heat effect [97,98]. In the Tibetan Plateau, due to the lower atmospheric energy transfer efficiency at high altitudes, heatwave events are rare and short-lived. The strong radiative cooling effect (manifested by rapid nighttime temperature drops) and frequent valley wind circulation accelerate heat dissipation, thus suppressing the persistence of high temperatures [27,99].

Although this study provides important insights into the spatiotemporal distribution and driving mechanisms of heatwaves, there are still certain limitations. First, the 11-year data period from 2013 to 2023 limits the analysis of long-term trends, and may not fully capture the impacts of climate events such as El Niño or the Atlantic Multidecadal Oscillation (AMO). As noted by Wei et al. [40], these climate events play a significant role in intensifying heatwaves. Climate research generally suggests using data periods of over 30 years to derive robust trends, so our results may reflect short-term anomalies rather than long-term climate patterns. Second, the applicability of ERA5 data in complex terrains (such as the Tibetan Plateau) warrants further examination. Zhao et al. [100] pointed out that ERA5 underestimates precipitable water vapor estimates, which may affect the representation of heatwaves in high-altitude areas. Moreover, the EHF index does not account for humidity, which may underestimate the heat stress of humid-heat compound events, particularly in southern regions where high humidity exacerbates heat stress [101]. This contrasts with the findings of Cheng et al. [98], who emphasized the important role of humidity in southern heatwaves, suggesting that our temperature-based index may overlook the critical impact of humidity on heatwaves.

Future research could address the short-term data limitation by extending the observation period to over 30 years and using CMIP6 models for multi-scenario simulations (SSP1-2.6 to SSP5-8.5) to explore the nonlinear changes in heatwave responses to rising greenhouse gas concentrations [102]. Additionally, combining Landsat 8 surface temperature data (30 m resolution) with drone-based thermal infrared remote sensing could improve spatial accuracy in complex terrains [103]. Developing multi-physics coupled models that integrate temperature, humidity, and wind speed with human heat tolerance thresholds would help establish health risk-oriented heatwave warning systems, enhancing their practical utility. Interdisciplinary collaborations, such as epidemiological studies quantifying the heatwave mortality dose–response relationship [101], or agricultural models assessing crop yield risks [104], would contribute to enhancing the scientific rigor and practical effectiveness of heatwave response strategies.

5. Conclusions

This study is based on observational data from 388 meteorological stations in China from 2013 to 2023. Using EHF, HWN, HWF, HWD, and HWI indices, it systematically reveals the spatial distribution patterns and characteristics of heatwaves. Furthermore, the study explores the driving mechanisms of atmospheric circulation and soil moisture using ERA5 reanalysis data. The results show that: (1) In terms of spatial distribution, the eastern coastal regions and the Yangtze River Basin, dominated by the subtropical high, experience frequent heatwaves, with the number of heatwave days generally ranging from 8.5 to 12.5 days per year, averaging 2 to 3.5 days. However, the intensity of these heatwaves is lower than that in the northern and northwestern regions, with intensities ranging from 5 to 15 °C^2. In the northwest inland areas (e.g., the Tarim Basin and the Hexi Corridor), due to strong evaporation and dry soil, the duration of heatwaves is the longest, lasting 10.5 to 16.5 days, with higher intensities ranging from 30 to 120 °C^2. On the Tibetan Plateau's low-elevation river valleys (e.g., the Yarlung Tsangpo River Valley), driven by local circulations, the occurrence of heatwave events is relatively low, averaging 0.5 to 1.75 times per year, with short durations (6.5 to 10.5 days) and low intensities (5 to 30 °C^2). (2) A driving mechanism, atmospheric circulation, particularly the dynamic variations of the subtropical high, plays a dominant role in the formation and persistence of heatwaves. The abnormal intensification of the subtropical high leads to prevailing subsidence airflow, suppressing cloud and precipitation formation while enhancing surface radiative heating, thereby contributing to the heat dome effect. This significantly prolongs the duration and intensifies the severity of heatwaves. Additionally, soil moisture indirectly influences heatwave development by regulating surface energy balance and water cycles, exerting a modulating effect on heatwave formation and progression. (3) This study has certain limitations in terms of data and methodology. First, the wind field and soil moisture data in the ERA5 reanalysis dataset used in this study lack validation against ground-based meteorological observations, which may affect the regional applicability and accuracy of the data. Second, the heatwave index (EHF) used in the study is based solely on air temperature, without incorporating key meteorological variables such as relative humidity, which results in a lack of comprehensiveness in the heat stress assessment. These limitations may, to some extent, impact the reliability of the study's conclusions. (4) Regarding regional adaptation strategies, for the eastern urban clusters, enhancing heat island mitigation technologies (e.g., high-albedo materials) and health warning systems is recommended. In arid northwestern regions, optimizing irrigation strategies to improve soil moisture retention is crucial. Meanwhile, on the Tibetan Plateau, establishing an eco-adaptive management framework based on local circulation characteristics is necessary to mitigate heatwave impacts.

Author Contributions: Conceptualization, J.L.; methodology, J.L.; software, J.L.; validation, J.L.; formal analysis, J.L.; investigation, J.L.; resources, J.L.; data curation, M.L.; writing—original draft preparation, J.L. and M.L.; writing—review and editing, J.L. and M.L.; visualization, J.L.; supervision, J.L.; project administration, J.L.; funding acquisition, J.L. All authors have read and agreed to the published version of the manuscript.

Funding: This research was jointly funded by the Key Laboratory of Mine Spatio-Temporal Information and Ecological Restoration, MNR (Grant No. KLM202301), Henan Provincial Science and Technology Research (Grant No. 242102320017), and Henan Province Joint Fund Project of Science and Technology (Grant No. 222103810097).

Institutional Review Board Statement: Not applicable.

Informed Consent Statement: Not applicable.

Data Availability Statement: The raw data supporting the conclusions of this article will be made available by the authors on request.

Conflicts of Interest: The authors declare no conflicts of interest.

References

1. *IPCC Climate Change 2023: Synthesis Report. Contribution of Working Groups I, II and III to the Sixth Assessment Report of the Intergovernmental Panel on Climate Change*; The Australian National University: Canberra, Australia, 2023; ISBN 92-9169-164-X.
2. Perkins-Kirkpatrick, S.E.; Gibson, P.B. Changes in Regional Heatwave Characteristics as a Function of Increasing Global Temperature. *Sci. Rep.* **2017**, *7*, 12256.

3. Dosio, A.; Mentaschi, L.; Fischer, E.M.; Wyser, K. Extreme Heat Waves Under 1.5 °C and 2 °C Global Warming. *Environ. Res. Lett.* **2018**, *13*, 54006. [CrossRef]
4. Perkins-Kirkpatrick, S.E.; Lewis, S.C. Increasing Trends in Regional Heatwaves. *Nat. Commun.* **2020**, *11*, 3357. [CrossRef]
5. García-Herrera, R.; Díaz, J.; Trigo, R.M.; Luterbacher, J.; Fischer, E.M. A Review of the European Summer Heat Wave of 2003. *Crit. Rev. Environ. Sci. Technol.* **2010**, *40*, 267–306. [CrossRef]
6. Monteiro, A.; Carvalho, V.; Oliveira, T.; Sousa, C. Excess Mortality and Morbidity During the July 2006 Heat Wave in Porto, Portugal. *Int. J. Biometeorol.* **2013**, *57*, 155–167. [CrossRef]
7. Rebetez, M.; Dupont, O.; Giroud, M. An Analysis of the July 2006 Heatwave Extent in Europe Compared to the Record Year of 2003. *Theor. Appl. Climatol.* **2009**, *95*, 1–7. [CrossRef]
8. Lee, S.-M.; Min, S.-K. Heat Stress Changes over East Asia Under 1.5 and 2.0 °C Global Warming Targets. *J. Clim.* **2018**, *31*, 2819–2831. [CrossRef]
9. Hoy, A.; Hänsel, S.; Skalak, P.; Ustrnul, Z.; Bochníček, O. The Extreme European Summer of 2015 in a Long-Term Perspective. *Int. J. Climatol.* **2017**, *37*, 943–962. [CrossRef]
10. Krzyżewska, A.; Dyer, J. The August 2015 Mega-Heatwave in Poland in the Context of Past Events. *Weather* **2018**, *73*, 207–214. [CrossRef]
11. Song, X.; Zhang, Z.; Chen, Y.; Wang, P.; Xiang, M.; Shi, P.; Tao, F. Spatiotemporal Changes of Global Extreme Temperature Events (ETEs) since 1981 and the Meteorological Causes. *Nat. Hazard.* **2014**, *70*, 975–994. [CrossRef]
12. Grumm, R.H. The Central European and Russian Heat Event of July–August 2010. *Bull. Am. Meteorol. Soc.* **2011**, *92*, 1285–1296. [CrossRef]
13. Barriopedro, D.; Fischer, E.M.; Luterbacher, J.; Trigo, R.M.; García-Herrera, R. The Hot Summer of 2010: Redrawing the Temperature Record Map of Europe. *Science* **2011**, *332*, 220–224. [CrossRef]
14. Mishra, V.; Mukherjee, S.; Kumar, R.; Stone, D.A. Heat Wave Exposure in India in Current, 1.5 °C, and 2.0 °C Worlds. *Environ. Res. Lett.* **2017**, *12*, 124012. [CrossRef]
15. Suarez-Gutierrez, L.; Müller, W.A.; Li, C.; Marotzke, J. Hotspots of Extreme Heat Under Global Warming. *Clim. Dyn.* **2020**, *55*, 429–447. [CrossRef]
16. Vogel, M.M.; Zscheischler, J.; Fischer, E.M.; Seneviratne, S.I. Development of Future Heatwaves for Different Hazard Thresholds. *J. Geophys. Res. Atmos.* **2020**, *125*, e2019JD032070. [CrossRef]
17. Liu, J.; Ren, Y.; Tao, H.; Liu, T.; Chen, H. The response of heatwave to carbon emission in China. *China Environ. Sci.* **2021**, *42*, 415–424.
18. Jiang, J.; Liu, Y.; Mao, J.; Wu, G. Extreme Heatwave over Eastern China in Summer 2022: The Role of Three Oceans and Local Soil Moisture Feedback. *Environ. Res. Lett.* **2023**, *18*, 44025. [CrossRef]
19. Qian, C.; Ye, Y.; Jiang, J.; Zhong, Y.; Zhang, Y.; Pinto, I.; Huang, C.; Li, S.; Wei, K. Rapid Attribution of the Record-Breaking Heatwave Event in North China in June 2023 and Future Risks. *Environ. Res. Lett.* **2024**, *19*, 14028. [CrossRef]
20. Xu, W.; Yuan, W.; Wu, D.; Zhang, Y.; Shen, R.; Xia, X.; Ciais, P.; Liu, J. Impacts of Record-Breaking Compound Heatwave and Drought Events in 2022 China on Vegetation Growth. *Agric. For. Meteorol.* **2024**, *344*, 109799. [CrossRef]
21. Wu, X.; Jiang, D.; Zhang, F. Increasing Impact of Compound Agricultural Drought and Hot Events on Maize Yield in China. *Clim. Res.* **2023**, *90*, 17–29.
22. Zhao, L.; Li, X.; Zhang, Z.; Yuan, M.; Sun, S.; Qu, S.; Hou, M.; Lu, D.; Zhou, Y.; Lin, A. Developing a Novel Framework to Re-Examine Half a Century of Compound Drought and Heatwave Events in Mainland China. *Sci. Total Environ.* **2023**, *874*, 162366. [CrossRef] [PubMed]
23. Shi, P.; Li, Y.; Biswas, A.; Wei, K.; Hou, M. Spatial-Temporal Evolution and Intrinsic Drivers of Compound Drought and Heatwave Events in Mainland China. *Sci. Total Environ.* **2024**, *948*, 174834. [CrossRef]
24. Chen, H.; Zhao, L.; Cheng, L.; Zhang, Y.; Wang, H.; Gu, K.; Bao, J.; Yang, J.; Liu, Z.; Huang, J. Projections of Heatwave-Attributable Mortality Under Climate Change and Future Population Scenarios in China. *Lancet Reg. Health—West. Pac.* **2022**, *28*, 100582. [CrossRef]
25. Chen, Y.; Liao, Z.; Shi, Y.; Tian, Y.; Zhai, P. Detectable Increases in Sequential Flood-Heatwave Events across China during 1961–2018. *Geophys. Res. Lett.* **2021**, *48*, e2021GL092549. [CrossRef]
26. Wang, P.; Tang, J.; Wang, S.; Dong, X.; Fang, J. Regional Heatwaves in China: A Cluster Analysis. *Clim. Dyn.* **2018**, *50*, 1901–1917.
27. Ji, H.; Feng, A.; Zhao, Y.; Liao, J.; Zhang, Z.; Gu, C.; Feng, A. Characteristics of Heat Waves in Mainland China since 1961 Based on Absolute and Relative Methods. *Atmosphere* **2023**, *14*, 544. [CrossRef]
28. Mukherjee, S.; Ashfaq, M.; Mishra, A.K. Compound Drought and Heatwaves at a Global Scale: The Role of Natural Climate Variability-Associated Synoptic Patterns and Land-Surface Energy Budget Anomalies. *J. Geophys. Res. Atmos.* **2020**, *125*, e2019JD031943. [CrossRef]

29. Yang, J.; Yin, P.; Sun, J.; Wang, B.; Zhou, M.; Li, M.; Tong, S.; Meng, B.; Guo, Y.; Liu, Q. Heatwave and Mortality in 31 Major Chinese Cities: Definition, Vulnerability and Implications. *Sci. Total Environ.* **2019**, *649*, 695–702. [CrossRef]
30. Luo, M.; Wu, S.; Liu, Z.; Lau, N. Contrasting Circulation Patterns of Dry and Humid Heatwaves over Southern China. *Geophys. Res. Lett.* **2022**, *49*, e2022GL099243. [CrossRef]
31. Marghidan, C.P.; Van Aalst, M.; Blanford, J.; Guigma, K.; Pinto, I.; Maure, G.; Marrufo, T. Heatwaves in Mozambique 1983–2016: Characteristics, Trends and City-Level Summaries Using High-Resolution CHIRTS-Daily. *Weather Clim. Extrem.* **2023**, *40*, 100565.
32. Nairn, J.R.; Fawcett, R.J. The Excess Heat Factor: A Metric for Heatwave Intensity and Its Use in Classifying Heatwave Severity. *Int. J. Environ. Res. Public Health* **2015**, *12*, 227–253. [CrossRef] [PubMed]
33. Nairn, J.; Ostendorf, B.; Bi, P. Performance of Excess Heat Factor Severity as a Global Heatwave Health Impact Index. *Int. J. Environ. Res. Public Health* **2018**, *15*, 2494. [CrossRef]
34. Langlois, N.; Herbst, J.; Mason, K.; Nairn, J.; Byard, R.W. Using the Excess Heat Factor (EHF) to Predict the Risk of Heat Related Deaths. *J. Forensic Legal Med.* **2013**, *20*, 408–411. [CrossRef]
35. Jegasothy, E.; McGuire, R.; Nairn, J.; Fawcett, R.; Scalley, B. Extreme Climatic Conditions and Health Service Utilisation Across Rural and Metropolitan New South Wales. *Int. J. Biometeorol.* **2017**, *61*, 1359–1370. [CrossRef]
36. Nairn, J.R.; Fawcett, R.G. *Defining Heatwaves: Heatwave Defined as a Heat-Impact Event Servicing All Community and Business Sectors in Australia*; Centre for Australian Weather and Climate Research: Melbourne, Australia, 2013; ISBN 1-922173-11-8.
37. Hatvani-Kovacs, G.; Belusko, M.; Pockett, J.; Boland, J. Can the Excess Heat Factor Indicate Heatwave-Related Morbidity? A Case Study in Adelaide, South Australia. *Ecohealth* **2016**, *13*, 100–110. [CrossRef] [PubMed]
38. Zhou, Q.-Y.; Gao, M.-N.; Yang, J.; Sun, X.-Y.; Lu, Y.-Y.; Jiang, T.; Su, B.-D.; Zhu, T. Future Changes in Population Exposure to Intensified Heatwaves over Three Major Urban Agglomerations in China Based on Excess Heat Factor. *Adv. Clim. Chang. Res.* **2025**, *16*, 12–24. [CrossRef]
39. Perkins, S.E. A Review on the Scientific Understanding of Heatwaves—Their Measurement, Driving Mechanisms, and Changes at the Global Scale. *Atmos. Res.* **2015**, *164*, 242–267. [CrossRef]
40. Wei, J.; Han, W.; Wang, W.; Zhang, L.; Rajagopalan, B. Intensification of Heatwaves in China in Recent Decades: Roles of Climate Modes. *NPJ Clim. Atmos. Sci.* **2023**, *6*, 98. [CrossRef]
41. Luo, M.; Lau, N.-C.; Liu, Z. Different Mechanisms for Daytime, Nighttime, and Compound Heatwaves in Southern China. *Weather Clim. Extrem.* **2022**, *36*, 100449. [CrossRef]
42. Deng, K.; Yang, S.; Ting, M.; Zhao, P.; Wang, Z. Dominant Modes of China Summer Heat Waves Driven by Global Sea Surface Temperature and Atmospheric Internal Variability. *J. Clim.* **2019**, *32*, 3761–3775. [CrossRef]
43. Wei, J.; Wang, W.; Shao, Q.; Yu, Z.; Chen, Z.; Huang, Y.; Xing, W. Heat Wave Variations across China Tied to Global SST Modes. *J. Geophys. Res. Atmos.* **2020**, *125*, e2019JD031612. [CrossRef]
44. Pun, I.-F.; Hsu, H.-H.; Moon, I.-J.; Lin, I.; Jeong, J.-Y. Marine Heatwave as a Supercharger for the Strongest Typhoon in the East China Sea. *NPJ Clim. Atmos. Sci.* **2023**, *6*, 128. [CrossRef]
45. Liang, L.; Yu, L.; Wang, Z. Identifying the Dominant Impact Factors and Their Contributions to Heatwave Events over Mainland China. *Sci. Total Environ.* **2022**, *848*, 157527. [CrossRef] [PubMed]
46. Jiang, S.; Lee, X.; Wang, J.; Wang, K. Amplified Urban Heat Islands during Heat Wave Periods. *J. Geophys. Res. Atmos.* **2019**, *124*, 7797–7812. [CrossRef]
47. He, B.-J.; Wang, J.; Zhu, J.; Qi, J. Beating the Urban Heat: Situation, Background, Impacts and the Way Forward in China. *Renew. Sustain. Energy Rev.* **2022**, *161*, 112350. [CrossRef]
48. Shi, Z.; Jia, G. Changes in Urban Heat Island Intensity During Heatwaves in China. *Environ. Res. Lett.* **2024**, *19*, 74061.
49. Wu, S.; Luo, M.; Liu, Z.; Wang, X.; Huang, Z.; Li, X. Longer- and Slower-Moving Contiguous Heatwaves Linked to El Niño. *Geophys. Res. Lett.* **2024**, *51*, e2024GL109067. [CrossRef]
50. He, X.; Wang, J.; Feng, J.; Yan, Z.; Miao, S.; Zhang, Y.; Xia, J. Observational and Modeling Study of Interactions Between Urban Heat Island and Heatwave in Beijing. *J. Clean. Prod.* **2020**, *247*, 119169. [CrossRef]
51. Freychet, N.; Tett, S.; Wang, J.; Hegerl, G. Summer Heat Waves over Eastern China: Dynamical Processes and Trend Attribution. *Environ. Res. Lett.* **2017**, *12*, 24015. [CrossRef]
52. Wang, W.; Zhou, W.; Li, X.; Wang, X.; Wang, D. Synoptic-Scale Characteristics and Atmospheric Controls of Summer Heat Waves in China. *Clim. Dyn.* **2016**, *46*, 2923–2941. [CrossRef]
53. Wang, P.; Zhang, Q.; Yang, Y.; Tang, J. The Sensitivity to Initial Soil Moisture for Three Severe Cases of Heat Waves over Eastern China. *Front. Environ. Sci.* **2019**, *7*, 18. [CrossRef]
54. Miralles, D.G.; Gentine, P.; Seneviratne, S.I.; Teuling, A.J. Land–Atmospheric Feedbacks during Droughts and Heatwaves: State of the Science and Current Challenges. *Ann. N. Y. Acad. Sci.* **2019**, *1436*, 19–35. [CrossRef]

55. Han, R.; Li, Z.; Li, Z.; Han, Y. Spatial–Temporal Assessment of Historical and Future Meteorological Droughts in China. *Atmosphere* **2021**, *12*, 787. [CrossRef]
56. Liu, Y.; Tian, J.; Liu, R.; Ding, L. Influences of Climate Change and Human Activities on NDVI Changes in China. *Remote Sens.* **2021**, *13*, 4326. [CrossRef]
57. Lei, X.; Wang, Z.; Lin, G.; Lai, C. Response of Vegetation Dynamics to Drought at the Eco-Geographical Region Scale Across China. *Arabian J. Geosci.* **2021**, *14*, 2649. [CrossRef]
58. Wei, W.; Zhang, J.; Zhou, J.; Zhou, L.; Xie, B.; Li, C. Monitoring Drought Dynamics in China Using Optimized Meteorological Drought Index (OMDI) Based on Remote Sensing Data Sets. *J. Environ. Manag.* **2021**, *292*, 112733. [CrossRef]
59. Zheng, Z.; Lin, X.; Chen, L.; Yan, C.; Sun, T. Effects of Urbanization and Topography on Thermal Comfort During a Heat Wave Event: A Case Study of Fuzhou, China. *Sustain. Cities Soc.* **2024**, *102*, 105233. [CrossRef]
60. Yue, T.-X.; Zhao, N.; Ramsey, R.D.; Wang, C.-L.; Fan, Z.-M.; Chen, C.-F.; Lu, Y.-M.; Li, B.-L. Climate Change Trend in China, with Improved Accuracy. *Clim. Chang.* **2013**, *120*, 137–151. [CrossRef]
61. Song, M.; Wang, J.; Zhao, J. Effects of rising and extreme temperatures on production factor efficiency: Evidence from China's cities. *Int. J. Prod. Econ.* **2023**, *260*, 108847. [CrossRef]
62. Yang, X.; Zhou, B.; Xu, Y.; Han, Z. CMIP6 Evaluation and Projection of Temperature and Precipitation over China. *Adv. Atmos. Sci.* **2021**, *38*, 817–830. [CrossRef]
63. Liu, J.; Zhang, T.; Ren, Y.; Willems, P.; Mirchi, A.; Arshad, A.; Liu, T.; Pham, Q.B. Three-Dimensional Evaluation Framework of Hazard–Exposure–Vulnerability for Mapping Heatwave Risk and Associated Dominant Dimensions in China. *Int. J. Climatol.* **2024**, *44*, 5136–5155. [CrossRef]
64. General Administration of Quality Supervision, Inspection and Quarantine of the People's Republic of China; Standardization Administration of the People's Republic of China. GB/T 35221-2017 Specifications for Surface Meteorological Observation—General. Available online: https://www.cma.gov.cn/zfxxgk/gknr/flfgbz/bz/202209/t20220921_5099079.html (accessed on 5 April 2025).
65. Wu, Z.; Feng, H.; He, H.; Zhou, J.; Zhang, Y. Evaluation of Soil Moisture Climatology and Anomaly Components Derived from ERA5-Land and GLDAS-2.1 in China. *Water Resour. Manag.* **2021**, *35*, 629–643. [CrossRef]
66. Liu, H.; Dong, L.; Yan, R.; Zhang, X.; Guo, C.; Liang, S.; Tu, J.; Feng, X.; Wang, X. Evaluation of Near-Surface Wind Speed Climatology and Long-Term Trend over China's Mainland Region Based on ERA5 Reanalysis. *Clim. Environ. Res.* **2021**, *26*, 299–311.
67. Liu, Y.; Yang, Y. Detecting a Declining Trend of Multidepth Soil Moisture over the Mongolian Plateau from 1950 to 2020 Using ERA5-Land Reanalysis Datasets. *IEEE Access* **2022**, *10*, 95509–95526. [CrossRef]
68. You, Q.; Jiang, Z.; Kong, L.; Wu, Z.; Bao, Y.; Kang, S.; Pepin, N. A Comparison of Heat Wave Climatologies and Trends in China Based on Multiple Definitions. *Clim. Dyn.* **2017**, *48*, 3975–3989. [CrossRef]
69. Perkins, S.E.; Alexander, L.V. On the Measurement of Heat Waves. *J. Clim.* **2013**, *26*, 4500–4517. [CrossRef]
70. Jun, Y.; Liu, H.Z.; Ou, C.Q.; Lin, G.Z.; Yan, D.; Qin, Z.; Shen, J.C.; Chen, P.Y. Impact of Heat Wave in 2005 on Mortality in Guangzhou, China. *Biomed. Environ. Sci.* **2013**, *26*, 647–654.
71. Tong, S.; FitzGerald, G.; Wang, X.-Y.; Aitken, P.; Tippett, V.; Chen, D.; Wang, X.; Guo, Y. Exploration of the Health Risk-Based Definition for Heatwave: A Multi-City Study. *Environ. Res.* **2015**, *142*, 696–702. [CrossRef]
72. Xu, Z.; FitzGerald, G.; Guo, Y.; Jalaludin, B.; Tong, S. Impact of Heatwave on Mortality Under Different Heatwave Definitions: A Systematic Review and Meta-Analysis. *Environ. Int.* **2016**, *89*, 193–203. [CrossRef]
73. Xu, Z.; Cheng, J.; Hu, W.; Tong, S. Heatwave and Health Events: A Systematic Evaluation of Different Temperature Indicators, Heatwave Intensities and Durations. *Sci. Total Environ.* **2018**, *630*, 679–689. [CrossRef]
74. Tolika, K. Assessing Heat Waves over Greece Using the Excess Heat Factor (EHF). *Climate* **2019**, *7*, 9. [CrossRef]
75. Oliveira, A.; Lopes, A.; Soares, A. Excess Heat Factor Climatology, Trends, and Exposure across European Functional Urban Areas. Weather Clim. *Extremes* **2022**, *36*, 100455. [CrossRef]
76. Borg, M.; Nitschke, M.; Williams, S.; McDonald, S.; Nairn, J.; Bi, P. Using the Excess Heat Factor to Indicate Heatwave-Related Urinary Disease: A Case Study in Adelaide, South Australia. *Int. J. Biometeorol.* **2019**, *63*, 435–447. [CrossRef] [PubMed]
77. Perkins-Kirkpatrick, S.; White, C.; Alexander, L.; Argüeso, D.; Boschat, G.; Cowan, T.; Evans, J.; Ekström, M.; Oliver, E.; Phatak, A. Natural Hazards in Australia: Heatwaves. *Clim. Chang.* **2016**, *139*, 101–114. [CrossRef]
78. Nicholls, N.; Skinner, C.; Loughnan, M.; Tapper, N. A Simple Heat Alert System for Melbourne, Australia. *Int. J. Biometeorol.* **2008**, *52*, 375–384. [CrossRef] [PubMed]
79. Saez, M.; Sunyer, J.; Castellsagué, J.; Murillo, C.; Antó, J.M. Relationship between Weather Temperature and Mortality: A Time Series Analysis Approach in Barcelona. *Int. J. Epidemiol.* **1995**, *24*, 576–582. [CrossRef] [PubMed]
80. Hajat, S.; Kovats, R.S.; Atkinson, R.W.; Haines, A. Impact of Hot Temperatures on Death in London: A Time Series Approach. *J. Epidemiol. Community Health* **2002**, *56*, 367–372. [CrossRef]

81. Ding, T.; Qian, W. Geographical Patterns and Temporal Variations of Regional Dry and Wet Heatwave Events in China During 1960–2008. *Adv. Atmos. Sci.* **2011**, *28*, 322–337. [CrossRef]
82. Luo, M.; Ning, G.; Xu, F.; Wang, S.; Liu, Z.; Yang, Y. Observed Heatwave Changes in Arid Northwest China: Physical Mechanism and Long-Term Trend. *Atmos. Res.* **2020**, *242*, 105009. [CrossRef]
83. Tan, Z.; Liu, Y.; Tang, W.; Li, D.; Gao, J.; Luo, M. Impact of the Tibetan Plateau Heat Source on Heatwaves in China. *Clim. Dyn.* **2025**, *63*, 2. [CrossRef]
84. Dong, C.; Wang, X.; Ran, Y.; Nawaz, Z. Heatwaves Significantly Slow the Vegetation Growth Rate on the Tibetan Plateau. *Remote Sens.* **2022**, *14*, 2402. [CrossRef]
85. Change, I.P.O.C. Climate Change 2007: The Physical Science Basis. *Agenda* **2007**, *6*, 333.
86. Sun, Y.; Zhang, X.; Zwiers, F.W.; Song, L.; Wan, H.; Hu, T.; Yin, H.; Ren, G. Rapid Increase in the Risk of Extreme Summer Heat in Eastern China. *Nat. Clim. Chang.* **2014**, *4*, 1082–1085. [CrossRef]
87. Wu, W.; Liu, Q.; Li, H.; Huang, C. Spatiotemporal Distribution of Heatwave Hazards in the Chinese Mainland for the Period 1990–2019. *Int. J. Environ. Res. Public Health* **2023**, *20*, 1532. [CrossRef] [PubMed]
88. Ming, Y.; Shuqing, S. A Study on the Response of Subtropical High over the Western Pacific on the SST Anomaly. *Chin. J. Atmos. Sci. (Chin. Ed.)* **2000**, *24*, 206–213.
89. Yimin, L.; Guoxiong, W. Progress in the Study on the Formation of the Summertime Subtropical Anticyclone. *Adv. Atmos. Sci.* **2004**, *21*, 322–342. [CrossRef]
90. Gong, H.; Ma, K.; Hu, Z.; Dong, Z.; Ma, Y.; Chen, W.; Wu, R.; Wang, L. Attribution of the August 2022 Extreme Heatwave in Southern China: Role of Dynamical and Thermodynamical Processes. *Bull. Am. Meteorol. Soc.* **2024**, *105*, E193–E199. [CrossRef]
91. Entekhabi, D.; Rodriguez-Iturbe, I.; Castelli, F. Mutual Interaction of Soil Moisture State and Atmospheric Processes. *J. Hydrol.* **1996**, *184*, 3–17. [CrossRef]
92. Dong, D.; Tao, H.; Zhang, Z. Historic Evolution of Population Exposure to Heatwaves in Xinjiang Uygur Autonomous Region, China. *Sci. Rep.* **2023**, *13*, 7401. [CrossRef]
93. Liu, J.; Wang, A.; Zhang, T.; Pan, P.; Ren, Y. Projected Increase in Heatwaves under 1.5 and 2.0 C Warming Levels Will Increase the Socio-Economic Exposure Across China by the Late 21st Century. *Atmosphere* **2024**, *15*, 900. [CrossRef]
94. Dong, B.; Sutton, R.T.; Shaffrey, L. Understanding the Rapid Summer Warming and Changes in Temperature Extremes Since the Mid-1990s over Western Europe. *Clim. Dyn.* **2017**, *48*, 1537–1554. [CrossRef]
95. Miralles, D.G.; Teuling, A.J.; Van Heerwaarden, C.C.; Vilà-Guerau de Arellano, J. Mega-Heatwave Temperatures Due to Combined Soil Desiccation and Atmospheric Heat Accumulation. *Nat. Geosci.* **2014**, *7*, 345–349. [CrossRef]
96. Chen, L.; Chen, H.; Qi, Y. Regional Differences in the Effects of Land–Atmosphere Coupling on Summer High-Temperature Events in Eastern China and Their Possible Mechanisms. *Chin. J. Atmos. Sci.* **2024**, *49*, 107–122. [CrossRef]
97. Seneviratne, S.I.; Lüthi, D.; Litschi, M.; Schär, C. Land–Atmosphere Coupling and Climate Change in Europe. *Nature* **2006**, *443*, 205–209. [CrossRef]
98. Cheng, S.; Wang, S.; Li, M.; He, Y. Summer Heatwaves in China During 1961–2021: The Impact of Humidity. *Atmos. Res.* **2024**, *304*, 107366. [CrossRef]
99. Lu, G.; Li, Q.; Sun, X.; Zhao, M.; Dong, L.; Wu, Q.; Wang, L.; Zhao, L.; Duan, C.; Yin, Y. Comparative Analysis of Peak-Summer Heatwaves in the Yangtze–Huaihe River Basin of China in 2022 and 2013: Thermal Effects of the Tibetan Plateau. *Atmos. Res.* **2024**, *300*, 107222. [CrossRef]
100. Zhao, J.; Li, T.; Shi, K.; Qiao, Z.; Xia, Z. Evaluation of ERA-5 Precipitable Water Vapor Data in Plateau Areas: A Case Study of the Northern Qinghai-Tibet Plateau. *Atmosphere* **2021**, *12*, 1367. [CrossRef]
101. Zhang, J.; Ren, G.; You, Q. Assessing the Escalating Human-Perceived Heatwaves in a Warming World: The Case of China. *Weather Clim. Extrem.* **2024**, *43*, 100643. [CrossRef]
102. Guo, X.; Huang, J.; Luo, Y.; Zhao, Z.; Xu, Y. Projection of Heat Waves over China for Eight Different Global Warming Targets Using 12 CMIP5 Models. *Theor. Appl. Climatol.* **2017**, *128*, 507–522. [CrossRef]
103. Jiang, Y.; Yang, K.; Shao, C.; Zhou, X.; Zhao, L.; Chen, Y.; Wu, H. A Downscaling Approach for Constructing High-Resolution Precipitation Dataset over the Tibetan Plateau from ERA5 Reanalysis. *Atmos. Res.* **2021**, *256*, 105574. [CrossRef]
104. Jones, J.W.; Hoogenboom, G.; Porter, C.H.; Boote, K.J.; Batchelor, W.D.; Hunt, L.; Wilkens, P.W.; Singh, U.; Gijsman, A.J.; Ritchie, J.T. The DSSAT Cropping System Model. *Eur. J. Agron.* **2003**, *18*, 235–265. [CrossRef]

atmosphere

MDPI

Article

Advancing Heat Health Risk Assessment: Hotspot Identification of Heat Stress and Risk Across Municipalities in Algiers, Algeria

Dyna Chourouk Zitouni [1,2,3,*], Djihed Berkouk [3,4], Mohamed Elhadi Matallah [1,5], Mohamed Akram Eddine Ben Ratmia [2,3] and Shady Attia [1,*]

1 Sustainable Building Design Lab, Department UEE, Applied Sciences, University of Liège, 4000 Liège, Belgium; elhadi.matallah@univ-biskra.dz
2 Laboratory of Design and Modeling of Architectural and Urban Forms and Ambiances (LACOMOFA), University of Biskra, Biskra 07000, Algeria; akrameddine.benratmia@univ-biskra.dz
3 Department of Architecture, Mohamed Khider University, Biskra 07000, Algeria; dberkouk@dah.edu.sa
4 Department of Architecture, School of Engineering, Computing & Design, Dar Al-Hekma University, Jeddah 22246, Saudi Arabia
5 Civil Engineering and Hydraulics Laboratory, Sustainable Development and Environment (LARGHYDE), University of Biskra, Biskra 07000, Algeria
* Correspondence: dczitouni@uliege.be (D.C.Z.); shady.attia@uliege.be (S.A.)

Abstract: With accelerating surface warming trends in urban regions, cities like Algiers are increasingly exposed to extreme heat, contributing to a growing concern over heat-related illnesses. For a comprehensive long-term assessment (2001–2023) of heat-related risks in Algiers, multi-decade satellite, meteorological, and census data were used in this study to map and assess spatial patterns of the Heat Health Risk Index (HHRI) within the framework established by the Intergovernmental Panel on Climate Change (IPCC) incorporating hazard, exposure and vulnerability components. The Universal Thermal Climate Index (UTCI) was then calculated to assess thermal stress levels during the same period. Following this, the study addressed a critical research gap by coupling the HHRI and UTCI and identified hotspots using the Getis-Ord Gi* statistical analysis tool. Our findings reveal that the intensity of HHRI has increased over time since "very-low" risk areas had an outstanding decrease (93%) and a 6 °C UTCI rise over 23 years reaching the "very strong heat stress" level. The coupled index demonstrated greater and different risk areas compared to the HHRI alone, suggesting that the coupling of both indicators enhances the sensitivity of heat risk assessment. Finally, persistently identified hotspots in central and eastern regions call for localized, targeted interventions in those areas and highlight the value of remote sensing in informing policymakers and enhancing climate resilience.

Keywords: heat vulnerability; thermal comfort; risk assessment; environmental resilience; remote sensing; climate change

1. Introduction

Exposure to heat poses a significant threat to human health and is increasingly impacting the sustainability of urban areas, including cities in North Africa such as Algiers [1]. Ad hoc indicators have recently been introduced to track the consequences that global warming is inducing on human health [2]. The indicators are based on Crichton's [3] three key risk components: (1) hazard, which is a natural physical event that has the potential to cause discomfort or damage to vulnerable and exposed components; (2) exposure, which refers to the individual or buildings situated in locations where a potential hazard might occur; (3) vulnerability refers to how likely exposed elements are to be affected by a hazard (known as sensitivity), and also includes the ability of individuals or systems to adapt to and recover from that hazard.

By combining the three risk components via impact models, they can capture the heterogeneous set of factors and relationships linking climate change with health impacts, including heat-related mortality and changes in labor capacity [4,5]. Similarly, the Intergovernmental Panel on Climate Change (IPCC)'s Sixth Assessment Report (AR6) [4] used the three components to quantitatively analyze the spatial distribution of heat health risk (HHR). While other frameworks, such as the Climate Change and Disaster Indicators (CCDIs) developed by the Expert Team on Weather and Climate Extremes Monitoring and Assessment (ETWMO), offer alternative approaches to risk assessment, the present study adopts the IPCC's structure due to its suitability for spatial integration of environmental and socio-demographic indicators, and its compatibility with thermal stress models.

Although there have been relatively few studies following the hazard–exposure–vulnerability framework [6], the number has been increasing in recent years, with a significant concentration of research being conducted in Asia [7–11].

Geographic Information Systems (GISs) and remote-sensing technologies have an essential role in HHR assessment by providing accurate spatial and temporal data, which is crucial for mapping and analyzing the pattern of heat-related hazard, exposure, and vulnerability [10,12–15]. The literature on HHR assessment using the IPCC's framework has increasingly used remote-sensing tools to capture spatial variations in risk patterns across different regions. In 2019, the MODIS Land Surface Temperature (LST) and vegetation indices were combined with socioeconomic indicators to map HHR for the elderly in a mountainous area in China [16]. A 2020 study in the Philippines applied remotely sensed surface temperature data and social–ecological indicators to assess the current HHR [17]. Subsequently, a 2023 study in Singapore integrated remote sensing to quantify UHI intensity and coupled it with demographic data to assess the HHR for the elderly population [18]. Another 2023 study in Australia employed remote-sensing data and GIS tools to perform a comprehensive HHR assessment in Australian capital cities, incorporating health-related indicators to evaluate population sensitivity to extreme heat [19]. More recently, in 2024, a study conducted a spatially explicit assessment of the HHR in the Yangtze River Delta, China, using multi-sensor remote-sensing images [7]. Finally, a 2025 study leveraged the Sustainable Development Science Satellite 1 (SDGSAT-1) to further enhance our understanding of health-integrated heat risk assessment for Karachi, Pakistan, and inform targeted interventions [11]. These studies evidence the ability to integrate GIS and remote-sensing techniques to visualize the HHR's spatial and temporal patterns as well as to identify its hotspots, i.e., areas where vulnerable populations are disproportionately affected by extreme heat events. Despite that, only a limited number of studies have undertaken long-term evaluations of HHR [20,21]. Further, they mostly have represented heat by means of air temperature, neglecting the effect of other physiologically relevant parameters on the human body, as it neglects other meteorological factors contributing to heat risk, such as humidity, air movement, and solar radiation [10].

Heat, in a human health context, is thermal energy transferred to the human body from the surrounding environment. Conditions of air temperature, humidity, wind speed, and radiation can be such that the heat transferred to the body is excessive [22]. This condition, called heat stress, may lead to heat stroke and dehydration and additionally trigger asthma attacks and respiratory and cardiovascular health diseases [23]. This combination of factors, along with the increased need for reliable evaluation methods of the outdoor environment [24], underscores the importance of studying human exposure to extreme heat rather than air temperature alone and developing effective heat mitigation strategies aimed at enhancing heat stress [10,25,26]. The Universal Thermal Climate Index (UTCI) is an internationally recognized human biometeorological indicator used to evaluate the connections between the outdoor environment, heat, and human well-being [27]. UTCI values are categorized into ten levels of thermal stress, ranging from "extreme cold stress" to "extreme heat stress" providing a comprehensive framework for evaluating thermal stress in various climatic conditions and increasing preparedness against extreme heat disasters [28].

To the best of our knowledge, there has been only one study that has incorporated heat, as per its proper definition, into the HHR assessment and used the UTCI for this purpose [6]. The study, however, was based on a heat stress–exposure–vulnerability framework and did not include hazard components. In contrast with traditional hazard indicators that directly measure external heat-related events such as heatwaves, the UTCI evaluates the effects of those events on human physiology by construction, and it does so by considering multiple factors that influence an individual's heat stress rather than evaluating the severity of a potential hazard. Therefore, it appears more suitable to use alongside other hazard–exposure–vulnerability indicators rather than as a standalone hazard indicator in the assessment of the heat risk to human health. The absence of a direct connection between HHR and thermal stress has resulted in a research gap, particularly in the accuracy and reliability of spatial and temporal patterns of the Heat Health Risk Indicator (HHRI) as a human health indicator during extreme heat events. This gap is further compounded by the significant lack of data on heat-related mortality and morbidity, making it challenging to understand and address the impacts of extreme heat fully.

In recent years, the Mediterranean region has emerged as a changing climate hotspot, with rising temperatures exceeding the global average rate [29]. Heatwaves have become more frequent, persistent, and deadly across the region [30]. Algeria, which has a 1200 km coastline along the Mediterranean Sea, has been listed by the IPCC as one of the vulnerable countries to climate change [31]. Every year, more than 260 people are estimated to die from heat-related illnesses in the country [32]. With annual temperatures predicted to increase by 2.5 to 3 °C and the temperature of the warmest month by 1.9 °C by 2100, prolonged exposure to extreme heat, which ultimately results in heat stress, is expected to occur at least once in the next five years with potentially deadly consequences [33]. On 26–28 April 2023, local temperatures in many regions in Algeria were up to 20 °C higher than the climatological average (1991–2020) for that time of year. This was made 100 times more likely by climate change, according to attribution studies [34].

Within this context, there is a clear need to consider environmental, social, and public health information and deploy it to develop effective, locally relevant climate-adaptive health solutions. Our paper responds to this need by deploying remote sensing, meteorological, and census data at the municipality level across 23 years (2001–2023). This research aims at (a) conducting a long-term spatio-temporal analysis (2001–2023) of the HHRI and the UTCI within an integrated hazard–exposure–vulnerability framework, (b) couple the HHRI with the UTCI to construct a comprehensive, multidimensional indicator that is able to enhance the understanding of heat-related health risks, and (c) identify and compare hotspots, i.e., areas that exhibit significant clustering of both high HHRI and the new coupled indicator values, particularly when these values are categorized as "high" or "very high". This research is the first in the Mediterranean context to assess HHR using the IPCC AR6 framework, providing a methodologically robust foundation for targeted adaptation plans and improved public health resilience in regions highly vulnerable to heat extremes.

2. Materials and Methods

The proposed methodology for the present research integrates hazard, exposure, and vulnerability components with the UTCI, a heat stress indicator that integrates air temperature (Ta), wind speed (V), vapor pressure (e), and the mean radiant temperature (Tmrt) to provide comprehensive spatio-temporal analysis and hotspot identification within the study area. The HHR assessment follows the framework defined by the IPCC's AR6 [4]. Data availability and previous studies guided the selection of these indicators. Table 1 presents each indicator, which either increases or lowers the HHR.

The methodology is divided into four main steps as outlined in Figure 1. It is designed to be reproducible, depending on open data published in a dataset repository to assure accuracy and transparency [35]. Moreover, the ArcGIS Pro 3.2.0 software [36] was used throughout the process to perform spatial analysis and visualization. We visualized maps by splitting the study period into four main intervals: (a) 2001–2006, (b) 2007–2011, (c) 2012–2016, and (d) 2017–2023.

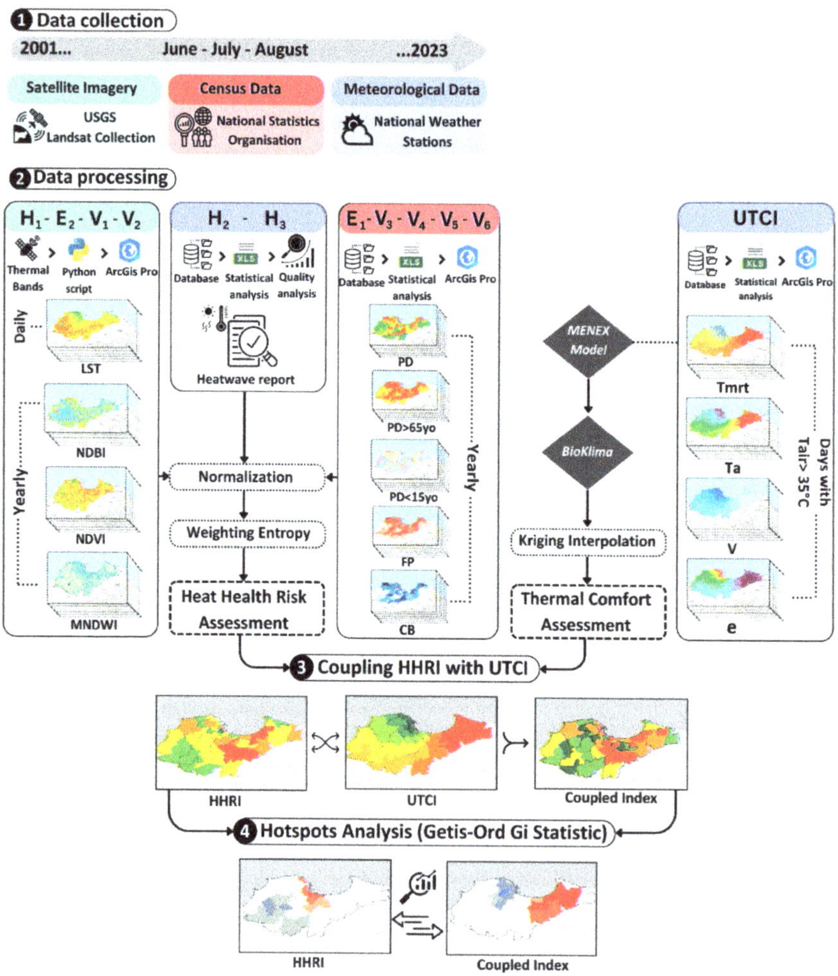

Figure 1. Study conceptual framework.

Table 1. Detailed description and calculation methods for each indicator used in the HHR assessment.

Category	Indicator	Description	Formula	Impact	References
Hazard	H1: LST Land surface temperature	The temperature of the Earth's surface is measured using satellite images of thermal bands	$LST = \frac{BT}{1+(w\times\rho BT)\times ln(\varepsilon)}$	(+)	[17,20,37]
	H2: Hot days	Days with maximum air temperatures greater than 35 °C	Count of days with Tair Max > 35 °C	(+)	[9,13,16]
	H3: Heatwave frequency	The number of events where the max daily average temperature exceeds 35 °C for more than 3 days	Count of days with Tair max > 35 °C for more than 3 consecutive days	(+)	[9,13,38]
Exposure	E1: (PD) Population density	Number of people per square kilometer	$\frac{Population}{Area}$	(+)	[15,37,38]
	E2: (NDBI) Normalized difference building index	An index used to measure built-up areas and urbanization	$\frac{(SWIR-NIR)}{(SWIR+NIR)}$	(+)	[39,40]
Vulnerability	V1: (NDVI) Normalized difference vegetation index	An index used to measure the density of vegetation	$\frac{(NIR-Red)}{(NIR+Red)}$	(−)	[37,40]
	V2: (MNDWI) Modified normalized difference water index	An index used to identify water bodies	$\frac{(NIR-SWIR)}{(NIR+SWIR)}$	(−)	[16,37,39]
	V3: (PD > 65 yo) Elderly population	The number of people aged 65 and older per square kilometer	$\frac{\text{Elderly population}}{Area}$	(+)	[14,16,18]
	V4: (PD < 15 yo) Young population	The number of people aged 15 and younger per square kilometer	$\frac{\text{Young population}}{Area}$	(+)	[14,17,41]
	V5: (FP) Female population	Number of females per square kilometer	$\frac{\text{Female population}}{Area}$	(+)	[6,15,39,42]
	V6: (CB) Care beds	The number of available care beds available per municipality	Total Care Beds per municipality	(−)	[6,9,15]

SWIR: reflectance in the shortwave infrared band; NIR: reflectance in the near-infrared band; Red: reflectance in the red visible band.

The indicators (H1, H2, H3, E1, E2, V1, V2, V3, V4, V5, V6) are detailed in Table 1.

2.1. Data Collection

The study focuses on Algiers, the capital of Algeria, situated at 36°46′34″ N and 3°03′36″ E (Figure 2). With its 80 km coastline along the Mediterranean Sea and varied topography, Algiers has a Mediterranean climate (Csa) according to the Köppen climate classification. It typically has long, hot, and dry summers, with average daily maximum temperatures exceeding 30 °C, heatwaves occasionally pushing temperatures above 35 °C, and minimal to no precipitation. Considering it is the most densely populated province in Algeria, with an estimated population of 3,309,896 divided into 57 municipalities according to the 2020 statistical yearbook, it serves as a great case study for assessing heat health concerns and heat stress.

Three types of data were used in the study: satellite, meteorological and census data.

Satellite data were obtained from the United States Geological Survey (USGS) [43] during March and April 2024, leveraging Landsat Collection 2. The data covered a series of sensors across different Landsat missions, specifically Landsat 5 Thematic Mapper (TM), Landsat 7 Enhanced Thematic Mapper Plus (ETM+), and Landsat 8/9 Operational Land Imager and Thermal Infrared Sensor (OLI/TIRS). Scenes with a cloud cover of 5% or less were selected, with a spatial resolution of 30 m. The temporal resolution of the dataset spans the summer months of June, July, and August from 2001 to 2023. The associated Metadata (MTL) files were used to perform atmospheric corrections, ensuring the accuracy of derived hazard, exposure, and vulnerability indices, namely the LST, normalized difference building index (NDBI), normalized difference vegetation index (NDVI), and modified normalized difference water index (MNDWI).

Meteorological data were acquired from four national weather stations within the study area, covering the same study period (summer months from 2001 to 2023). The data include hourly air temperature, relative humidity, and wind speed parameters. This meteorological dataset is required to calculate the UTCI [27] and the hazard components of the HHRI (hot days and heatwave frequency).

Census data were obtained from the statistical yearbook established by the National Statistical Organization for the study area from 2001 to 2023. This dataset covers population density, population under 15 and above 65, female population, and the number of available public care beds in each municipality to measure both exposure and vulnerability components.

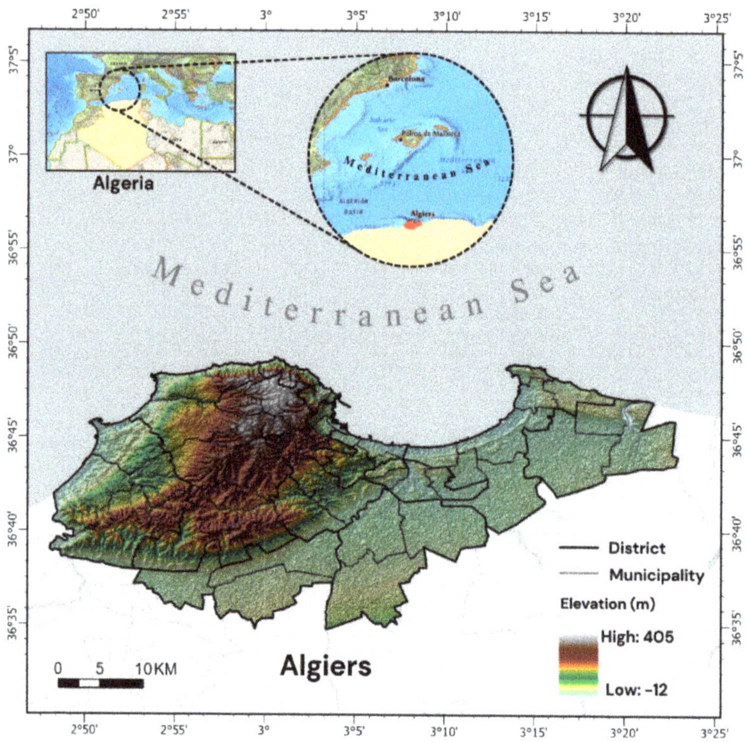

Figure 2. Study area location and elevation map.

2.2. Data Processing

2.2.1. Heat Health Risk Assessment

The spatial and temporal resolution of the collected data provided a solid foundation for our investigation, allowing comprehensive analysis and mapping of the indicators, and ensuring accurate spatial representation. The assessment of heat hazards included LST (H1) to quantify the surface-level heat intensity that directly impacts the HHR [44], as well as hot days (H2) and heatwave frequency (H3) to highlight prolonged periods of extreme heat that can increase the HHR [45]. Population density (E1) was selected for its high concentration of at-risk individuals, while NDBI (E2) reflects built-up areas' contribution to heat risk [46]. Vulnerability assessment included NDVI (V1) and MNDWI (V2) as positive indicators, as vegetation and water bodies help to reduce thermal stress [47]. The elderly population (V3) and young population (V4) were negative indicators due to their higher sensitivity to heat-related illnesses [42,48]. The female population (V5) was included due to potential gender disparities in heat tolerance [49]. Care beds (V6) were a positive indicator, reflecting healthcare capacity to handle HHR emergencies.

Satellite data were used to compute four key indicators (H1, E1, V2, V3) with atmospheric corrections. Python scripts in ArcGIS Pro 3.2.0 generated 148 daily LST maps, with 43% linked to heatwaves. Annual NDBI, NDVI, and MNDWI averages produced 23 maps per indicator (2001–2023). Meteorological analysis identified 770 hot days and 23 heatwave events, while census data provided municipality-level maps of population, age groups, female ratio, and care beds.

Because the HHR assessment integrates multiple environmental and demographic indicators, each with different units, the indicators were normalized to a common scale using the Min–Max normalization method to guarantee that each indicator contributes equally to the final index by bringing all variables to a uniform scale based on Equation (1):

$$\text{Xnorm}\begin{cases} = \frac{\text{X}-\text{Xmin}}{\text{Xmax}-\text{Xmin}}\ (for\ positive\ indicators) \\ = \frac{\text{Xmax}-\text{X}}{\text{Xmax}-\text{Xmin}}\ (for\ negative\ indicators) \end{cases} \tag{1}$$

where Xnorm is the normalized value, X is the original value, Xmin is the minimum value, and Xmax is the maximum value.

Next, the entropy weighting approach was performed to assign weights to each indicator based on their variability and contribution to heat risk. This approach was used in several studies in the HHR assessment [39,50] and includes four main steps as presented in Figure 3.

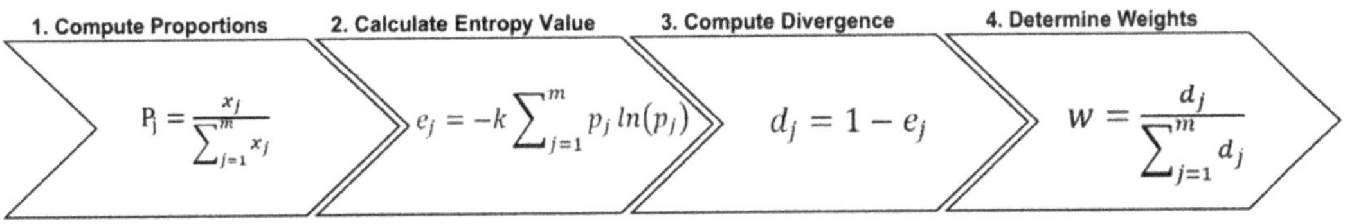

Figure 3. Entropy weighting process for indicator weight calculation.

Where Pj is the proportion of the normalized indicator, ej is the entropy value of the indicator j, dj is the degree of diversification for the indicator j, Wj is the weight assigned to the indicator j, xj is the normalized value of the indicator j, m is the total number of indicators, and K is a constant defined as 1/ln (m).

These indicators were then combined into a single index and finally mapped to visualize the overall HHRI using the Crichton risk triangle equation shown in Equation (2), and the levels of risk were then divided into five classes from "very high" to "very low" using the natural breaks (Jenks) option in ArcGIS Pro:

$$\text{HHRI} = \text{H} \times \text{E} \times \text{V} \tag{2}$$

where Hrepresents hazard, E represents exposure, and V represents vulnerability.

2.2.2. Heat Stress Assessment

Unlike simpler indices, UTCI integrates factors such as air temperature (Ta, °C), wind speed (V, m/s), vapor pressure (e, hPa), and the mean radiant temperature (Tmrt, °C), to provide a detailed understanding of thermal stress on the human body [51]. UTCI is expressed in terms of an assessment scale, which is composed of ten stress categories. Each category represents specific physiological and thermoregulatory responses to environmental conditions [52]. As this study focuses on the summer season, five stress levels ranging from no thermal stress to extreme heat stress were considered (Table 2).

Table 2. UTCI heat stress assessment scale (adapted from Bröde et al., 2012 [27]).

UTCI (°C)	Stress Category
UTCI ≥ 46	Extreme heat stress
38 ≤ UTCI < 46	Very strong heat stress
32 ≤ UTCI < 38	Strong heat stress
26 ≤ UTCI < 32	Moderate heat stress
9 ≤ UTCI < 26	No thermal stress

Daytime meteorological parameters (from 6:00 am to 6:00 pm) were extracted for each day within the study period. The kriging interpolation tool was used to represent the spatial distribution of these parameters across the study area. The UTCI was calculated for hot days, defined as days when the air temperature exceeded 35 °C, in order to better capture the extreme heat risk conditions which typically occurred between 1:00 pm and 3:00 pm, to better reflect peak thermal stress periods and avoid underestimation due to lower morning values. This same time window was applied consistently across all input parameters used in the UTCI computation. Additionally, wind speed measurements were adjusted from 10 m to 1.2 m using the logarithmic wind profile equation (Equation (3)):

$$v(z) = v_{\text{ref}} \frac{\ln(z/z_0)}{\ln(z_{\text{ref}}/z_0)} \tag{3}$$

where $v(z)$ is the wind speed at 1.2 m, $v(z_{\text{ref}})$ is the measured wind speed at the reference height, z is the target height, z_{ref} is the reference height, and z_0 is the surface roughness length.

The UTCI was calculated using the collected data and the Tmrt via the operational procedure by [27]. The calculation of UTCI was carried out by the software package BioKlima 2.6 [53] using Equation (4):

$$\text{UTCI} = \text{Ta} + \text{Offset (Ta,V,e,Tmrt)} \tag{4}$$

The MENEX model [54] was used to calculate Tmrt [6,55], through Equations (5)–(8):

$$\text{Tmrt} = \left[(\text{Rprim} + 0.5\text{Lg} + 0.5\text{La}) / \left(0.95 \times 5.667 \times 10^{-8}\right)\right]^{1/4} - 273.16 \tag{5}$$

$$\text{La} = 5.5 \times 10^{-8}(273.16 + \text{Ta})^4 \left[0.82 - 0.25 \times 10^{-0.094 \times 0.75\text{e}}\right] \tag{6}$$

$$\text{Lg} = 5.5 \times 10^{-8}(273.16 + \text{Tg})^4 \tag{7}$$

$$\text{Rprim} = 1.4[-100.428 + 73.981\ln(\text{h})](1 - 0.01\text{a}_\text{c})\text{Irc} \tag{8}$$

where La and Lg represent the longwave radiation emitted by the air and the ground, respectively, both measured in (W/m^2); Tg refers to LST (°C); Rprim (W/m^2) represents the absorbed solar radiation; the terms h and a_c denote the sun altitude (degree) and albedo of skin (%), respectively; and Irc represents the reduction coefficient of heat transfer through clothing.

2.3. Coupling HHRI and UTCI

The major step in our methodology involves the coupling of the HHRI with the UTCI to provide a comprehensive view of heat-related risks and thermal stress in Algiers. This process integrates the spatial distributions of both indicators to identify areas experiencing the highest combined risk. To achieve this, both the HHRI and UTCI values were then normalized to a common scale (0–1) to ensure compatibility and comparability, as well as the integration of the two indicators into a single map. The normalization was performed using the Min–Max normalization method expressed in Equation (1). The normalized HHRI and UTCI were then coupled using the raster calculator tool in ArcGIS Pro 3.2.0 software following Equation (9):

$$\text{Coupled Index} = \frac{\text{HHRI} + \text{UTCI}}{2} \tag{9}$$

We used equal weights to ensure that both HHRI and the UTCI contribute equally to the combined risk assessment. The resulting coupled index was then mapped to provide a visual representation of areas where high HHRI and high UTCI values intersected.

2.4. Hotspots Analysis (Getis-Ord Gi*)

The final step of our methodology aimed at identifying spatial clusters of extreme risk areas within the study area. To further refine our understanding, Getis-Ord Gi*—a spatial analysis tool that identifies areas of significant clustering, including those that are most vulnerable to risk [56]—was applied to both the HHRI and the coupled index maps. The analysis identifies both hot and cold spots, which are areas of statistically significant clustering of "very high" and "high" or "very low" and "low" values. They were classified into seven categories: cold spot with 99% confidence, cold spot with 95% confidence, cold spot with 90% confidence, not significant, hotspot with 99% confidence, hotspot with 95% confidence, and hotspot with 90% confidence. By identifying hotspots of both HHRI and the new coupled index, we created detailed maps highlighting the areas that require targeted interventions to heat stress, considering both HHR risk and thermal stress.

3. Results

The results are structured into four sections: (1) HHR assessment, covering hazard, exposure, vulnerability, and HHRI; (2) UTCI analysis; (3) the coupled HHRI–UTCI index; (4) hotspot analysis for refined risk evaluation. The study spans four periods, (a) 2001–2006, (b) 2007–2012, (c) 2013–2018, and (d) 2019–2023, presented at the municipal level in Algiers (57 municipalities).

3.1. Heat Health Risk Assessment

3.1.1. Hazard Index

Figure 4 shows the spatial distribution of the hazard index. Throughout (a), the hazard index mainly ranged from "moderate" to "very high", with 58% of total municipalities (e.g., Shaoula). Moving to (b), the extent of "very high" areas expanded to 19% in the southern districts (e.g., Birtouta). In contrast, in (c) there was a consistent expansion in the hazard index in the eastern regions (e.g., Dar El Beida). Finally, (d) indicates a continuation of these patterns, with 39% of areas classified as "high" and "very high" due to the rising LST values there.

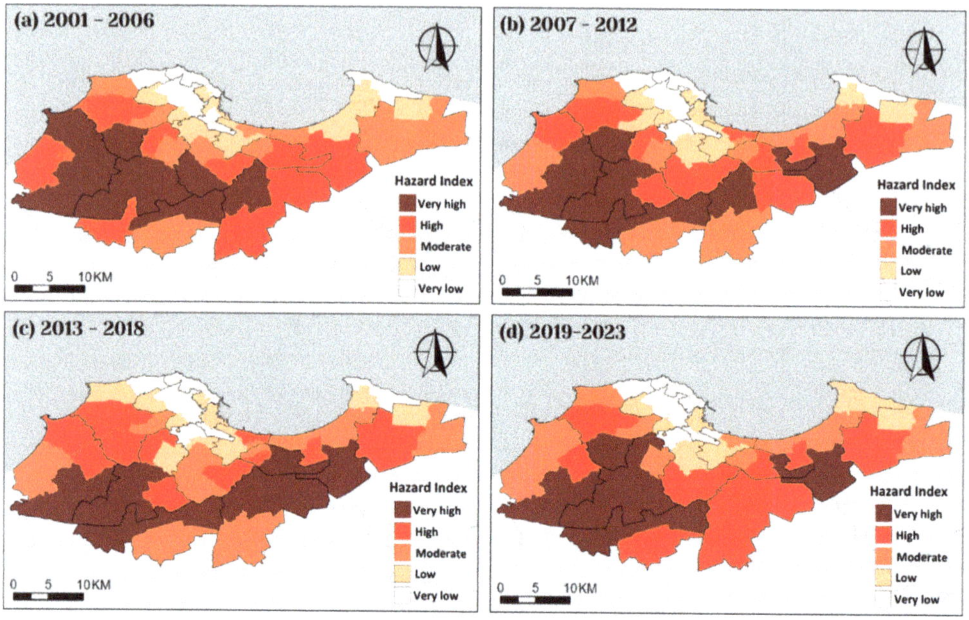

Figure 4. Spatial distributions of the hazard index in Algiers. (**a**) 2001–2006, (**b**) 2007–2012, (**c**) 2013–2018, (**d**) 2019–2023.

3.1.2. Exposure Index

Figure 5 illustrates the spatial distribution of the exposure index. In (a) and (b), 30% of the total municipalities had the highest levels of exposure, classified as "high" and "very high", mainly in the north (e.g., The Casbah), while 50% experienced "very low" levels (e.g., Mahelma). In (c), there was a further expansion of "high" levels, with 26% of municipalities identified in the central regions (e.g., El Achour). This trend continued, with "very low" levels nearly disappearing in (d) with only 11% of the total municipalities. At the same time, new areas were categorized as "high" and "very high" (e.g., Mahelma and Baba Hassen) due to increasing population and urbanization in the last 10 years in those areas.

3.1.3. Vulnerability Index

Figure 6 demonstrates the spatial distribution of the vulnerability index. In (a), "very high" levels were dispersed across 12% of the total municipalities in the central and eastern regions (e.g., Bordj El Kiffan). In contrast, "low" and "very low" levels were primarily found in the southern municipalities (e.g., Zeralda) with 35%. In (b), although vulnerability levels increased and "very low" areas nearly disappeared, the proportion of "very high" municipalities decreased to 8%, which is lower than in the previous period. However, (c) showed a shift in the spatial pattern, with lower vulnerability levels in the central districts (e.g., Shaoula). Finally, in (d), the pattern was similar to (c), but with only 5% of areas classified as at the "very high" vulnerability level and only one municipality (Sidi M'hamed) at the "very low" level. The increase in vulnerability over time is likely due to the growing population of vulnerable groups, such as the elderly and young, which intensifies the overall risk.

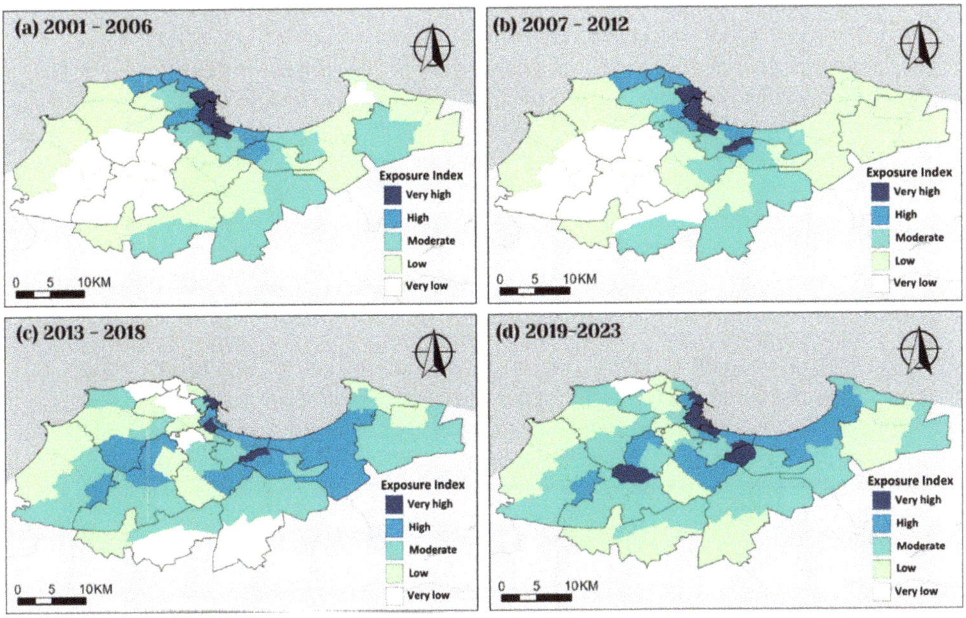

Figure 5. Spatial distributions of the exposure index in Algiers. (**a**) 2001–2006, (**b**) 2007–2012, (**c**) 2013–2018, (**d**) 2019–2023.

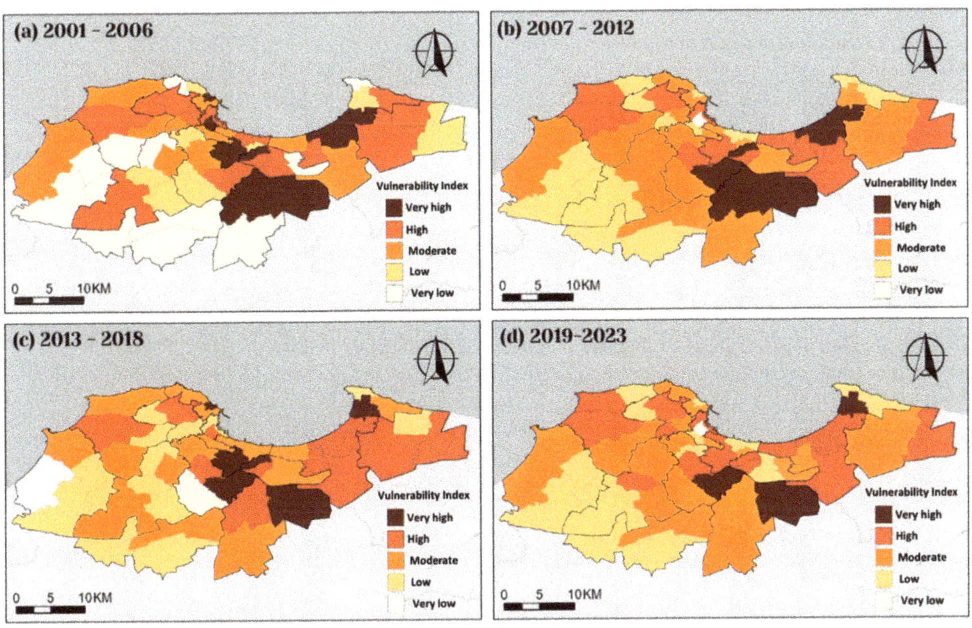

Figure 6. Spatial distributions of the vulnerability index in Algiers. (**a**) 2001–2006, (**b**) 2007–2012, (**c**) 2013–2018, (**d**) 2019–2023.

3.1.4. Heat Health Risk Index (HHRI)

Figure 7 demonstrates the spatial distribution of the HHRI. In (a), the HHRI was higher in the central regions as well as several scattered municipalities in both the east and west, with 18% of municipalities identified as "very high" risk and 19% as "high". However, "moderate" risk was more common, with 28% of municipalities across the province. The areas with a majority of "low" and "very low" risk tended to be in the southern and western parts (e.g., Birtouta). Moving into (b), an increase in HHRI intensity was observed across the province, with areas that were classified as having "low" and "very low" risk levels increasing to "moderate" risk in nearly half of the total municipalities (e.g., Draria); however, the "very high" risk level identified municipalities slightly decreased to 11%. Algiers witnessed a decrease in HHRI intensity throughout period (c), with only two municipalities remaining in the "very high" risk category (Bab El Oued and Bachedjerah). Both "high" and "moderate" risks reduced equally, while "low" risk levels remained unchanged, and "very low" levels were identified in only one single municipality (Shaoula). Lastly, (d) exhibited a rise in intensity, as some central municipalities (e.g., Kouba) experienced a shift from "moderate" to "high" risk, and 7% of municipalities (e.g., Eucalyptus) reverted to the "very high" category. The changes in risk levels over time highlight the varying degrees of resilience and vulnerability within different municipalities, particularly in the central regions where persistent high-risk areas reflect ongoing challenges in managing HHR.

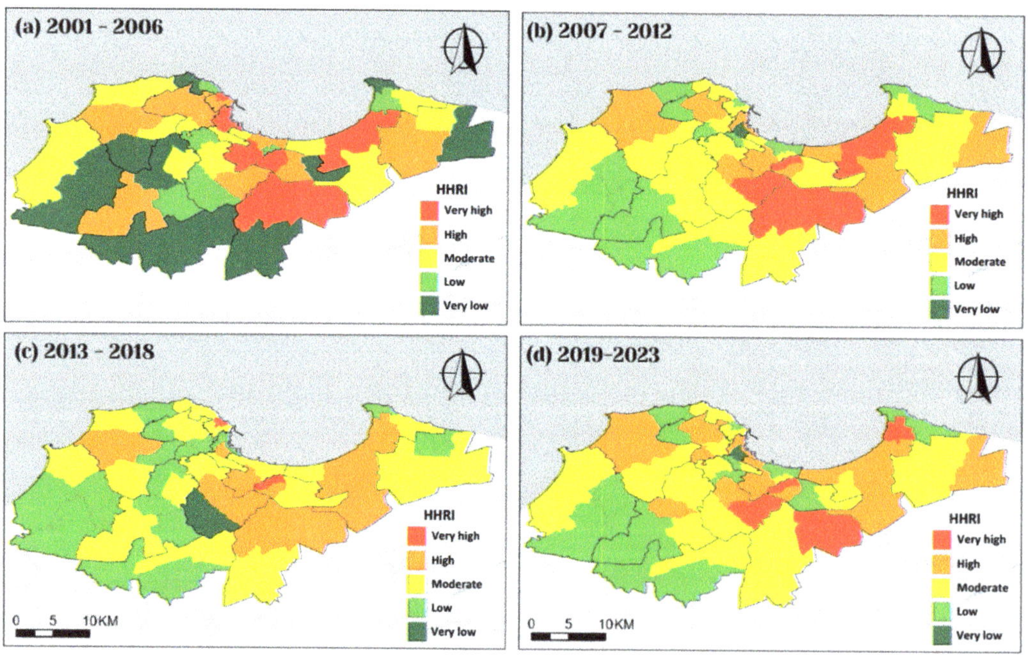

Figure 7. Spatial distributions of HHRI in Algiers. (**a**) 2001–2006, (**b**) 2007–2012, (**c**) 2013–2018, (**d**) 2019–2023.

3.2. Universal Thermal Climate Index (UTCI)

The UTCI maps (Figure 8) illustrate the spatial distribution of thermal stress across the study area, identifying regions experiencing varying levels of heat stress over the four distinct periods. In (a), UTCI values ranged from 24.1 °C to 32.0 °C, with the majority of areas in the "moderate heat stress" category. Notably, the highest levels of thermal stress were concentrated in the eastern and southern municipalities (e.g., Ain Taya). In contrast, the western and northern ones (e.g., Rais Hamidou) exhibited more favorable conditions, falling within the "no thermal stress" to "moderate heat stress" categories. The thermal stress further intensified in (b), with the highest values reaching 37.0 °C. This shift places the central and eastern regions (e.g., Oued Smar) into the "strong heat stress" category. At the same time, the northern areas (e.g., El Biar) continued to experience lower stress levels, mostly within the "moderate heat stress" range. In (c), the thermal stress decreased slightly, with the highest UTCI recorded at 36.1 °C, maintaining most of the area inside the "strong heat stress" category. However, the central regions (e.g., Sidi M'hamed) exhibited a slight decrease in stress. In contrast, the northern regions (e.g., Oued Koriche) continued to benefit from comparatively thermally favorable conditions, with the lowest UTCI recorded at 28.8 °C, inserting them in the "moderate heat stress" category. Lastly, (d) experienced a small rise in thermal stress, with the highest UTCI value identified during the four periods rising to 38.1 °C and with a minimum temperature of 29.3 °C, reaching the "very strong heat stress" category. Despite these moderate shifts, the overall trend of thermal stress across the study area and for all four periods remained stable, with the northern areas constantly experiencing lower stress levels compared to the central and eastern ones.

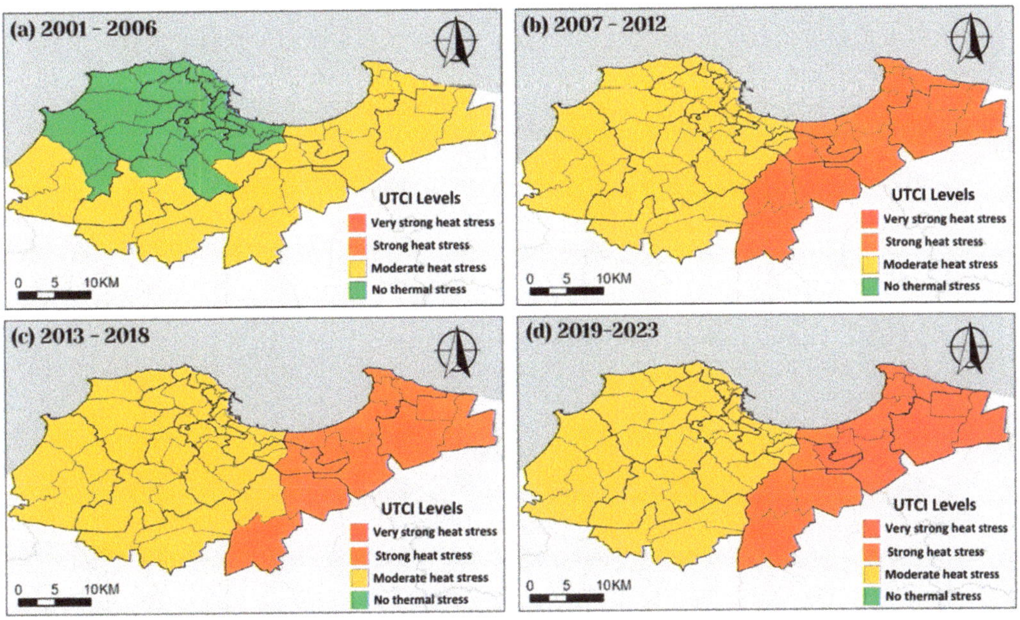

Figure 8. Spatial distributions of UTCI in Algiers. (**a**) 2001–2006, (**b**) 2007–2012, (**c**) 2013–2018, (**d**) 2019–2023.

3.3. Coupled Index

The coupling of the HHRI with the UTCI indicates considerable variations in risk throughout the study area and during the four periods (Figure 9). In (a), the combined index was higher in the eastern regions (e.g., Rouiba) compared to other municipalities, with 7% of them classified as "very high" risk and 14% as "high." The "moderate" risk level was identified in a few scattered municipalities (e.g., Kouba) across the province. However, "low" and "very low" levels were more common across the study area, with 37% and 25% of municipalities (e.g., Ouled Chbel and Sidi Moussa), respectively. Moving into (b), the intensity of the combined index increased to 42% of municipalities under "high" and "very high" levels in the central, eastern, and western regions of the city (e.g., Kouba and Heraoua). Moreover, fewer "low" and "very low" risk areas (42%) were identified in the south (e.g., Souidania). A continuing escalation in intensity was observed throughout (c), with just one municipality (Saoula) remaining in the "very low" risk category. However, the "moderate" risk level saw an increase, with almost half of the total municipalities spread across the study area in a disconnected pattern, and only 7% of total municipalities (e.g., Bab Ezzouar) identified in the "very high" category. Finally, (d) saw a spike in intensity, as the majority (47%) of central municipalities shifted from "low" to "moderate" levels (e.g., Dely Brahim). In comparison, the "very low" level was maintained in only 5% of municipalities. Despite this increase, some municipalities (e.g., Bab Ezzouar) shifted from the "moderate" to the "high" risk category, demonstrating some overall instability in the broader region.

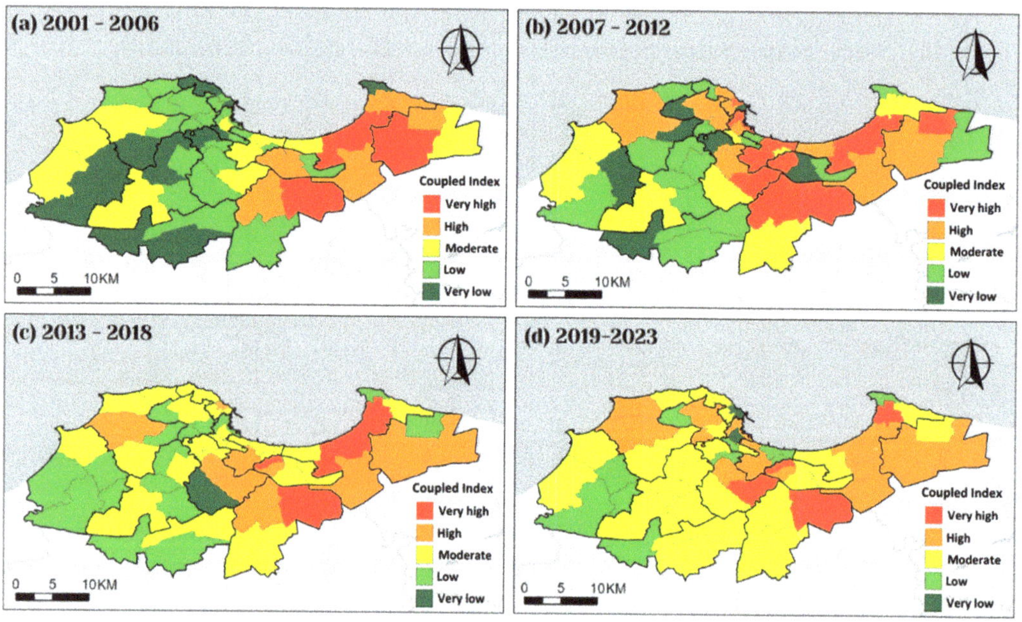

Figure 9. Spatial distributions of the coupled index in Algiers. (**a**) 2001–2006, (**b**) 2007–2012, (**c**) 2013–2018, (**d**) 2019–2023.

The coupling of the HHRI and UTCI provided an improved understanding of the spatial and temporal dynamics of heat health risks in Algiers. The variations in risk levels recorded across different periods reflect the evolving interplay between heat stress and population vulnerability. This pattern notably highlights the increased risks in the central and eastern regions, which may be influenced by factors such as urban density and demographic composition.

3.4. Hotspots Analysis

Figure 10 represents the spatial distribution of the HHRI hotspots (a, b, c, d) and the coupled index hotspots (a′, b′, c′, d′) over the four study periods using the Getis-Ord Gi* statistical method resulting in three levels for both hotspots and cold spots along with a not-significant area.

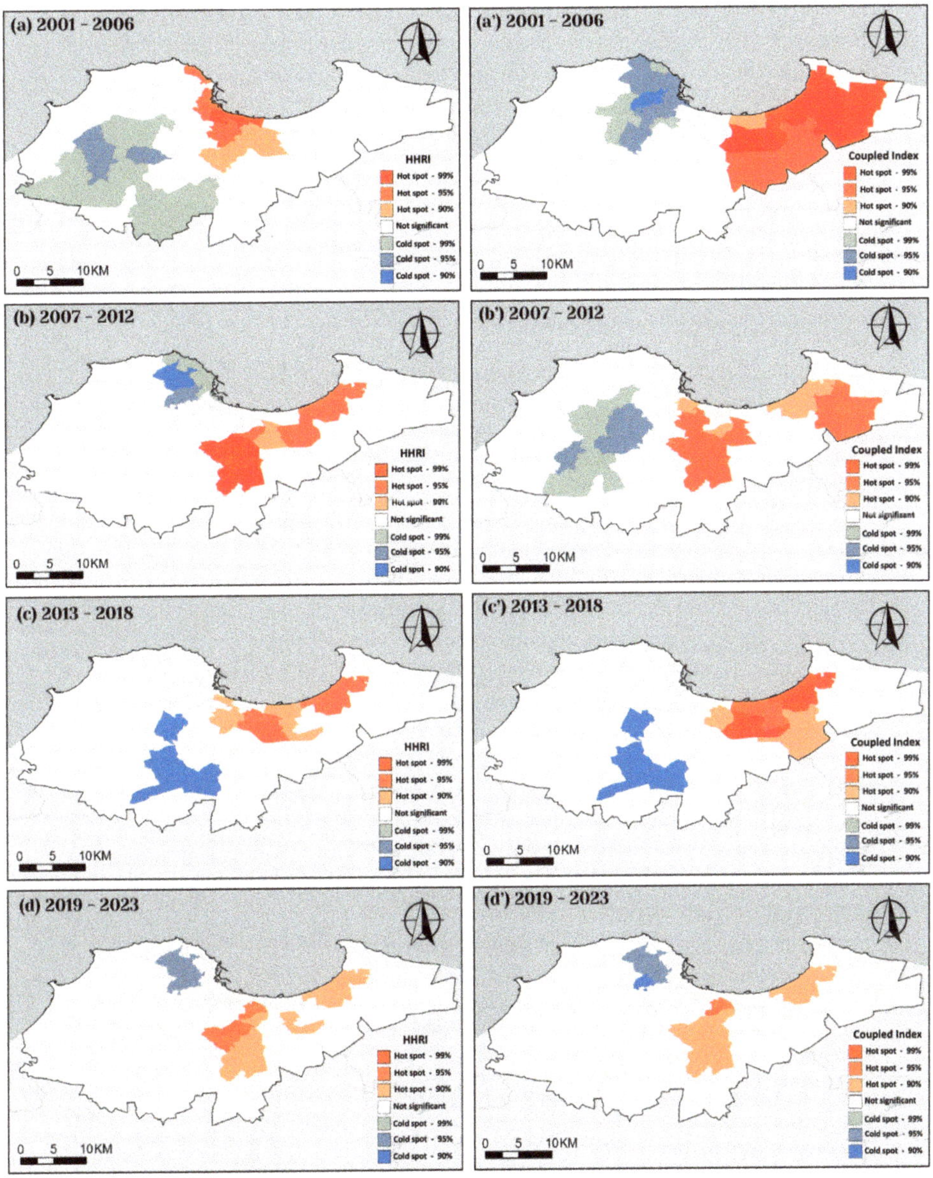

Figure 10. Spatial distributions of HHRI hotspots and cold spots in Algiers. (**a**) 2001–2006, (**b**) 2007–2012, (**c**) 2013–2018, (**d**) 2019–2023. Spatial distributions of the coupled index hotspots and cold spots in Algiers. (**a′**) 2001–2006, (**b′**) 2007–2012, (**c′**) 2013–2018, (**d′**) 2019–2023.

In the first period, HHRI hotspots were identified primarily in the extreme northern and eastern regions, as shown in (a), with 1.18% of the total area classified as a hotspot with 99% confidence. Most of the regions (66.5%) were not significant, and no cold spot with 99% confidence was identified, while 21% of the total area was classified as a cold spot with 95% confidence. Alternatively, (a′) shows a significantly larger area (14.22%) under the hotspot category at 99% confidence in the coupled index, mainly in the eastern regions. This suggests a broader risk when human thermal stress is integrated, as shown in Figure 11, while the cold spot area was smaller compared to the HHRI, at 3.4% with 99% confidence.

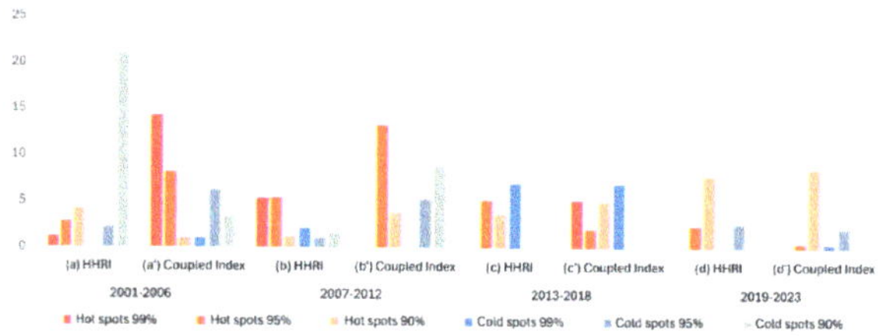

Figure 11. Temporal analysis of hotspots and cold spots in the HHRI and the coupled index in Algiers municipalities (2001–2023).

During the second period, both the intensity and spatial distribution of the HHRI and coupled index hotspots changed. As shown in (b), HHRI hotspots with 99% confidence increased to 5.3% of the total area, with concentration in the central regions, alongside a notable rise in "not significant" areas, reaching 83.3%. Cold spots decreased to 2.09% with 99% confidence, primarily in the northwestern regions. In contrast, (b′) shows a decrease in coupled index hotspot intensity, with no area classified at 99% confidence but 13.23% at 95% confidence, mostly in the eastern and central regions. The "not significant" areas remained constant compared to the previous period. At the same time, cold spots decreased to 5.23% at 95% confidence in the central to western regions, with no cold spots at 99% confidence, as observed in Figure 11.

The third period went through another shift in both indicators, with practically the same distribution across the study area. Figure 10c reveals a large drop in HHRI hotspots, with no areas categorized at 99% confidence and only 5.2% of the total area categorized at 95% confidence in eastern regions, while cold spots increased to 6.98% at 99% confidence in the south, leaving most of the city categorized as "not significant" area with 84.1%. Moreover, (c″) shows coupled index hotspots with 5.17% at 99% confidence in the eastern regions once again, while it shares the same cold spot intensity and distribution with the HHRI as shown in Figure 11.

Finally, there was a further change for both indicators in the fourth period, although with a different intensity but the same spatial pattern. Figure 10d indicates a shift of HHRI hotspots from the extreme north to the central regions, with no areas categorized at 99% confidence, while 2.46% of the total area was categorized at 95% confidence. Furthermore, cold spots experienced an entire migration towards northern regions with just 2.62% at 95% confidence. The combined index hotspots, however, decreased to only one municipality in the east and a few other ones in the central region with only 0.62% at 95% confidence as indicated in (d′). At the same time, there were no hotspots categorized at 99% confidence. Cold spots classified at 99% confidence were identified within a smaller area with just 0.48%, while a "not significant" area remained dominant with 88.1% as summarized in Figure 11.

Pearson correlation analysis was conducted for both indicators of cold/hotspots across the four periods (Figure 12). The p-value increased from -0.25 in the first period to 0.96 in the last period, highlighting how linked environmental, urban dynamics, and demographic factors are in increasing heat risk. This upward trend indicates a strengthening relationship between the distribution of hotspots and the underlying risk components, supporting the spatial patterns observed in the HHRI analysis.

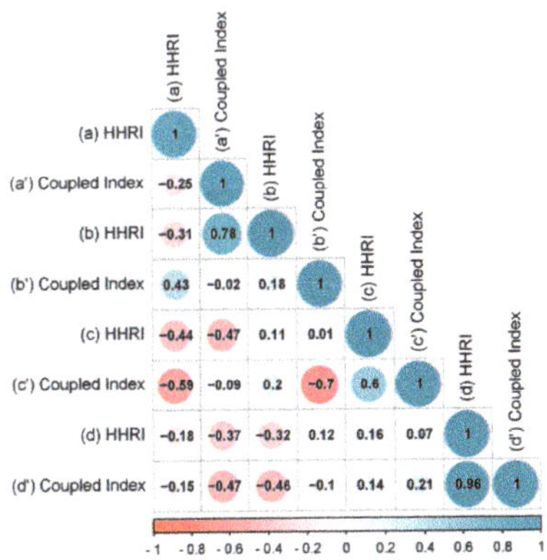

Figure 12. Matrix of the Pearson correlation coefficients between the HHRI and the coupled index (2001–2023).

Throughout the study period, persistent hotspots (Figure 13) were identified in Bordj El Kiffan, El Harrach, and Bourouba, while consistent cold spots were observed in Draria, Bouzareah, and Ben Aknoun.

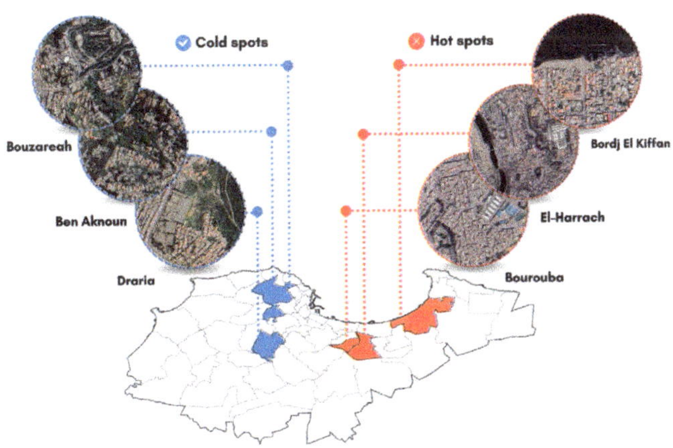

Figure 13. Persistent hotspots and cold spots of the coupled index in Algiers (2001–2023).

4. Discussion

4.1. HHR Assessment

To our knowledge, this paper is the first to assess HHR in Algiers and, more broadly, the first to undertake a long-term multi-decade (2001–2023) analysis of HHRI in any Mediterranean, African, or MENA country. While there is limited prior research in this area, one previous study has conducted a long-term HHR assessment, although in a different climatic context [20]; another study focused on a shorter timeframe [17]; and several others have examined HHRI during summer days and months [8,15,19], as well as during heatwave periods [6,12]. Our HHR assessment framework followed the IPCC AR6 employing the three risk components, hazard, exposure, and vulnerability, as defined from decades-spanning satellite imagery, census data, and meteorological data.

Our findings reveal that the intensity of HHRI has increased over time, since "very-low" risk areas had an outstanding decrease (93%) and almost disappeared—likely due to the global and regional climate change in rising temperatures and the frequency of extreme heat events [57]. Moreover, Algiers had a massive and rapid urbanization with a substantial increase (113.67%) in built-up areas from 1995 to 2023 [58]. People from all over the country are looking for better life conditions in the capital, leading to a massive expansion of built-up areas. Our analysis also suggests that the shifts in HHRI hotspots are due to changes in land use and the development of new infrastructure areas that were once less developed and urbanized, causing higher risk levels [59].

However, our results differ from previous studies that identified the most populated areas as the highest risk for heat-related health issues [19,20]. This is due to several factors. First, Algiers possesses significantly varied geographic and environmental features, such as elevation and proximity to the sea, where the most populated areas are located. Second, those areas are found within old urban fabrics, characterized by narrow, often covered streets, and colonial-era buildings with arcades providing solar shading. Furthermore, according to a pedestrians' thermal comfort investigation [60], these urban fabrics demonstrate better thermal conditions compared to the other layouts found in the remaining municipalities. This is further supported by [61,62], who indicated that these covered passages and streets consistently offer the best thermal comfort performance across all measurements.

4.2. UTCI Assessment

Algeria, particularly Algiers, is representative of the Mediterranean climate zone, where urban heat dynamics are influenced not only by global climate trends but also by local geographic and environmental factors. Furthermore, it has a very diverse urban layout and morphology along with various land use and building materials all together influencing thermal stress parameters.

The UTCI results from this research showed that during the summer from 2001 to 2023, Algiers was classified in the first period (Figure 8a) as both "no thermal heat stress" and "moderate thermal stress", while in the remaining periods, (Figure 8b–d), it was classified as "moderate thermal stress" and "strong thermal stress", indicating a clear increase in heat stress across Algiers throughout 23 years, which aligns with the global trends due to climate change. Similar findings have been reported in other regions, showing that UTCI values are increasing with time, leading to intense thermal stress experienced by urban populations during extreme heat events [63,64]. However, unlike HHRI, the spatial distribution of UTCI remained stable because the environmental conditions across different districts have not varied significantly over the decades. Moreover, topography has a crucial influence on altering thermal stress levels [65]. The northwestern areas of Algiers, which are situated at higher elevations (Figure 2) had lower UTCI values. This inverse correlation between altitude and thermal stress explains the stability of the UTCI spatial pattern, as the topography of the city has not changed over the study period. The elevated areas have cooler temperatures, which assist in mitigating the impact of heat stress [66], resulting in the observed consistency in the distribution of UTCI within these regions.

4.3. Coupled Index Hotspots

Previous studies [17,19,20,38] have used LST as a proxy for human-perceived thermal conditions. However, the way humans perceive LST is different from their perception of air temperature. In light of this, [19] suggested incorporating air temperatures in heat risk assessment studies. Later, [15] found that it is more appropriate to integrate a combined indicator of air temperature and relative humidity for a better human thermal perception. According to our knowledge, only one study [6] tried to incorporate the UTCI within the HHR assessment by integrating heat stress, social vulnerability, and human exposure, thus neglecting the hazard indicator as a direct measure of the intensity and frequency of extreme heat events. To overcome this limitation, we coupled the HHRI with the UTCI for a more comprehensive identification of high-risk areas, and we then determined hotspots using the Getis-Ord Gi* statistics.

Over the considered period (2001–2023), the coupled index demonstrated greater and different risk areas compared to the HHRI alone. This suggests that the coupling of both indicators enhances the sensitivity of risk assessment, particularly where thermal stress becomes severe in areas affected by heat-related vulnerabilities. Furthermore, synchronization between HHRI and the coupled index hotspots became apparent over time, especially during the last period, indicating that areas with high HHRI levels are also experiencing high levels of thermal stress. The urban structure and population distribution may have reached an amount of balance, where external climate factors are now the main driving factors of further rises in risk. The Pearson correlation matrix (Figure 12) supports this observation, showing a significant positive correlation between HHRI and coupled index with time, particularly during the last period.

The persistent hotspots in central and eastern regions throughout all periods (Figure 13) demonstrate the ongoing risk in these areas. The identification of persistent cold spots in the northern regions, especially at higher altitudes, indicates areas that maintain a certain level of resistance. However, their varying intensity demonstrates that even these sites are not completely protected from increasing temperatures and greater thermal stress.

4.4. Implications and Recommendations

As cities expand, achieving sustainable urban development requires integrating health considerations into urban planning, yet these interconnections remain underexplored in global urban science [67]. The insights gained from coupling HHR with thermal stress and the associated cold/hotspots have significant implications for urban planning and public health. According to Wu et al. [8], the complexity of HHR is influenced by both regional and intra-regional factors and requires policy decisions to be made at multiple levels. Our work addresses these requirements by suggesting a multi-scale strategy. At the national level, this strategy aims to (a) create country-wide guidelines for climate-resilient infrastructure that cities and municipalities across Algeria can adopt including heat mitigation strategies; (b) prepare a national heat action plan involving early-warning systems and emergency response strategies [68–70]; (c) develop a national database that combines the real-time monitoring of meteorological data with statistics regarding heat mortality and morbidity during summer and heatwaves in particularly. At the city scale, the strategy aims to (a) encourage the renovation of existing buildings to enhance thermal comfort and energy efficiency [71]; (b) establish an equal distribution of healthcare services across districts, particularly focusing on areas with high HHRI and low care-bed availability.

Furthermore, the identification of persistent hotspots calls for localized, targeted interventions, such as increasing greenspaces and tree cover [72], with more tree-lined streets [73], implementing green roofs [74], and enhancing urban fabrics with their building materials [75]. Moreover, our strategy improves the municipality's infrastructure ability against extreme heat by enhancing water delivery systems and guaranteeing a consistent energy supply for air conditioning [17]. Finally, it raises public awareness of the symptoms of heat-related illnesses and the necessary behavioral adjustments that might minimize the risk of exposure and vulnerability.

By addressing both environmental and health-related urban vulnerabilities, these insights support the goals of Sustainable Development Goal 3 (Good Health and Well-Being), Goal 11 (Sustainable Cities and Communities), and Goal 13 (Climate Action), reinforcing the role of heat-health risk research in global sustainability efforts.

4.5. Limitations and Future Perspectives of the Study

While this study sets an example of how to combine the IPCC's HHR assessment framework with thermal stress, it introduces several novel contributions, including the coupling of HHRI with UTCI, the use of a high-temporal-resolution dataset spanning over two decades, and the application of hotspot-mapping techniques. Together, these provide a more comprehensive understanding of vulnerable areas during extreme heat conditions that could lead to heat-related mortality and morbidity. It is important to acknowledge certain limitations that may affect the interpretation of the results. First, although Algeria is experiencing rising temperatures, there is a considerable lack of public weather stations regarding its large surface, particularly in the capital, Algiers. Therefore, expanding the meteorological spatial resolution in the province would have improved our scientific understanding regarding improvements in thermal, heatwave, and heat risk mitigation. Furthermore, and in agreement with previous research, there is a lack of city-level data on heat-related mortality and morbidity during summer in general and heatwaves in particular [6,9,13,15,17]. If these data can be gathered at a finer scale in the future, studies should prioritize investigating heat-related mortality and morbidity to validate and improve the accuracy of HHR assessment findings. This could provide further insights into the vulnerabilities among different populations. Finally, many variables, such as data on air-conditioning use and the health conditions of the population, are challenging to gather yet might significantly impact the vulnerability assessment [17,20]. Future research should explore various weighting strategies to enhance the precision and relevance of risk estimates in diverse urban contexts.

In addition, the current study was conducted at the municipal level, which limits the ability to capture micro-scale variations in thermal stress. As a result, local-scale variation in UTCI may be underestimated.

Future studies should consider the integration of microclimatic simulation tools such as ENVI-met or SOLWEIG to more accurately assess intra-urban variability and improve our understanding of heat exposure at finer spatial resolutions.

5. Conclusions

This study conducted a long-term assessment (2001–2023) of heat-related risks in Algiers by integrating the Heat Health Risk Index (HHRI) and the Universal Thermal Climate Index (UTCI). Using the IPCC's risk framework, we evaluated hazard, exposure, and vulnerability, and coupled this with thermal stress analysis through UTCI to improve the sensitivity of heat risk identification. This approach offers a more comprehensive understanding of the relationship between environmental heat and human health vulnerability in urban contexts. Key findings from this study include the following:

- HHRI showed a significant increase over time, reflecting the growing vulnerability of Algiers to extreme heat events

- UTCI increased in Algiers between 2001 and 2023, reaching the strong heat stress category.
- Coupling HHRI with UTCI enhanced the sensitivity and accuracy of heat-related risk assessment.
- The identification of persistent hotspots and cold spots offers crucial insights for targeted climate resilience interventions in the most vulnerable areas.

This work addressed a critical research gap, demonstrating the need to include thermal stress into the traditional HHR framework for a better understanding of the vulnerable areas during extreme heat conditions that could cause heat-related mortality and morbidity. It highlights the value of combining HHR indicators with thermal stress data in informing targeted interventions and policy decisions aimed at enhancing climate resilience in rapidly urbanizing regions like Algiers.

Author Contributions: Conceptualization, D.C.Z., D.B., M.E.M. and S.A.; data curation, D.C.Z.; methodology, D.C.Z. and M.A.E.B.R.; software, D.C.Z. and M.A.E.B.R.; supervision, D.B., M.E.M. and S.A.; visualization, D.C.Z. and M.A.E.B.R.; writing—original draft, D.C.Z.; writing—review and editing, D.C.Z., D.B., M.E.M. and S.A. All authors have read and agreed to the published version of the manuscript.

Funding: This research received no external funding.

Institutional Review Board Statement: Not applicable.

Informed Consent Statement: Not applicable.

Data Availability Statement: The data supporting the findings of this study have been deposited in the Harvard Dataverse repository and can be accessed at https://doi.org/10.7910/DVN/D9WGVN.

Acknowledgments: The authors would like to express their sincere gratitude to Claudia Di Napoli for her valuable insights into heat stress assessment and her expertise in the calculation of the Universal Thermal Climate Index (UTCI), which contributed to the methodological framework of this study.

Conflicts of Interest: The authors declare no conflicts of interest.

Abbreviations

The following abbreviations are used in this manuscript:

E	Exposure
GIS	Geographic Information System
H	Hazard
HHRI	Heat Health Risk Index
IPCC	Intergovernmental Panel on Climate Change
LST	Land Surface Temperature
MENA	Middle East and North Africa Region
MNDWI	Modified Normalized Difference Water Index
MODIS	Moderate Resolution Imaging Spectroradiometer
NDBI	Normalized Difference Building Index
NDVI	Normalized Difference Vegetation Index
NIR	Near Infrared
RED	Red Visible Band
SWIR	Shortwave Infrared
UTCI	Universal Thermal Climate Index
V	Vulnerability

References

1. Semahi, S.; Benbouras, M.A.; Mahar, W.A.; Zemmouri, N.; Attia, S. Development of spatial distribution maps for energy demand and thermal comfort estimation in algeria. *Sustainability* **2020**, *12*, 6066. [CrossRef]
2. Di Napoli, C.; McGushin, A.; Romanello, M.; Ayeb-Karlsson, S.; Cai, W.; Chambers, J.; Dasgupta, S.; Escobar, L.E.; Kelman, I.; Kjellstrom, T.; et al. Tracking the impacts of climate change on human health via indicators: Lessons from the Lancet Countdown. *BMC Public Health* **2022**, *22*, 1–8. [CrossRef] [PubMed]
3. Crichton, D. The risk triangle. *Nat. Disaster Manag.* **1999**, *102*, 102–103.
4. IPCC 6th Assessment Report. Available online: https://www.ipcc.ch/assessment-report/ar6/ (accessed on 28 August 2024).
5. Di Napoli, C.; Romanello, M.; Minor, K.; Chambers, J.; Dasgupta, S.; Escobar, L.E.; Hang, Y.; Hänninen, R.; Liu, Y.; Batista, M.L.; et al. The role of global reanalyses in climate services for health: Insights from the Lancet Countdown. *Meteorol. Appl.* **2023**, *30*, e2122. [CrossRef]
6. Dong, J.; Peng, J.; He, X.; Corcoran, J.; Qiu, S.; Wang, X. Heatwave-induced human health risk assessment in megacities based on heat stress-social vulnerability-human exposure framework. *Landsc. Urban. Plan.* **2020**, *203*, 103907. [CrossRef]
7. Sun, H.; Chen, Y.; Li, K.; Gao, S. Spatio-temporal assessment of heat health risk in Chinese metropolitan cities based on the modified multi-indicators coupled risk framework. *Sustain. Cities Soc.* **2024**, *108*, 105451. [CrossRef]
8. Wu, H.; Xu, Y.; Zhang, M.; Su, L.; Wang, Y.; Zhu, S. Spatially explicit assessment of the heat-related health risk in the Yangtze River Delta, China, using multisource remote sensing and socioeconomic data. *Sustain. Cities Soc.* **2024**, *104*, 105300. [CrossRef]
9. Wu, H.; Zhao, C.; Zhu, Y.; Pan, Y. A multiscale examination of heat health risk inequality and its drivers in mega-urban agglomeration: A case study in the Yangtze River Delta, China. *J. Clean. Prod.* **2024**, *458*, 142528. [CrossRef]
10. Ye, J.; Yang, F. Towards multi-scale and context-specific heat health risk assessment—A systematic review. *Sustain. Cities Soc.* **2025**, *119*, 106102. [CrossRef]
11. Baqa, M.F.; Lu, L.; Guo, H.; Song, X.; Alavipanah, S.K.; Nawaz-Ul-Huda, S.; Li, Q.; Chen, F. Investigating heat-related health risks related to local climate zones using SDGSAT-1 high-resolution thermal infrared imagery in an arid megacity. *Int. J. Appl. Earth Obs. Geoinf.* **2025**, *136*, 104334. [CrossRef]
12. Tomlinson, C.J.; Chapman, L.; E Thornes, J.; Baker, C.J. Including the urban heat island in spatial heat health risk assessment strategies: A case study for Birmingham, UK. *Int. J. Health Geogr.* **2011**, *10*, 42. [CrossRef] [PubMed]
13. Hu, K.; Yang, X.; Zhong, J.; Fei, F.; Qi, J. Spatially Explicit Mapping of Heat Health Risk Utilizing Environmental and Socioeconomic Data. *Environ. Sci. Technol.* **2017**, *51*, 1498–1507. [CrossRef] [PubMed]
14. Chen, Q.; Ding, M.; Yang, X.; Hu, K.; Qi, J. Spatially explicit assessment of heat health risk by using multi-sensor remote sensing images and socioeconomic data in Yangtze River Delta, China. *Int. J. Health Geogr.* **2018**, *17*, 1–15. [CrossRef]
15. Zha, F.; Lu, L.; Wang, R.; Zhang, S.; Cao, S.; Baqa, M.F.; Li, Q.; Chen, F. Understanding fine-scale heat health risks and the role of green infrastructure based on remote sensing and socioeconomic data in the megacity of Beijing, China. *Ecol. Indic.* **2024**, *160*, 111847. [CrossRef]
16. Zhang, W.; Zheng, C.; Chen, F. Mapping heat-related health risks of elderly citizens in mountainous area: A case study of Chongqing, China. *Sci. Total. Environ.* **2019**, *663*, 852–866. [CrossRef]
17. Estoque, R.C.; Ooba, M.; Seposo, X.T.; Togawa, T.; Hijioka, Y.; Takahashi, K.; Nakamura, S. Heat health risk assessment in Philippine cities using remotely sensed data and social-ecological indicators. *Nat. Commun.* **2020**, *11*, 1581. [CrossRef] [PubMed]
18. Zhu, W.; Yuan, C. Urban heat health risk assessment in Singapore to support resilient urban design—By integrating urban heat and the distribution of the elderly population. *Cities* **2023**, *132*, 104103. [CrossRef]
19. Wang, S.; Sun, Q.C.; Huang, X.; Tao, Y.; Dong, C.; Das, S.; Liu, Y. Health-integrated heat risk assessment in Australian cities. *Environ. Impact Assess. Rev.* **2023**, *102*, 107176. [CrossRef]
20. Wang, C.; Ren, Z.; Guo, Y.; Zhang, P.; Hong, S.; Ma, Z.; Hong, W.; Wang, X. Assessing urban population exposure risk to extreme heat: Patterns, trends, and implications for climate resilience in China (2000–2020). Sustain. *Cities Soc.* **2024**, *103*, 105260. [CrossRef]
21. Zhang, C.; Yang, Y.; Yu, L. Assessing urban surface thermal environment and heat health risk in Chinese cities: A twenty-year study. *Urban Clim.* **2025**, *59*, 102304. [CrossRef]
22. McGregor, G.R.; Vanos, J.K. Heat: A primer for public health researchers. *Public Health* **2018**, *161*, 138–146. [CrossRef] [PubMed]
23. Crimmins, A.; Balbus, J.; Gamble, J.L.; Beard, C.B.; Bell, J.E.; Dodgen, D.; Eisen, R.J.; Fann, N.; Hawkins, M.D.; Herring, S.C. The Impacts of Climate Change on Human Health in the United States: A Scientific Assessment. 2016. Available online: http://www.preventionweb.net/publication/impacts-climate-change-human-health-united-states-scientific-assessment (accessed on 28 August 2024).

24. Bröde, P.; Błazejczyk, K.; Fiala, D.; Havenith, G.; Holmér, I.; Jendritzky, G.; Kuklane, K.; Kampmann, B. The universal thermal climate index UTCI compared to ergonomics standards for assessing the thermal environment. *Ind. Health* **2013**, *51*, 16–24. [CrossRef] [PubMed]
25. Hu, J.; Zhou, Y.; Yang, Y.; Chen, G.; Chen, W.; Hejazi, M. Multi-city assessments of human exposure to extreme heat during heat waves in the United States. *Remote Sens. Environ.* **2023**, *295*, 113700. [CrossRef]
26. Mahia, L.; Berkouk, D.; Bouzir, T.A.K.; Pigliautile, I.; Pisello, A.L. Investigating the Relationship between Spatial Morphology, Meteorological Factors, and Elderly People Responses in a Traditional Algerian Village, Sustain. *Cities Soc.* **2025**, *121*, 106212. [CrossRef]
27. Bröde, P.; Fiala, D.; Błażejczyk, K.; Holmér, I.; Jendritzky, G.; Kampmann, B.; Tinz, B.; Havenith, G. Deriving the operational procedure for the Universal Thermal Climate Index (UTCI). *Int. J. Biometeorol.* **2012**, *56*, 481–494. [CrossRef]
28. Coccolo, S.; Kämpf, J.; Scartezzini, J.-L.; Pearlmutter, D. Outdoor human comfort and thermal stress: A comprehensive review on models and standards. *Urban Clim.* **2016**, *18*, 33–57. [CrossRef]
29. Lewis, S.C.; King, A.D.; Perkins-Kirkpatrick, S.E.; Mitchell, D.M. Regional hotspots of temperature extremes under 1.5 °C and 2 °C of global mean warming. *Weather. Clim. Extrem.* **2019**, *26*, 100233. [CrossRef]
30. Wedler, M.; Pinto, J.G.; Hochman, A. More frequent, persistent, and deadly heat waves in the 21st century over the Eastern Mediterranean. *Sci. Total Environ.* **2023**, *870*, 161883. [CrossRef]
31. Ali, E.; Cramer, W.; Georgopoulou, E.; Hilmi, N.J.M.; Le Cozannet, G.; Lionello, P. Cross-Chapter Paper 4: Mediterranean Region. In *Climate Change 2022: Impacts, Adaptation and Vulnerability. Contribution of Working Group II to the Sixth Assessment Report of the Intergovernmental Panel on Climate Change*; Cambridge University Press: Cambridge, UK, 2022. [CrossRef]
32. Hajat, S.; Proestos, Y.; Araya-Lopez, J.-L.; Economou, T.; Lelieveld, J. Current and future trends in heat-related mortality in the MENA region: A health impact assessment with bias-adjusted statistically downscaled CMIP6 (SSP-based) data and Bayesian inference. *Lancet Planet. Health* **2023**, *7*, e282–e290. [CrossRef]
33. GFDRR. Global Facility for Disaster Reduction and Recovery (GFDRR). 2024. Available online: https://www.gfdrr.org/en (accessed on 11 September 2024).
34. Philip, S.; Kew, S.; Vautard, R.; Vahlberg, M.; Singh, R.; Driouech, F.; Lguensat, R.; Barnes, C.; Otto, F. Extreme April Heat in Spain, Portugal, Morocco & Algeria Almost Impossible Without Climate Change. 2023 (pp. 1–11). Available online: http://hdl.handle.net/10044/1/103833 (accessed on 28 August 2024).
35. Zitouni, D.C.; Berkouk, D.; Matallah, M.E.; Attia, S. Dataset for Heat Health Risk and Thermal Comfort Assessment in Algiers (2001–2023). *Harv. Dataverse* **2024**. [CrossRef]
36. Esri. ArcGis Pro. 2023. Available online: https://www.esri.com/fr-fr/arcgis/products/arcgis-pro/overview (accessed on 20 December 2024).
37. Ma, L.; Huang, G.; Johnson, B.A.; Chen, Z.; Li, M.; Yan, Z.; Zhan, W.; Lu, H.; He, W.; Lian, D. Investigating urban heat-related health risks based on local climate zones: A case study of Changzhou in China. *Sustain. Cities Soc.* **2023**, *91*, 104402. [CrossRef]
38. Dong, W.; Liu, Z.; Zhang, L.; Tang, Q.; Liao, H.; Li, X. Assessing Heat Health Risk for Sustainability in Beijing's Urban Heat Island. *Sustainability* **2014**, *6*, 7334–7357. [CrossRef]
39. Huang, H.; Ma, J.; Yang, Y. Spatial heterogeneity of driving factors for urban heat health risk in Chongqing, China: A new identification method and proposal of planning response framework. *Ecol. Indic.* **2023**, *153*, 110449. [CrossRef]
40. Pramanik, S.; Punia, M.; Yu, H.; Chakraborty, S. Is dense or sprawl growth more prone to heat-related health risks? Spatial regression-based study in Delhi, India. *Sustain. Cities Soc.* **2022**, *81*, 103808. [CrossRef]
41. Hua, J.; Zhang, X.; Ren, C.; Shi, Y.; Lee, T.-C. Spatiotemporal assessment of extreme heat risk for high-density cities: A case study of Hong Kong from 2006 to 2016. Sustain. *Cities Soc.* **2021**, *64*, 102507. [CrossRef]
42. Mahia, L.; Berkouk, D.; Bouzir, T.A.K.; Pigliautile, I.; Pisello, A.L. Exploring the Relationship between Elderly Women Responses, Multi-domain Factors, and Street Geometry Indicators: A Case Study in a Mediterranean Context. *Build. Environ.* **2025**, *273*, 112647. [CrossRef]
43. USGS. Earth Explorer. 2024. Available online: https://earthexplorer.usgs.gov (accessed on 1 March 2024).
44. Kestens, Y.; Brand, A.; Fournier, M.; Goudreau, S.; Kosatsky, T.; Maloley, M.; Smargiassi, A. Modelling the variation of land surface temperature as determinant of risk of heat-related health events. *Int. J. Health Geogr.* **2011**, *10*, 7. [CrossRef]
45. Huang, H.; Liu, X.; Ren, L. Analysis of the spatiotemporal mechanism of high temperature on residents' irritability in Beijing based on multiscale geographically weighted regression model. *Front. Ecol. Evol.* **2022**, *10*, 973365. [CrossRef]
46. Krüger, E.; Gobo, J.P.A.; Tejas, G.T.; Souza, R.M.d.S.d.; Neto, J.B.F.; Pereira, G.; Mendes, D.; Di Napoli, C. The impact of urbanization on heat stress in Brazil: A multi-city study. *Urban Clim.* **2024**, *53*, 101827. [CrossRef]
47. Matallah, M.E.; Ahriz, A.; Zitouni, D.C.; Arrar, H.F.; Ben Ratmia, M.A.E.; Attia, S. A methodological approach to evaluate the passive cooling effect of Oasis palm groves. *Sustain. Cities Soc.* **2023**, *99*, 104887. [CrossRef]

48. Khan, A.A. Heat related illnesses: Review of an ongoing challenge. *Saudi Med. J.* **2019**, *40*, 1195. [CrossRef] [PubMed]
49. Aboubakri, O.; Khanjani, N.; Jahani, Y.; Bakhtiari, B. Thermal comfort and mortality in a dry region of Iran, Kerman; a 12-year time series analysis. *Theor. Appl. Clim.* **2020**, *139*, 403–413. [CrossRef]
50. Wu, X.; Liu, Q.; Huang, C.; Li, H. Mapping heat-health vulnerability based on remote sensing: A case study in karachi. *Remote Sens.* **2022**, *14*, 1590. [CrossRef]
51. Blazejczyk, K.; Epstein, Y.; Jendritzky, G.; Staiger, H.; Tinz, B. Comparison of UTCI to selected thermal indices. *Int. J. Biometeorol.* **2012**, *56*, 515–535. [CrossRef]
52. Błażejczyk, K.; Jendritzky, G.; Bröde, P.; Fiala, D.; Havenith, G.; Epstein, Y.; Psikuta, A.; Kampmann, B. An introduction to the universal thermal climate index (UTCI). *Geogr. Pol.* **2013**, *86*, 5–10. [CrossRef]
53. Błażejczyk, K.; Błażejczyk, M. BioKlima—Institute of Geography and Spatial Organization PAS. 2010. Available online: https://www.igipz.pan.pl/bioklima-crd.html (accessed on 11 September 2024).
54. Blazejczyk, K. New climatological and physiological model of the Human Heat Balance outdoor (MENEX) and its applications in bioclimatological studies in different scales. *Zeszyty Instytutu Geografii i Przestrzennego Zagospodarowania PAN* **1994**, *28*, 27–58.
55. Wang, C.; Zhan, W.; Liu, Z.; Li, J.; Li, L.; Fu, P.; Huang, F.; Lai, J.; Chen, J.; Hong, F.; et al. Satellite-based mapping of the Universal Thermal Climate Index over the Yangtze River Delta urban agglomeration. *J. Clean. Prod.* **2020**, *277*, 123830. [CrossRef]
56. Manepalli, U.R.; Bham, G.H.; Kandada, S. Evaluation of hotspots identification using kernel density estimation (K) and Getis-Ord (Gi*) on I-630. In Proceedings of the 3rd International Conference on Road Safety and Simulation, National Academy of Sciences, Indianapolis, IN, USA, 14–16 September 2011; pp. 14–16. Available online: https://onlinepubs.trb.org/onlinepubs/conferences/2011/RSS/2/Manepalli,UR.pdf (accessed on 22 August 2024).
57. Wang, Y.; Chan, A.; Lau, G.N.; Li, Q.; Yang, Y.; Yim, S.H.L. Effects of urbanization and global climate change on regional climate in the Pearl River Delta and thermal comfort implications. *Int. J. Clim.* **2019**, *39*, 2984–2997. [CrossRef]
58. Alioua, N.E.H.; Kemmouche, A. Assessing Urban Expansion of Algiers with Random Forest-Based Multi-Index Using Landsat Imagery. In Proceedings of the 2024 IEEE Mediterranean and Middle-East Geoscience and Remote Sensing Symposium (M2GARSS), Oran, Algeria, 15–17 April 2024; pp. 139–143.
59. Heidari, H.; Mohammadbeigi, A.; Khazaei, S.; Soltanzadeh, A.; Asgarian, A.; Saghafipour, A. Assessment of land use–land cover changes using GIS, remote sensing, and CA–Markov model: A case study of Algiers. *Algeria Appl. Geomat* **2020**. [CrossRef]
60. Talhi, A.; Barlet, A.; Bruneau, D.; Aichour, B. Towards a prediction of outdoor human thermal comfort adapted for designers of urban spaces: Examining UTCI and APCI in the context of Algiers (Algeria). *Int. J. Biometeorol.* **2020**, *64*, 651–662. [CrossRef] [PubMed]
61. Arrar, F.H.; Kaoula, D.; Matallah, M.E.; Abdessemed-Foufa, A.; Taleghani, M.; Attia, S. Quantification of Outdoor Thermal Comfort Levels under Sea Breeze in the Historical City Fabric: The Case of Algiers Casbah. *Atmosphere* **2022**, *13*, 575. [CrossRef]
62. Smail, S.A.; Zemmouri, N.; Djenane, M.; Nikolopoulou, M. Investigating the transient conditions of "Sabat" space and its influence on pedestrian sensations during thermal walks. Algiers' Casbah case study. *Build. Environ.* **2024**, *261*, 111760. [CrossRef]
63. Błażejczyk, K.; Twardosz, R. Secular changes (1826–2021) of human thermal stress according to UTCI in Kraków (southern Poland). *Int. J. Clim.* **2023**, *43*, 4220–4230. [CrossRef]
64. Kuchcik, M.; Błażejczyk, K.; Halaś, A. Long-term changes in hazardous heat and cold stress in humans: Multi-city study in Poland. *Int. J. Biometeorol.* **2021**, *65*, 1567–1578. [CrossRef]
65. Zheng, Z.; Lin, X.; Chen, L.; Yan, C.; Sun, T. Effects of urbanization and topography on thermal comfort during a heat wave event: A case study of Fuzhou, China. *Sustain. Cities Soc.* **2024**, *102*, 105233. [CrossRef]
66. Zhou, D.; Zhao, S.; Liu, S.; Zhang, L.; Zhu, C. Surface urban heat island in China's 32 major cities: Spatial patterns and drivers. *Remote Sens. Environ.* **2014**, *152*, 51–61. [CrossRef]
67. Tonne, C.; Adair, L.; Adlakha, D.; Anguelovski, I.; Belesova, K.; Berger, M.; Brelsford, C.; Dadvand, P.; Dimitrova, A.; Giles-Corti, B.; et al. Defining pathways to healthy sustainable urban development. *Environ. Int.* **2021**, *146*, 106236. [CrossRef]
68. Lowe, D.; Ebi, K.L.; Forsberg, B. Heatwave Early Warning Systems and Adaptation Advice to Reduce Human Health Consequences of Heatwaves. *Int. J. Environ. Res. Public Health* **2011**, *8*, 4623–4648. [CrossRef]
69. Kien, N.D.; My, N.H.D.; Thu, D.T.A.; Tri, T.T.C.; Son, N.H.; Phong, T.K.; Tin, H.C.; Lan, N.H.; Thang, T.B.; The, B.D.; et al. Valuation of a Heatwave Early Warning System for Mitigating Risks Associated with Heat-Related Illness in Central Vietnam. *Sustainability* **2023**, *15*, 15342. [CrossRef]
70. Trahan, A.; Walshe, R.; Mehta, V. Extreme heat, gender, and access to preparedness measures: An analysis of the heatwave early warning system in Ahmedabad, India. *Int. J. Disaster Risk Reduct.* **2023**, *99*, 104080. [CrossRef]
71. Duan, Z.; de Wilde, P.; Attia, S.; Zuo, J. Prospect of energy conservation measures (ECMs) in buildings subject to climate change: A systematic review. *Energy Build.* **2024**, *322*, 114739. [CrossRef]

72. Fahmy, M.; Mahdy, M.; Mahmoud, S.; Abdelalim, M.; Ezzeldin, S.; Attia, S. Influence of urban canopy green coverage and future climate change scenarios on energy consumption of new sub-urban residential developments using coupled simulation techniques: A case study in Alexandria, Egypt. *Energy Rep.* **2020**, *6*, 638–645. [CrossRef]
73. Necira, H.; Matallah, M.E.; Bouzaher, S.; Mahar, W.A.; Ahriz, A. Effect of Street Asymmetry, Albedo, and Shading on Pedestrian Outdoor Thermal Comfort in Hot Desert Climates. *Sustainability* **2024**, *16*, 1291. [CrossRef]
74. Arrar, H.F.; Kaoula, D.; Santamouris, M.; Foufa-Abdessemed, A.; Emmanuel, R.; Matallah, M.E.; Ahriz, A.; Attia, S. Coupling of different nature base solutions for pedestrian thermal comfort in a Mediterranean climate. *Build. Environ.* **2024**, *256*, 111480. [CrossRef]
75. Ben Ratmia, F.Z.; Ahriz, A.; Santi, G.; Bouzaher, S.; Mahar, W.A.; Ben Ratmia, M.A.E.; Matallah, M.E. Street Design Strategies Based on Spatial Configurations and Building External Envelopes in Relation to Outdoor Thermal Comfort in Arid Climates. *Sustainability* **2024**, *16*, 221. [CrossRef]

MDPI AG
Grosspeteranlage 5
4052 Basel
Switzerland
Tel.: +41 61 683 77 34

Atmosphere Editorial Office
E-mail: atmosphere@mdpi.com
www.mdpi.com/journal/atmosphere

www.ingramcontent.com/pod-product-compliance
Lightning Source LLC
Chambersburg PA
CBHW061511210926
51850CB00008B/1884